World Human Rights Guide

WORLD HUMAN RIGHTS GUIDE

THIRD EDITION

Originated and compiled by
CHARLES HUMANA

New York   Oxford
OXFORD UNIVERSITY PRESS
1992

Oxford University Press

Oxford New York Toronto
Delhi Bombay Calcutta Madras Karachi
Kuala Lumpur Singapore Hong Kong Tokyo
Nairobi Dar es Salaam Cape Town
Melbourne Auckland
and associated companies in
Berlin Ibadan

Published by Oxford University Press, Inc.,
200 Madison Avenue, New York, New York 10016

Oxford is a registered trademark of Oxford University Press

Library of Congress Cataloging-in-Publication Data

Humana, Charles.
World human rights guide/
originated and compiled by Charles Humana,—3rd ed.
p. cm. Includes bibliographical references and index.
ISBN 0-19-507674-5   ISBN 0-19-507926-4 (pbk.)
1. Civil rights—Handbooks, manuals, etc.
2. Human rights—Handbooks, manuals, etc.
I. Title.
JC571.H788   1992   323′.02′.02—dc20   92-9758

9 8 7 6 5 4 3 2 1

Printed in United States of America
on acid-free paper

Preface

This is the third edition of the guide since it was first published in 1983.

The original intention was to update the assessment of human rights every 3/4 years, but the political and social changes that were destabilizing Eastern Europe, southern Africa, and parts of Asia, and which at any moment could overthrow governments and systems, made it advisable to delay the revision.

This hesitation was resolved, however, by two events. The first was the prominence given to the 1986 edition of the guide by the Human Development Report 1991, published by the United Nations Development Programme. The guide was used in the report as the basis for its Human Freedom Index, an exercise in the classification of countries by their human rights performance. The HFI, as it was called, was the first to be compiled by the UNDP and the report described the guide as the "most systematic and extensive coverage" of all attempts to classify and measure human rights.

But it was soon apparent to the compilers of the HFI that information that was 5 years out of date seriously diminished the authority of the index, a fact that significantly reinforced the opposition of the many member-states of the United Nations resisting any form of measuring, scrutinizing, or even referring to human rights.

The second source of encouragement for beginning a 1991 update was the incredible suddenness with which countries under one-party rule adopted multiparty democracy. It was a category which included Poland, East Germany, Czechoslovakia, Hungary, Bulgaria, and Romania, and other countries in Africa and Asia. But what of the USSR, where internal developments were making the breakup of the state seem inevitable?

The compiler's reluctance to undertake the next update had also been influenced by the conviction that the USSR would disintegrate into a number, perhaps a large number, of independent states, and that this would probably happen when it was too late to correct the revision. It was therefore fortuitous that the transitional period ended in August 1991, when the Soviet Union ceased to be a unified state.

But the welcome change created a new problem. How did one assess the human rights performances of newly independent countries? They could hardly create "instant" constitutions, they could not establish "instant" human rights laws, or at least have proved over a period of time that those rights were being honored. The difficulty has therefore been avoided by updating the USSR at the moment of that state's dissolution.

A similar decision concerning a seemingly insoluble problem has been applied to Yugoslavia. As of November 1991, it was no longer a single independent country, and what will follow is beyond prediction.

To turn to the significant human rights changes of the last 5 years, the extent of these is clearly illustrated by the statistics set out in the Progress Report 1986–91 on page xi. A study of the figures and the magnitude of the shift from one-party government with little respect for human rights to multiparty democracy suggest that history may well refer to the last 10 or 15 years of the 20th century as the Age of Human Rights. The Magna Carta, the American Bill of Rights, the French Déclaration des Droits de l'Homme et du Citoyen, and the Universal Declaration of Human Rights adopted in 1948 by the United Nations did not immediately produce universal changes comparable with those of the recent past.

The swiftness with which this major transformation has taken place, the numbers of people who are now free to choose their own governments, and, in the case of Europe and excepting Yugoslavia, the avoidance of bloody revolution, must be considered one of the most uplifting achievements of our social evolution. The compiler has therefore been fortunate that the succeeding updates of his work have coincided with such an era.

C. H.

Acknowledgments

A single compiler could not have completed this work without the authoritative and generous help of many human rights and other organizations, experts, research workers, and loyal friends. I would have liked to have mentioned the major contributors individually but this would have meant an injustice to others almost as important. One must also acknowledge the debt owed to those who helped with previous editions of the guide, contributions which have become a permanent feature of the work.

In the circumstances I set out this distinguished assembly in alphabetical order, and thank them all.

Dr. Jose Aitken
Sheila Annesley
Professor W.H.G. Armytage
Dr. David L. Banks (Carnegie-Mellon University, USA)
Martin Breese International
Dr. Harry Cargas (Webster University, USA)
Dr. Georges Chapouthier
Peter Davies
Paul LaRose Edwards
Ralph Estling
INDEX on Censorship
    (Philip Spender, Lek Hor Tan, Andrew Graham-Yooll, Adewale Maja Pearce,
    and Adel Darwish)
Dr. Eva Komaromi
Professor K.-Y. Liang
Jeremy McBride (Faculty of Law, University of Birmingham, UK)
Aryeh Neier (Human Rights Watch)
Peter Pack (Amnesty International)
Graciela Plouin
Robert Pullen (University of Sheffield)
Michael Rubinstein
George Schöpflin (London School of Economics)
John and Ching Yee Smithback
Professor Herbert Spirer
Louise Spirer
Dr. Etsuro Totsuka (London School of Economics)
United Nations Association
United Nations Information Centre (London)
Marilyn Wilson (International Planned Parenthood Federation)
Wang Zhen

I must also state my debt to those who have ensured that the appearance of the guide complements the contents. A special word, then, for my colleague Jim Dening of Archive Editions and his son James. Their professional and technical expertise has involved them in research, computer assistance, and the preparation of a projected human rights database. And my appreciation to Ann Buchan for a major contribution to the presentation of the work.

Lastly, I affirm that those whose help and support I have acknowledged bear no responsibility for possible mistakes or errors of judgment that may be contained in this work.

Many people have attempted to classify human rights and to measure each country against that classification. . . .

The one that offers the most systematic and extensive coverage is the index designed by Charles Humana. He examined various UN conventions and international treaties, and from them distilled 40 distinct criteria for judging freedom. These include freedom of movement, the rights of assembly and free speech, the rights to ethnic and gender equality, the rule of law, and other democratic freedoms. Humana's index is thus more than a political freedom index, more than a human rights index. It is a *human freedom* index.

From the Human Development Report 1991
of the United Nations Development Programme

Contents

Progress report 1986–91

Set out below is a comparison of statistics from the *World Human Rights Guide* for the years 1986 and 1991. They show a significant trend toward greater respect for human rights worldwide.

	1991	1986
Average human rights rating (Afghanistan to Zimbabwe)	**62%**	55%
Form of government:	*(Population in millions)*	
Multiparty democracy or similar	**2,530 or 48%**	1,900 or 40%
One-party or one-person rule	**2,340 or 44%**	2,400 or 51%
Military or effective military rule	**410 or 8%**	440 or 9%
Total world population	**5,280**	4,740

The statistics indicate an improvement over a 5-year period which is unparalleled in history. The average ratings have increased from 55% to 62% or, in terms of population, multiparty democracy is now enjoyed by 630,000,000 more people. As a percentage this represents the repudiation of one-party or one-person rule by nearly one in eight of the world's population since 1986.

The scale of this progression, and the suddenness and unexpectedness of the events that caused it, permits the observation that if three countries of the Far East with major populations adopt multiparty democracy, then the proportion of people living under a freer political and economic system would rise from 48% to 75%. Two of these countries, China with a population of 1,139 million and Vietnam with 67 million, are under Communist party rule, while Indonesia, with a population of 184 million, is under effective military rule.

Note: The 1991 total for multiparty democracies includes countries under short-term interim governments which are genuinely preparing for early multiparty elections.

Source of the population figures:
Human Development Reports (United Nations Development Programme). Figures from the 1991 report are actually for 1990.

The basis of the questionnaire

The questionnaire used in the *World Human Rights Guide* is drawn totally from UN instruments. The indicators – or questions – are listed below together with the appropriate parts of the relevant articles.

The instruments are abbreviated as follows:

UDHR: Universal Declaration of Human Rights
ICESCR: International Covenant on Economic, Social and Cultural Rights
ICCPR: International Covenant on Civil and Political Rights

Question 1. Article 13(1) UDHR: Everyone has the right to freedom of movement and residence within the borders of each state.

Question 2. Article 13(2) UDHR: Everyone has the right to leave any country, including his own, and to return to his country.

Question 3. Article 21 ICCPR: The right of peaceful assembly shall be recognized. No restrictions may be placed on the exercise of this right other than those imposed in conformity with the law and which are necessary in a democratic society in the interests of national security or public safety, public order (*ordre public*), the protection of public health or morals or the protection of the rights and freedoms of others.

Question 4. Article 19 ICCPR: Everyone shall have the right to hold opinions without interference.
Everyone shall have the right to freedom of expression; this right shall include freedom to seek, receive and impart information and ideas of all kinds, regardless of frontiers, either orally, in writing or in print, in the form of art, or through any other media of his choice.

Question 5. Article 19 UDHR: Everyone has the right to freedom of opinion and expression; this right includes freedom to hold opinions without interference and to seek, receive and impart information and ideas through any media and regardless of frontiers.

Question 6. Article 27 ICCPR: In those States in which ethnic, religious or linguistic minorities exist, persons belonging to such minorities shall not be denied the right, in community with the other members of their group, to enjoy their own culture, to profess and practise their own religion, or to use their own language.

Question 7. Article 10(3) ICESCR: Children and young persons should be protected from economic and social exploitation . . . states should also set age limits below which the paid employment of child labour should be prohibited and punishable by law.

Question 8. Article 6(1) ICCPR: Every human being has the inherent right to life. This right shall be protected by law. No one shall be arbitrarily deprived of his life.

Question 9. Article 5 UDHR: No one shall be subjected to torture or to cruel, inhuman or degrading treatment or punishment.

Question 10. Article 23(1) UDHR: Everyone has the right to work, to free choice of employment, to just and favourable conditions of work and to protection against unemployment.

Question 11. Article 6(6) ICCPR: Nothing in this article shall be invoked to delay or to prevent the abolition of capital punishment by any State Party to the present Covenant.

Question 12. Article 7 ICCPR: No one shall be subjected to torture or to cruel, inhuman or degrading treatment or punishment. In particular, no one shall be subjected without his free consent to medical or scientific experimentation.

Question 13.   Article 9(1) ICCPR: Everyone has the right to liberty and security of person. No one shall be subjected to arbitrary arrest or detention. No one shall be deprived of his liberty except on such grounds and in accordance with such procedures as are established by law.

Question 14.   Article 20(2) UDHR: No one may be compelled to belong to an association.

Question 15.   Article 18 UDHR: Everyone has the right to freedom of thought, conscience and religion; this right includes freedom to change his religion or belief, and freedom, either alone or in community with others and in public or private, to manifest his religion or belief in teaching, practice, worship and observance.

Question 16.   See Article 19 ICCPR above.

Question 17.   See Article 19 UDHR above.

Question 18.   Article 12 UDHR: No one shall be subjected to arbitrary interference with his privacy, family, home or correspondence, nor to attacks upon his honour and reputation. Everyone has the right to the protection of the law against such interference or attacks.

Question 19.   Article 25 ICCPR: Every citizen shall have the right and the opportunity . . . (a) To take part in the conduct of public affairs, directly or through freely chosen representatives; (b) To vote and to be elected at genuine periodic elections which shall be by universal and equal suffrage and shall be held by secret ballot, guaranteeing the free expression of the will of the electors; (c) To have access, on general terms of equality, to public service in his country.

Question 20.   See Article 25 ICCPR above.

Question 21.   Article 2 UDHR: Everyone is entitled to all the rights and freedoms set forth in this Declaration, without distinction of any kind, such as race, colour, sex, language, religion, political or other opinion, national or social origin, property, birth or other status. . .

Question 22.   Article 23(2) UDHR: Everyone, without any discrimination, has the right to equal pay for equal work.

Question 23.   See Article 23(2) UDHR above.

Question 24.   See Article 19 UDHR above.

Question 25.   See Article 19 UDHR above.

Question 26.   See Article 19 UDHR above.

Question 27.   Article 10 UDHR: Everyone is entitled in full equality to a fair and public hearing by an independent and impartial tribunal, in the determination of his rights and obligations and of any criminal charge against him.

Question 28.   Article 8 ICESCR: The States Parties to the present Covenant undertake to ensure (a) The right of everyone to form trade unions and join the trade union of his own choice, subject only to the rules of the organisation concerned, for the promotion and protection of his economic and social interests.

Question 29.   Article 15 UDHR: (1) Everyone has the right to a nationality. (2) No one shall be arbitrarily deprived of his nationality nor denied the right to change his nationality.

Question 30.   Article 11(1) UDHR: Everyone charged with a penal offence has the right to be presumed innocent until proved guilty according to law in a public trial at which he has had all the guarantees necessary for his defence.

Question 31.   Article 14(3) ICCPR: In the determination of any criminal charge against him, everyone shall be entitled to the following minimum guarantees, in full equality: (d) To be tried in his presence, and to defend himself in person or through legal assistance of his own choosing; to be informed, if he does not have legal assistance, of this right; and to have legal assistance assigned to him, in any case where the interests of justice so require, and without payment by him in any such case if he does not have sufficient means to pay for it.

Question 32.   See Article 10 UDHR above.

Question 33. Article 9(3) ICCPR: Anyone arrested or detained on a criminal charge shall be brought promptly before a judge or other officer authorized by law to exercise judicial power and shall be entitled to trial within a reasonable time or to release. . .

Question 34. Article 17(1) ICCPR: No one shall be subjected to arbitrary or unlawful interference with his privacy, family, home or correspondence, nor to unlawful attacks on his honour and reputation.

Question 35. Article 17 UDHR: (1) Everyone has the right to own property alone as well as in association with others. (2) No one shall be arbitrarily deprived of his property.

Question 36. Article 16(1) UDHR: Men and women of full age, without any limitation due to race, nationality or religion, have the right to marry and to found a family. They are entitled to equal rights as to marriage, during marriage and at its dissolution.

Question 37. See Article 16(1) UDHR above.

Question 38. Article 18(1) ICCPR: Everyone shall have the right to freedom of thought, conscience and religion. This right shall include freedom to have or to adopt a religion or belief of his own choice, and freedom, either individually or in community with others and in public or private, to manifest his religion or belief in worship, observance, practice and teaching.

Question 39. Article 15(1b) ICESCR: The States Parties to the present Covenant recognize the right of everyone: (a) To take part in cultural life; (b) To enjoy the benefits of scientific progress and its applications.

Question 40. See Article 12 UDHR above.

Notes on data and statistics

Population: all figures 1990
Life expectancy: all figures 1990
Infant mortality: all figures 1989
GNP per capita: all figures 1988
Health, military, and education spending as a percentage of GNP (gross national product): 1986

Income per head   In quoting the gross national product (GNP) per head it must be stressed that this may be a misleading indicator of how the wealth of a country is shared. Distinctions between rich and poor and between urban and rural, to mention only two extremes, can be very great. The day-to-day variations in the value of the US dollar in international financial markets mean that it is constantly changing in value against the local currencies.

Ethnic/ethnic minorities   Questions 6 and 23 use the word "ethnic" to cover a possible diversity of race, tribe, clan, or minority.

Corporal punishment   Article 7 of the International Covenant on Civil and Political Rights states that "no one shall be subjected to . . . degrading . . . punishment." In countries where corporal punishment may not strictly be a permitted court sentence but where torture and physical coercion are practiced with the clear complicity of the government, it is brought into the assessment of question 12.

Capital punishment   This is not outlawed by the United Nations but Article 6 of the International Covenant on Civil and Political Rights, after mentioning a number of safeguards for those sentenced to death, states: "Nothing in this article shall be invoked to delay or to prevent the abolition of capital punishment by any State Party to the present Covenant." The careful wording of Article 6 is complemented by the Second Optional Protocol to the International Covenant on Civil and Political Rights aiming at the abolition of the death penalty.

Limited terrorism   In certain countries such as Spain, UK, Irish Republic, France, and Italy, where the human rights situation is relatively satisfactory, area restrictions to combat local terrorist groups are taken into consideration when there is a clear need to protect the population and where the security forces of those countries do not abuse their powers.

Abbreviations   The following have been used: UDHR (Universal Declaration of Human Rights), ICCPR (International Covenant on Civil and Political Rights), ICESCR (International Covenant on Economic, Social and Cultural Rights), WHO (World Health Organization), ILO (International Labour Organization).

Quotation marks   The use of quotation marks under COMMENTS to the country reports usually indicates that the references are from official or important sources.

Question 21   Although women enjoy political and legal equality in many countries, they are always underrepresented in governments and parliaments, and frequently not represented at all. For the purpose of the rating of this question, constitutional equality may not prevent the reality of an overall discrimination.

Question 32   The term "civilian trials" has been used instead of "civil and criminal trials" for the sake of brevity.

Question 36   There are many variations of accepted forms of marriage in Muslim countries, and it has been impossible to offer a brief and standard answer to this question. Intermarriage is always forbidden. Some Muslim countries also require a civil marriage, and there are further distinctions between Arab and non-Arab Muslim countries.

Crime of crimes   An Armenian correspondent in Marseilles has correctly stated that this guide does not try to assess the worst human rights crime of all, *genocide*. True, many instances today are described as such but nothing outside an actual war between nations, which is in a different category, compares with the millions who died in the mass extermination of Armenians, Jews, Cambodians, and the kulaks of the USSR in this century. Crimes of the past, which predate the compiling of this guide, must be judged by history.

Convention on equality for women   In the interests of brevity this refers to the Convention on the Elimination of All Forms of Discrimination against Women.

UN nonmembers   Switzerland and Hong Kong, which is a British crown colony, are not member-states of the United Nations.

Countries omitted   A population cutoff of one million has been applied to the countries covered by the guide. In some instances, however, where countries are in a state of turmoil or boundaries are in the process of being redrawn, a human rights assessment has not been attempted. Such countries include Ethiopia, Liberia, Lebanon, Haiti, and Somalia. A few other countries above the limit have been omitted because the information or the sources of that information did not satisfy the criteria set by the compiler.

Assessment of countries

Information cutoff date

The information in this work covers the period to November 1991

Page	Country	Capital	Human rights rating (%)		
	World average		*(above)*	**62**	*(below)*
	Afghanistan	Kabul			28
	Algeria	Algiers	66		
	Angola	Luanda			27
	Argentina	Buenos Aires	84		
	Australia	Canberra	91		
	Austria	Vienna	95		
	Bangladesh	Dacca			59
	Belgium	Brussels	96		
	Benin	Porto Novo	90		
	Bolivia	Sucre	71		
	Botswana	Gaborone	79		
	Brazil	Brasília	69		
	Bulgaria	Sofia	83		
	Burma (Myanmar)	Rangoon			17
	Cambodia	Phnom Penh			33
	Cameroon	Yaoundé			56
	Canada	Ottawa	94		
	Chile	Santiago	80		
	China	Beijing			21
	Colombia	Bogotá			60
	Costa Rica	San José	90		
	Cuba	Havana			30
	Czechoslovakia	Prague	97		
	Denmark	Copenhagen	98		
	Dominican Republic	Santo Domingo	78		
	Ecuador	Quito	83		
	Egypt	Cairo			50
	El Salvador	San Salvador			53
	Finland	Helsinki	99		
	France	Paris	94		
	Germany	Bonn	98		
	Ghana	Accra			53
	Greece	Athens	87		
	Guatemala	Guatemala City		62	

Honduras	Tegucigalpa	65	
Hong Kong	Victoria	79	
Hungary	Budapest	97	
India	New Delhi		54
Indonesia	Jakarta		34
Iran	Teheran		22
Iraq	Baghdad		17
Irish Republic	Dublin	94	
Israel	Jerusalem	76	
Italy	Rome	90	
Ivory Coast	Abidjan	75	
Jamaica	Kingston	72	
Japan	Tokyo	82	
Jordan	Amman	65	
Kenya	Nairobi		46
Korea, North	Pyongyang		20
Korea, South	Seoul		59
Kuwait	Kuwait		33
Libya	Tripoli		24
Malawi	Lilongwe		33
Malaysia	Kuala Lumpur		61
Mexico	Mexico City	64	
Morocco	Rabat		56
Mozambique	Maputo		53
Nepal	Kathmandu	69	
Netherlands	Amsterdam	98	
New Zealand	Wellington	98	
Nicaragua	Managua	75	
Nigeria	Abuja		49
Norway	Oslo	97	
Oman	Muscat		49
Pakistan	Islamabad		42
Panama	Panama City	81	
Papua New Guinea	Port Moresby	70	
Paraguay	Asunción	70	
Peru	Lima		54
Philippines	Quezon City	72	
Poland	Warsaw	83	
Portugal	Lisbon	92	
Romania	Bucharest	82	
Rwanda	Kigali		48
Saudi Arabia	Riyadh		29
Senegal	Dakar	71	
Sierra Leone	Freetown	67	
Singapore	Singapore		60
South Africa	Pretoria/Capetown		50
Spain	Madrid	87	
Sri Lanka	Colombo		47
Sudan	Khartoum		18

Country	City			
Sweden	Stockholm	98		
Switzerland	Berne	96		
Syria	Damascus			30
Tanzania	Dar es Salaam			41
Thailand	Bangkok		62	
Togo	Lomé			48
Trinidad	Port-of-Spain	84		
Tunisia	Tunis			60
Turkey	Ankara			44
Uganda	Kampala			46
Union of Soviet Socialist Republics (USSR)	Moscow			54
United Kingdom (UK)	London	93		
United States of America (USA)	Washington DC	90		
Uruguay	Montevideo	90		
Venezuela	Caracas	75		
Vietnam	Hanoi			27
Yemen	Sana'a			49
Yugoslavia	Belgrade			55
Zaire	Kinshasa			40
Zambia	Lusaka			57
Zimbabwe	Harare	65		

Information cutoff date
The information in this work covers the period to November 1991

Main sources

Amnesty International
Amnesty International Report 1991
Article 19 World Report 1991 (UK)
Anti-Slavery International (UK)
American Statistical Association (Human Rights Committee)
Reporters Sans Frontières (France)
Statesman's Yearbook 1990 (Macmillan, London)
Harvard Institute for International Development
International Boundaries Research Unit (University of Durham)
Human Rights Watch (USA)
Human Rights Watch Report 1990
US Department of State Country Reports on Human Rights Practices
Minority Rights Group (UK)
Index on Censorship (UK)
International Planned Parenthood Federation
International Labour Organization
United Nations Development Programme 1991 Report
World Bank Publications
The Economist (UK)
Le Monde (France)
The Times (UK)
The Washington Post (USA)
The New York Times (USA)
The Guardian (UK)
Far Eastern Economic Review (Hong Kong)
International PEN Writers in Prison Committee
Europa Yearbook
CH Network (monitors and researchers in certain countries)
Fédération Internationale des Droits de L'Homme (France)
Commonwealth Secretariat publications (London)

World Human Rights Guide

Introduction

Purpose of the guide

The questionnaires that follow cover 40 indicators from the major UN treaties. The selected human rights are those that are able to be accurately defined and measured. Not all of the treaties make this possible. In some the articles are general statements of the obligations of governments or parties to a just society or to higher principles, but the questionnaire of the guide adheres strictly to human rights which can be clearly assessed. The four graded evaluations of the YES/yes/no/NO method, which has proved in previous editions to be accurate and satisfactory, are again used.

The general reader may wonder why the United Nations, the organization best qualified to undertake such an exercise, does not issue for the public periodic assessments of the human rights performances of its member-states. Such an omission on the part of the international body is explained by the fact that most of the individual governments are against such an authoritative disclosure of human rights violations, the reasons for which will be evident on a closer study of the questionnaires.

From time to time other attempts to measure human rights performances have been undertaken, but these have been confined to distinct categories of rights, limited regions of the world, or specific causes such as the fate of oppressed minorities. A worldwide survey is regularly undertaken by Freedom House of the USA, but this, while assessing most countries, does not attempt a clear article-by-article evaluation. Major world reviews are also annually published by Amnesty International, Human Rights Watch, Article 19, and the US government, though without attempting a classification of countries.

The purpose of the *World Human Rights Guide* is to correct this omission. It is also intended, without lengthy explanations or wordiness, to present the facts in an easily understood form for those who may not be familiar with the subject. Hundreds of millions of the world's population, perhaps more, are unaware of their rights. The guide is therefore published with a second purpose in mind. An informed population is more likely to demand its rights, and to achieve this legitimate purpose, than those kept in ignorance by regimes which, though paying lip service to their obligations as members of the United Nations, seek to deny them.

The importance of monitoring

The compilation of the guide is part of the worldwide exercise of monitoring. The limited authority of the United Nations and its inability to impose on its member-states respect for its own treaties and principles have meant that public knowledge of human rights violations, atrocities, and crimes comes almost entirely from other sources. Apart from the better-known international monitors such as Amnesty International and Human Rights Watch, many countries have their own active monitoring groups, some effective enough to bring meaningful pressure on their governments. But there are countries where these local monitoring groups are not only forbidden, their members are in danger of being assassinated, tortured, or held in prison indefinitely.

As will be observed from the questionnaires that follow, human rights monitors may suffer this fate, either from the security forces of a country or from terrorists, in over 20 member-states of the United Nations. But there is an aspect of this "on-the-ground" or "frontline" form of monitoring that cannot be crushed or intimidated, and that is by the international corps of journalists and correspondents of the world's media. No tyrant, no regime, is safe from these witnesses to their violations, reports and pictures of which can appear almost simultaneously on the television screens and in the newspapers of the most distant parts of the world – as was seen in the information that came out of Baghdad during the recent gulf war.

Monitoring has therefore proved to be the most effective way of applying pressure to regimes perpetuating the most evil human rights crimes. Not even the most powerful of tyrants is indifferent to the details of his dark secrets becoming that day's feature of the world's press and television.

It is to this form of pressure that the guide adds its modest contribution.

Human rights – what are they?

Human rights are the laws, customs, and practices that have evolved over the centuries to protect ordinary people, minorities, groups, and races from oppressive rulers and governments. The greatest number of advances has probably come as the result of wars, rebellions, and violence of one kind or another, but at certain historically significant moments improvements in the area of human rights have been introduced, or at least codified, by charters that now form the basis of modern rights, particularly as they are framed by UN instruments.

The most important of these foundation charters were the Magna Carta imposed on King John of England in 1215 by his barons, the French Déclaration des Droits de l'Homme et du Citoyen of 1789, and the American Bill of Rights of 1791. Nearly 130 years later, after the First World War, human rights were given an international dimension when some were incorporated into the conventions of the League of Nations. Established mainly to prevent future wars, which it failed to do, the League of Nations nevertheless had a number of humanitarian successes in the areas of labor conditions, slavery, and health.

In 1945, after the Second World War, the League was superseded by the United Nations. From an initial membership of 51 countries, the international organization today has expanded to include 175 states, and one of its major concerns is respect for human rights. Beginning with the Charter of 1945, the instruments of the United Nations now cover all aspects of life, from civil and political rights to economic and cultural rights, and include specific conventions to further ensure the rights of children, of women, of married women, of exploited prostitutes, of the stateless, and of refugees.

There are three major UN human rights instruments. The first is the Universal Declaration of Human Rights (UDHR), which was adopted in 1948 without a dissenting vote and, although not considered to be binding at the time, it has now become part of customary international law. To reinforce it, however, two major covenants setting out more specifically certain categories of human rights were adopted in 1966 and came into force after they had been ratified by 35 countries in 1976.

Together with the UDHR, the International Covenant on Civil and Political Rights (ICCPR) and the International Covenant on Economic, Social and Cultural Rights (ICESCR) form the basis of the questionnaire used in the guide.

The structure of the guide

The guide covers 104 countries with populations over one million. The questions are laid out as a simple checklist. Can one travel freely in one's own country? Is there the risk of extrajudicial killing? Is there indefinite detention without trial? Do women enjoy equal rights? To ensure that the 40 indicators of the questionnaire are consistent with the purpose of the guide, and are not regarded as the arbitrary choice of the compiler, each is drawn from articles of the three major UN instruments listed above. The pages on the basis of the questionnaire set out the sources in detail.

The evaluation of each indicator of the questionnaire, which eventually establishes a country's rating, is in two stages. The first is to gather information on human rights in different countries. The main sources are listed on page xxi. These are supplemented by a network of international correspondents which regularly informs the compiler of developments in their areas. A limited measure of help has been received from embassies and high commissions in London but information from these sources, which are unlikely to be objective, is always treated with caution. In some instances completed questionnaires have been received which read like a description of paradise.

The assembled material for each indicator is then graded into four categories or levels. This number was chosen after many different alternatives were tried and evaluated, and has remained the compiler's preferred system since the first guide was published in 1983. The grades or categories are indicated on the questionnaires as YES, yes, no, and NO, details of which are explained below.

YES represents the category of unqualified respect for the freedoms, rights, or guarantees of the article or indicator of the questionnaire.

 qualifies otherwise satisfactory answers on the grounds of occasional breaches of respect for the freedoms, rights, or guarantees of the article or indicator of the questionnaire.

 indicates frequent violations of the freedoms, rights, or guarantees of the article or indicator of the questionnaire.

NO indicates a constant pattern of violations of the freedoms, rights, or guarantees of the article or indicator of the questionnaire.

After careful assessment, the collected information receives points on the basis of 3 points for YES, 2 for yes, 1 for no and none for NO. The maximum total for the 40 questions, disregarding the question of weighting which is discussed below, is therefore 120 points which, when converted, becomes a human rights rating of 100%.

Two other factors should be mentioned in determining the accuracy of the results. They are the extent of consultations with leading human rights organizations, and the importance to the compiler of their "on the ground" conclusions and, second, the fact that a spread of forty questions minimizes any possible distortions to the final country ratings.

The refinement of weighting

The equal treatment of all questions on the basis of the above ratings has, however, the obvious disadvantage of leveling all 40 of the human rights to a uniform seriousness. A system of weighting has therefore been adopted. This will help to correct – though hardly to a point of perfection – the differences between the gravity of the various human rights.

Any method of weighting 40 questions, and arranging them in order of degrees of oppression, inevitably raises the issue of subjective judgments. Individuals who are devoutly religious, for example, might place freedom of religion above censorship of the press. Women will naturally feel more strongly about equal rights for their sex; ethnic groups, about oppression by the dominant section of a country. The infinity of possible variations therefore made it necessary to find a simple form of weighting that would stand up to the test of what constitutes the worst atrocities "against the person.'

The problem – and challenge – has been resolved by making a clear distinction between physical suffering inflicted directly on the individual and the denial of political and social rights. An individual screaming while subjected to torture or locked for years in a black cell because of his or her opinions or dissent is enduring a degree of physical or mental suffering greater than the denial of a free vote or of having his or her newspapers censored.

A number of purists, usually statisticians who find that weighting introduces a further degree of subjectivity into a field that ideally should be free of such complexities, are against what the compiler regards as a straightforward exercise of common sense, but on balance his original conviction on the subject is unaltered.

The weighted questions are numbers 7–13. They were finally weighted by a factor of 3.0. Experiments proved that weighting values in the range 2.0–5.0 typically alter the final human rights rating by only ±3–4%. Results are therefore relatively independent of the exact figure used and of the compiler's value judgments. Further, the results are also robust in that, for the same range of weights (factors 2.0–5.0), the lists after numerous tests retain the same groupings of the 10 countries with the "highest" and "lowest" ratings; only their order within the groupings fluctuates.

The example below illustrates how the rating is reached. For reference, the maximum possible point score for the 40 questions *after weighting* but before conversion to a percentage is 162 or $(33 \times 3) + (7 \times 3 \times 3)$.

MALAYSIA
This is one of the middle band of countries and it has a rating of 61%. This is based on 69 points from the 33 nonweighted questions plus 30 points from the 7 weighted questions. The rating for Malaysia is therefore:

$$\frac{\text{Nonweighted total } 69 + (\text{total to be weighted } 10 \times 3) \times 100}{162} = 61\%$$

For and against

Past editions of the guide have evoked some controversy. Many feel that the measurement of human rights should not be attempted because it will always be a subjective exercise, relying on information that can never be totally reliable, and because countries are at different stages of development, a fact which will certainly influence their societies. As well as the standard of living or the level of prosperity of a country, the critics claim that classifications of human rights ignore traditions and customs, religion and history, all of which will affect the final results.

A further criticism is that the 40 indicators of the questionnaire are mainly from articles of the UDHR and the ICCPR rather than the more "collective" rights covered by the ICESCR. It is claimed, therefore, by those giving priority to the ICESCR, that one-party systems may be doing more for their people in the areas of health, education, and social matters at the present juncture of their development than could ever be achieved by a democratic system that might lack the power to mobilize and direct its people.

There is some substance to these criticisms. How does one answer them? One must begin by reminding critics that all member-states of the United Nations are obliged to uphold the Purposes and Principles of the UN Charter, of which Article 55(c) calls for "universal respect for, and observance of, human rights and fundamental freedoms for all without distinction as to race, sex, language or religion." To complain of different stages of development, of the unfairness of comparisons, overlooks the fact that the binding treaties may be criticized for appearing to favor the more prosperous countries but they form the "rules of the game" by which all states have chosen to play.

The guide simply sets their performance against their obligations. How do they compare? If the information gathered is accurate, and the method of measurement reliable, then criticism must be against the exercise itself rather than the results. But how tenable are these results? As has been mentioned elsewhere, the guide was chosen as the basis for the Human Freedom Index of the Human Development Report 1991 of the United Nations Development Programme. Among the technical notes set out in the report, the choice is justified as follows:

> The strengths of the Humana index are threefold. Each question is based on an internationally recognized human right. It has clear and reproducible computational procedures. And the final scale has good discrimination. The ranked countries are fairly evenly distributed all the way along the scale 0% to 100%, and not simply grouped into "good" at one end and "bad" at the other.

But the report of the UNDP is not alone in its approval of ratings and the system of measurement. In 1985, Dr. David L. Banks, now of the Carnegie-Mellon University, USA, presented a paper to the American Statistical Association on the measurement of human rights which featured an earlier edition of the Humana guide. In his conclusions, he stated:

> The first conclusion is that there is sufficient reliability in human rights scoring to disarm accusations that the subject is intrinsically incapable of measurement. No doubt there are strong cultural and individual differences in our working concept of human freedom but for practical purposes different people find substantial agreement in their ratings.

Similar approval of the purpose and method of the guide has been stated by other distinguished sources, and the accuracy of the ratings and the classification has usually been challenged only by those who are most guilty of human rights violations. The most common defense of such erring regimes is a legalistic one, namely the quoting of Article 2(7) of the UN Charter, which reads:

> Nothing contained in the present Charter shall authorize the United Nations to intervene in matters which are essentially within the domestic jurisdiction of any state.

In other words, human rights atrocities committed by governments in their own countries do not concern the rest of humanity. There is, however, increasing support for the concept of the droit d'ingérance or the right of the international community to intervene in the internal affairs of a state in exceptional humanitarian circumstances.

To return to another criticism of the composition of the questionnaire, the fact that it makes little use of the ICESCR, it should be explained that this covenant is concerned with broader social and economic questions and that the articles usually refer to vague guarantees such as "recognising the right of" or "taking steps

towards" respecting a particular human right.
  Article 12(2) of the ICESCR, for example, states:

> The steps to be taken by the States Parties to the present Covenant to achieve the full realisation of this right (health) shall include those necessary for
> (a) The provision for the reduction of the stillbirth-rate and of infant mortality and for the development of the child;
> (b) The improvement of all aspects of environmental and industrial hygiene;
> (c) The prevention, treatment and control of epidemic, endemic, occupational and other diseases;
> (d) the creation of conditions which would assure to all medical service and medical attention in the event of sickness.

Since promises and aspirations cannot be measured, and in practice each country will be reporting on its own progress, the questionnaire could make only limited use of the articles of the ICESCR.

Human rights, a Western liberal concept

Opposition to the measuring of rights and classifying countries also centers on the perception that the purpose has a Western liberal bias. This is undeniably true. The human rights treaties of the United Nations reflect Western liberal traditions and values because they were fundamental to the founding of the world organization. Its treaties are drawn from the precedents of earlier Western treaties, and what can be described as the international political and social morality of the second half of the 20th century therefore has a similar origin.

Since the formation of the United Nations a number of regional treaties have been drawn up. They include those of the Organization of African Unity, the Organization of American States, the Council of Europe, and a looser association of Arab states. Although their aims and articles are not identical, the fact remains that the model for all of them, to a lesser or greater degree, has been the same as that which formed the basis of the UN international treaties.

But there is a major contradiction within the United Nations. Many countries which enjoy full membership and participate in all its activities adhere to traditions, religions, and beliefs which cannot really coexist with Western liberal ideas. This conflict has been resolved by a much-practiced exercise in tolerance, obliviousness, diplomacy, and insincerity, which permits states that cannot and do not honor their obligations to UN instruments to continue in their own way. For example, 30 countries are regarded as totally Muslim in character or have considerable Muslim minorities which influence their governments. Religious law and their traditions take precedence over all other considerations, and their freedom to follow this course is never disputed. Other exceptions, which evoke little censure, apply to many African countries, one-party states, and, occasionally, even to UN members actually at war with each other.

The questionnaire of this guide has not been constructed to conform to standards and practices which might be most acceptable to countries which fail to honor their treaty obligations. If the indicators have to bear the label of being "Western liberal" or those suited to "prosperous countries," or even of being "unlawfully intrusive," then the guide will have to live with such criticisms.

Women, the oppressed half of humanity

It is widely assumed that the progress made toward women's equality in Western and Communist countries has achieved its purpose and that the two sexes enjoy relatively similar rights. But when is equality rather less than equality? Not a single country has escaped the fact that traditions cannot be legislated away, that customs do not change overnight, and that the lifelong conditioning of men and women to their respective roles defies quick solutions.

A clear example of this is the underrepresentation of women in government. In the countries of Scandinavia, where women's equality is undisputed, government posts and Parliaments are largely dominated by men. Even more obviously, English-speaking democracies have governing assemblies in which women rarely achieve a 10% representation. A woman president or prime minister in Western Europe is regarded as exceptional, perhaps a phenomenon, while in the USA, where women appear to claim greatest equality, such an elevation is virtually impossible.

The same contradiction in practice is evident in pay and employment. Even where "equality boards" and ombudsmen have been appointed, discrimination cannot be totally eliminated, and an overall figure for pay differential shows that in many Western democracies women may earn only 70% of men's pay. The situation is similar with marriage and divorce. Even where developed countries offer protection in marriage and divorce to nonwage-earning wives, the fact that the husband is the wage earner gives him a financial advantage that is a reality outside the competence of the law.

Most complex of all, however, is women's position under Islamic Shari'a law. The application of this varies from one Muslim country to another but nearly always deals with marriage, divorce, inheritance, and "family" matters. There is no possibility of UN treaties on women's rights prevailing against a religion and tradition that demand total obedience and conformity. And an explanation, indisputable to those who put it forward, simply asserts that in Muslim societies women enjoy the honored status of the divine role for which they were created, that of being wife and mother.

As has already been stated, the guide does not seek to judge such beliefs, whether religious or customary, but simply sets them against the international obligations of their governments and countries. It also records without comment that the Convention on the Elimination of All Forms of Discrimination against Women has been ratified by over 100 member-states, including Indonesia, Yemen, and Bangladesh, but not by the USA, India, or Israel (as of 31 December 1990).

Conclusion

The Preface and the Progress Report provide evidence of the major human rights improvements since the previous update of the guide in 1986. But the contribution of the United Nations to this transformation has only been forthcoming when events, pressures, or self-interest have forced it to act. Explanations are given that initiatives are circumscribed by principles, laws, and the mandates of its various treaties, but all these could be overcome if there was a general will to do so.

The limitations of the United Nations therefore underline the crucial part played in human rights improvements by nongovernmental monitoring organizations. Evidence of the power of the spread of information, of people becoming aware of what they are denied, can be most clearly illustrated by what the compiler of this guide calls regional contagion. Over the last few years, there has been a worldwide spilling over frontiers of a demand for free and democratic societies. Once a country has achieved its freedom, a neighboring oppressive regime is immediately under threat.

An example that illustrates the fear of such a threat is the surprising survival of the Saddam Hussein regime after the gulf war. If the end of his regime had been followed by a multiparty democracy, how welcome would such a development have been to the rulers of neighboring Saudi Arabia and Kuwait, Syria, Iran, and

Jordan? The resistance to the contagion of multiparty democracy – a campaign, however, conducted without publicity – was as determined as any mounted against the Iraqi invaders.

But regional contagion continues to make progress elsewhere. In South America, for example, in the 1980s, one country after another overturned military regimes. On another continent, in southern Africa, a limited period of time has seen Namibia gain independence, South Africa move away from apartheid in face of international condemnation, Angola and Mozambique end their civil wars, and, most recently of all, a free election in Zambia result in a multiparty democracy. But an even more remarkable example of regional contagion has been seen in Eastern Europe, including, one after another, the republics of the USSR.

To all these changes, monitoring has made its contribution. But nothing that is done by nongovernmental human rights bodies could equal the similarly active participation of the United Nations. At the 39th (1984) session of the General Assembly, the delegation from Ecuador pleaded that "universality should be shown with regard to all aspects of the concept (human rights law): in the compilation of information, in the gathering and dissemination of facts, in the judgement of situations and in the search for solutions to problems. . . . Therefore we believe that only universality can give validity and authority to human rights." This intervention by Ecuador went on to call for "an annual worldwide report on the progress made in the implementation of Human Rights in each and every country."

Seven years later, the plea is as urgent as ever. A minimum duty for the international body is to overcome the opposition within its membership and "gather and disseminate" the details of violations on a country-by-country basis. And then to make them public.

Afghanistan

Human rights rating: 28%

YES 4 yes 3 no 15 NO 18

Population: 16,600,000
Life expectancy: 42.5
Infant mortality (0–5 years)
  per 1,000 births: 296
United Nations covenants ratified:
Civil and Political Rights, Economic,
Social and Cultural Rights

Form of government: One-party system
GNP per capita (US$): 200
% of GNP spent by state:
On health: n/a
On military: n/a
On education: 1.8

FACTORS AFFECTING HUMAN RIGHTS
The country remains divided between a Communist-dominated government that was
kept in power by Soviet military forces and the *mujahideen* or rebel groups. The latter
are composed of Islamic factions ranging from fundamentalists to those with more
secular policies. With the departure of Soviet forces in 1989, after 10 years of
considerable slaughter, the fighting has lessened and moves toward restoring the unity
of the country are taking place. Human rights are widely violated by all sides, and
individuals regarded as hostile may expect summary treatment.

	FREEDOM TO:		COMMENTS
1	Travel in own country	no	As well as the dangers of traveling between areas controlled by government and rebel forces, other threats include minefields and roadblocks. These checkpoints may extract "payments" from travelers
2	Travel outside own country	no	Only for a favored few who have influence with the authorities in their respective regions of the country
3	Peacefully associate and assemble	no	Despite promises of constitutional reforms toward a more democratic society, the government suppresses all public protests and rallies against its authority. Constant surveillance of suspect associations
4	Teach ideas and receive information	no	Strictly within government guidelines. Most distinguished academics have fled the country
5	Monitor human rights violations	yes	The present central government permits some investigations by international bodies but the continuing civil war creates major problems for monitors
6	Publish and educate in ethnic language	YES	Rights respected

FREEDOM FROM:		**COMMENTS**
7	Serfdom, slavery, forced or child labor	no	Minimal government control. The civil war involves children of most ages, including some in military service
8	Extrajudicial killings or "disappearances"	**NO**	Many hundreds of killings. By both government and the *mujahideen* armed opposition, who control major areas of the country
9	Torture or coercion by the state	**NO**	Widespread torture by the warring sides of a bitterly divided country. Beatings, burnings, electric shocks, food deprivation, etc.
10	Compulsory work permits or conscription of labor	no	The war situation permits emergency conscription, the individuals so drafted being described as "volunteers"
11	Capital punishment by the state	**NO**	By firing squad
12	Court sentences of corporal punishment	yes	The extent of arbitrary violence by the authorities, and the difficulties of monitoring court sentences, make a qualified assessment necessary
13	Indefinite detention without charge	**NO**	By both government and the *mujahideen*. Thousands detained. Temporary camps supplement prisons
14	Compulsory membership of state organizations or parties	**YES**	Rights respected
15	Compulsory religion or state ideology in schools	no	Schools under both government and the *mujahideen* subject to indoctrination of children
16	Deliberate state policies to control artistic works	no	The need of the central government to maintain its authority, and the flight from the country of many of its creative individuals, restricts or inhibits free expression
17	Political censorship of press	**NO**	Total censorship though the government claims to be respecting the amended constitution. A token opposition press has limited circulation
18	Censorship of mail or telephone tapping	**NO**	Wide interception of correspondence. All forms of communication under surveillance

FREEDOM FOR OR RIGHTS TO:		**COMMENTS**
19	Peaceful political opposition	**NO**	Civil war since the Communist coup of 1978
20	Multiparty elections by secret and universal ballot	no	Progress toward multiparty elections limited to statements of intent and some constitutional amendments

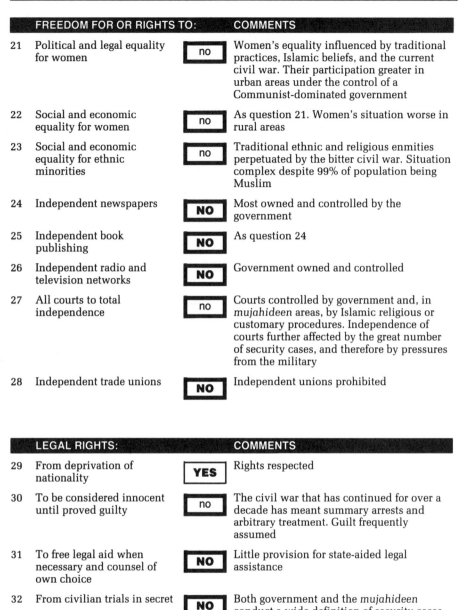

FREEDOM FOR OR RIGHTS TO:		COMMENTS
21	Political and legal equality for women	no	Women's equality influenced by traditional practices, Islamic beliefs, and the current civil war. Their participation greater in urban areas under the control of a Communist-dominated government
22	Social and economic equality for women	no	As question 21. Women's situation worse in rural areas
23	Social and economic equality for ethnic minorities	no	Traditional ethnic and religious enmities perpetuated by the bitter civil war. Situation complex despite 99% of population being Muslim
24	Independent newspapers	NO	Most owned and controlled by the government
25	Independent book publishing	NO	As question 24
26	Independent radio and television networks	NO	Government owned and controlled
27	All courts to total independence	no	Courts controlled by government and, in *mujahideen* areas, by Islamic religious or customary procedures. Independence of courts further affected by the great number of security cases, and therefore by pressures from the military
28	Independent trade unions	NO	Independent unions prohibited

LEGAL RIGHTS:		COMMENTS
29	From deprivation of nationality	YES	Rights respected
30	To be considered innocent until proved guilty	no	The civil war that has continued for over a decade has meant summary arrests and arbitrary treatment. Guilt frequently assumed
31	To free legal aid when necessary and counsel of own choice	NO	Little provision for state-aided legal assistance
32	From civilian trials in secret	NO	Both government and the *mujahideen* conduct a wide definition of security cases in secret
33	To be brought promptly before a judge or court	NO	The number of security cases makes respect for normal judicial procedures unlikely. The "state-of-war" situation has meant that the *mujahideen* continue to hold Soviet prisoners despite that country's ending of military support to the government
34	From police searches of home without a warrant	NO	The state-of-war conditions permit arbitrary searches

LEGAL RIGHTS:		COMMENTS
35	From arbitrary seizure of personal property	**NO**	The armed forces of both combatants seize, occupy, and expropriate. The only consistent pattern is its consistency

PERSONAL RIGHTS:		COMMENTS
36	To interracial, interreligious, or civil marriage	**NO**	In a country almost totally Muslim, interreligious marriage forbidden
37	Equality of sexes during marriage and for divorce proceedings	**NO**	Government changing constitution in favor of Muslims, whose support they now need. This means a return to Islamic traditions which favor husband in both marriage and divorce
38	To practice any religion	yes	But no proselytizing in a country 99% Muslim
39	To use contraceptive pills and devices	**YES**	Rights respected
40	To noninterference by state in strictly private affairs	no	Wide interference as required by the military situation. Villages destroyed, inhabitants transported to other areas

Algeria

Human rights rating: 66%

YES 13 **yes** 16 **no** 10 NO 1

Population: 25,000,000
Life expectancy: 65.1
Infant mortality (0–5 years)
  per 1,000 births: 102
United Nations covenants ratified:
Civil and Political Rights, Economic,
Social and Cultural Rights

Form of government: One-party system
GNP per capita (US$): 2,360
% of GNP spent by state:
On health: 2.2
On military: 1.9
On education: 6.1

FACTORS AFFECTING HUMAN RIGHTS

The intention to hold multiparty elections in 1991, after virtually one-party rule since the country gained independence in 1962 from France, was briefly affected by violent demonstrations by Islamic fundamentalists and the declaring of martial law. As of November 1991, the situation had been restored and the elections took place at the end of 1991. The victory of the fundamentalists, however, caused a state of emergency to be decreed and government was assumed by the High Council of State supported by the military.

	FREEDOM TO:		COMMENTS
1	Travel in own country	**YES**	Rights respected
2	Travel outside own country	yes	Travel abroad impeded by long delays in issuing passports but exit visas discontinued
3	Peacefully associate and assemble	yes	Swift police response to demonstrations that become a threat to law and order, a situation becoming more prevalent with the increase in militant Muslim fundamentalism. Police violence also against rallies of feminist movement
4	Teach ideas and receive information	yes	The increase of Muslim fundamentalism in universities has meant rise in tension and surveillance of academics and students
5	Monitor human rights violations	yes	Greater tolerance of human rights investigations. New monitoring groups being formed
6	Publish and educate in ethnic language	yes	Fewer complaints by Berber minority (25% of population) of language disadvantages and cultural discrimination

FREEDOM FROM:		COMMENTS
7	Serfdom, slavery, forced or child labor	yes	Regional pattern of child labor, particularly in rural areas
8	Extrajudicial killings or "disappearances"	**YES**	Rights respected
9	Torture or coercion by the state	no	Beatings and protracted brutalities. Recent reforms have reduced these but perpetrators are seldom charged. Human rights groups investigating violations
10	Compulsory work permits or conscription of labor	yes	Evidence of limited forced labor of prison inmates
11	Capital punishment by the state	**NO**	By firing squad
12	Court sentences of corporal punishment	**YES**	Rights respected
13	Indefinite detention without charge	yes	Position improved but the mid-1991 martial law declaration permitted detentions without charge
14	Compulsory membership of state organizations or parties	**YES**	Rights respected
15	Compulsory religion or state ideology in schools	**YES**	Rights respected
16	Deliberate state policies to control artistic works	**YES**	Rights respected
17	Political censorship of press	no	Progress toward a freer press was temporarily interrupted by declaration of martial law. A few leading foreign journals proscribed
18	Censorship of mail or telephone tapping	no	Proposed reforms of the intelligence services delayed by state of emergency. Surveillance continues, particularly of Muslim fundamentalists

FREEDOM FOR OR RIGHTS TO:		COMMENTS
19	Peaceful political opposition	yes	New constitution (1989) guaranteeing establishment of new parties temporarily affected by state of emergency
20	Multiparty elections by secret and universal ballot	no	The multiparty elections held at the end of 1991 were followed by the declaration of a one-year state of emergency
21	Political and legal equality for women	no	Underrepresentation at all levels of government and political life. The right of husbands to place proxy votes for wives declared unconstitutional in October 1991

FREEDOM FOR OR RIGHTS TO:		COMMENTS
22	Social and economic equality for women	**no**	Pay and employment inequalities. The patriarchal tradition toward women persists. Position worsened by growing Muslim fundamentalism
23	Social and economic equality for ethnic minorities	yes	Berber minority enjoying improved social status though areas of discrimination persist
24	Independent newspapers	yes	Progress toward an independent press includes transfer of government-owned newspapers to the leading political party and permission for many new newspapers to be published by recently formed political parties
25	Independent book publishing	yes	The limitations of book publishing usually apply to Islamic conformism. Some authors forced to publish in France
26	Independent radio and television networks	**no**	Stateowned and controlled but less overall bias with the government's move to democratization
27	All courts to total independence	yes	Occasional pressures on judiciary by senior ministers, particularly when cases involve fundamentalist violence. Some corruption
28	Independent trade unions	yes	New law permits formation of unions but no affiliation with political parties. Limitations on strike action

LEGAL RIGHTS:		COMMENTS
29	From deprivation of nationality	**YES**	Rights respected
30	To be considered innocent until proved guilty	**YES**	Rights respected
31	To free legal aid when necessary and counsel of own choice	**YES**	Free legal service by Algerian Bar Association
32	From civilian trials in secret	**no**	Situation varies with periodic proclamations of martial law
33	To be brought promptly before a judge or court	yes	Within 48 hours but many delays. Also affected by a degree of corruption
34	From police searches of home without a warrant	yes	Occasional abuses. More frequent during militant agitation by Muslim fundamentalists
35	From arbitrary seizure of personal property	**YES**	Rights respected

PERSONAL RIGHTS:		COMMENTS
36	To interracial, interreligious, or civil marriage	**no**	No interreligious marriage for practicing Muslims (99% of population). Those risking intermarriage in recent past have had passports confiscated
37	Equality of sexes during marriage and for divorce proceedings	**no**	Muslim traditions and practices favor husband both for marriage and divorce. Polygamy not illegal
38	To practice any religion	**YES**	Rights respected
39	To use contraceptive pills and devices	**YES**	State support
40	To noninterference by state in strictly private affairs	**YES**	Rights respected

Angola

Human rights rating: 27%

YES 6 yes 3 no 12 NO 19

Population: 10,000,000
Life expectancy: 45.5
Infant mortality (0–5 years)
  per 1,000 births: 292
United Nations covenants ratified:
Convention on Equality for Women

Form of government: One-party system
GNP per capita (US$): 870
% of GNP spent by state:
On health: 1.0
On military: 12
On education: 3.4

FACTORS AFFECTING HUMAN RIGHTS

Significant progress toward unifying the country, following 16 years of civil war, has been made. One-quarter of Angola has been controlled by UNITA, a strongly armed rebel movement, and advances toward the restoration of peace include plans for a multiparty political system. Elections are scheduled for 1992, after being postponed from 1991. The long period of fighting has meant little respect for human rights, but the continuous sequence of summary killings by all factions, of torture and general lawlessness, has moderated with the prospect of reconciliation.

	FREEDOM TO:		COMMENTS
1	Travel in own country	no	Limited by security considerations. The rebel movement, UNITA, designates the whole country as a "war zone"
2	Travel outside own country	no	Overseas travel only with approval of the authorities. Some easing of travel between Angola and Portugal, the previous colonial ruler
3	Peacefully associate and assemble	NO	Nothing that conflicts with the policies and security requirements of the government
4	Teach ideas and receive information	NO	As question 3
5	Monitor human rights violations	NO	No cooperation with international monitors. Inquiries and requests usually ignored
6	Publish and educate in ethnic language	YES	Five major language groups. The present division of the country due to civil war limits information. The official language, however, is Portuguese

	FREEDOM FROM:		COMMENTS
7	Serfdom, slavery, forced or child labor	no	Regional pattern of child labor. Both government and rebel soldiers, controlling over a quarter of the country, draft local labor when required

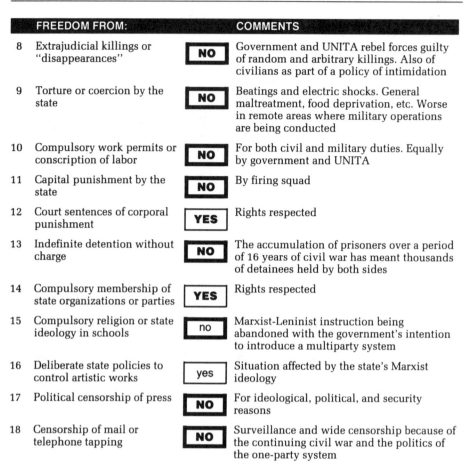

FREEDOM FROM:		COMMENTS
8	Extrajudicial killings or "disappearances"	**NO**	Government and UNITA rebel forces guilty of random and arbitrary killings. Also of civilians as part of a policy of intimidation
9	Torture or coercion by the state	**NO**	Beatings and electric shocks. General maltreatment, food deprivation, etc. Worse in remote areas where military operations are being conducted
10	Compulsory work permits or conscription of labor	**NO**	For both civil and military duties. Equally by government and UNITA
11	Capital punishment by the state	**NO**	By firing squad
12	Court sentences of corporal punishment	**YES**	Rights respected
13	Indefinite detention without charge	**NO**	The accumulation of prisoners over a period of 16 years of civil war has meant thousands of detainees held by both sides
14	Compulsory membership of state organizations or parties	**YES**	Rights respected
15	Compulsory religion or state ideology in schools	no	Marxist-Leninist instruction being abandoned with the government's intention to introduce a multiparty system
16	Deliberate state policies to control artistic works	yes	Situation affected by the state's Marxist ideology
17	Political censorship of press	**NO**	For ideological, political, and security reasons
18	Censorship of mail or telephone tapping	**NO**	Surveillance and wide censorship because of the continuing civil war and the politics of the one-party system

FREEDOM FOR OR RIGHTS TO:		COMMENTS
19	Peaceful political opposition	**NO**	Not permitted
20	Multiparty elections by secret and universal ballot	**NO**	An intention to introduce a multiparty system during 1991 has made little progress. Now planned for late 1992
21	Political and legal equality for women	no	Despite constitutional equality, underrepresentation at senior government levels and in the professions
22	Social and economic equality for women	no	Inequalities compounded by customary law. Women provide most agricultural labor
23	Social and economic equality for ethnic minorities	no	The two sides in the civil war draw support from different ethnic groups, which influences the degree of equality
24	Independent newspapers	**NO**	State control. Similar situation in rebel-held territories

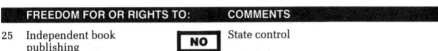

	FREEDOM FOR OR RIGHTS TO:		COMMENTS
25	Independent book publishing	**NO**	State control
26	Independent radio and television networks	**NO**	State control
27	All courts to total independence	no	Subject to political and military controls, despite guarantees in the constitution. Some corruption
28	Independent trade unions	**NO**	The trade union movement is controlled by the governing political party

	LEGAL RIGHTS:		COMMENTS
29	From deprivation of nationality	**YES**	Rights respected
30	To be considered innocent until proved guilty	no	The bitter civil war, begun in 1976, encourages arbitrary decisions by police and security forces
31	To free legal aid when necessary and counsel of own choice	no	Situation varies at local and national levels. Information limited but little evidence of concern for defendant's rights
32	From civilian trials in secret	**NO**	Situation arbitrary. Regional military councils empowered to hold secret trials
33	To be brought promptly before a judge or court	**NO**	Wide interpretation of security offenses permits 3 months' detention, which can be indefinitely renewed. After 6 months, however, the accused must be informed of the offense
34	From police searches of home without a warrant	**NO**	No restrictions on arbitrary searches, which are justified by security considerations
35	From arbitrary seizure of personal property	no	Despite constitutional guarantees, the continuing civil war permits the military and police forces to confiscate when necessary. Degree of corruption

	PERSONAL RIGHTS:		COMMENTS
36	To interracial, interreligious, or civil marriage	**YES**	Rights respected
37	Equality of sexes during marriage and for divorce proceedings	yes	Traditional practices rather than constitutional rights prevail, usually to husband's advantage
38	To practice any religion	yes	The previous hostility toward religion and the church is moderating as the government seeks support. Seized properties being returned
39	To use contraceptive pills and devices	**YES**	Rights respected

40 To noninterference by state   The continuing civil war, with harassment
    in strictly private affairs   no and intimidation by the authorities, presents
                                       a constant threat to family life and private
                                       affairs

Argentina

Human rights rating: 84%

YES 28 yes 8 no 4 NO 0

Population: 32,300,000
Life expectancy: 71.0
Infant mortality (0–5 years) per 1,000 births: 36
United Nations covenants ratified: Civil and Political Rights, Economic, Social and Cultural Rights, Convention on Equality for Women

Form of government: Parliamentary government with executive president
GNP per capita (US$): 2,520
% of GNP spent by state:
On health: 1.6
On military: 1.5
On education: 3.3

FACTORS AFFECTING HUMAN RIGHTS

The freely elected government that followed the military junta in 1983 quickly replaced the machinery of dictatorship with a comprehensive system of constitutional democracy. This has extended over the judiciary, the legislature, the media, and the security forces, with an even more impressive change in the respect for human rights. Economic problems, however, accompanied by considerable social unrest and violence from small terrorist groups, have meant that the security forces have resorted to unlawful excesses, including a few arbitrary killings, torture, and opposition to human rights monitoring.

FREEDOM TO: / COMMENTS

1	Travel in own country	**YES**	Rights respected
2	Travel outside own country	yes	A few restrictions on passports of those who have previously served prison sentences
3	Peacefully associate and assemble	**YES**	Rights respected
4	Teach ideas and receive information	**YES**	Rights respected
5	Monitor human rights violations	**YES**	Occasional threats by anonymous right-wing groups against monitoring of suspected offenses involving police and military
6	Publish and educate in ethnic language	**YES**	Rights respected

---

FREEDOM FROM: / COMMENTS

7	Serfdom, slavery, forced or child labor	yes	Regional child labor in rural areas

FREEDOM FROM:		COMMENTS
8	Extrajudicial killings or "disappearances"	**no**	A number of killings and disappearances occurred in 1990 with well-documented evidence of police complicity. In some areas military have taken over police duties which has led to deaths of guerrillas after capture
9	Torture or coercion by the state	**no**	Beatings, electric shocks, etc. Some police officers imprisoned after investigations and protests from international human rights groups
10	Compulsory work permits or conscription of labor	**YES**	Rights respected
11	Capital punishment by the state	**YES**	Abolished 1984 (for all ordinary crimes)
12	Court sentences of corporal punishment	**YES**	Rights respected
13	Indefinite detention without charge	**YES**	Rights respected
14	Compulsory membership of state organizations or parties	**YES**	Rights respected
15	Compulsory religion or state ideology in schools	**YES**	Rights respected
16	Deliberate state policies to control artistic works	**YES**	Rights respected
17	Political censorship of press	**YES**	Rights respected
18	Censorship of mail or telephone tapping	**no**	Police surveillance of human rights groups and some opposition politicians. Despite the end of the military dictatorship, a few covert activities by a small number of "old guard" officials

FREEDOM FOR OR RIGHTS TO:		COMMENTS
19	Peaceful political opposition	**YES**	Rights respected
20	Multiparty elections by secret and universal ballot	**YES**	Rights respected
21	Political and legal equality for women	**yes**	Despite constitutional equality, discrimination persists in politics and professions. Regional traditional factors
22	Social and economic equality for women	**yes**	Pay and employment inequalities continue. Position improving
23	Social and economic equality for ethnic minorities	**YES**	Rights respected
24	Independent newspapers	**YES**	Rights respected

FREEDOM FOR OR RIGHTS TO:		COMMENTS
25	Independent book publishing	**YES**	Rights respected
26	Independent radio and television networks	**YES**	State and private networks compete in all areas
27	All courts to total independence	**YES**	Judges threatened when investigating human rights violations by police and major crimes by the former military dictatorship. Independence also affected by presidential pardons of ex-military on murder charges
28	Independent trade unions	**YES**	Rights respected

LEGAL RIGHTS:		COMMENTS
29	From deprivation of nationality	**YES**	Rights respected
30	To be considered innocent until proved guilty	**YES**	Rights respected
31	To free legal aid when necessary and counsel of own choice	yes	Means test. "Defender of the poor" appointed by court
32	From civilian trials in secret	**YES**	Rights respected
33	To be brought promptly before a judge or court	yes	Within 8 days, but many long delays. Accusations of arbitrary detentions by judges of the "old guard" still holding office
34	From police searches of home without a warrant	yes	Police abuses. Evidence of unofficial activities by right-wing "parapolice" groups
35	From arbitrary seizure of personal property	**YES**	Rights respected

PERSONAL RIGHTS:		COMMENTS
36	To interracial, interreligious, or civil marriage	**YES**	Legal marriage at 18 for males, 16 for females
37	Equality of sexes during marriage and for divorce proceedings	yes	In practice, traditional status of women persists, to advantage of husbands
38	To practice any religion	**YES**	Rights respected
39	To use contraceptive pills and devices	**YES**	Limited state support
40	To noninterference by state in strictly private affairs	no	The large increase in crime due to the country's worsening economy has meant police actions against slum dwellers and excessive harassment of the family life of the poor

Australia

Human rights rating: 91%

YES 30 yes 10 no 0 NO 0

Population: 16,900,000
Life expectancy: 76.5
Infant mortality (0–5 years)
  per 1,000 births: 10
United Nations covenants ratified:
Civil and Political Rights, Economic,
Social and Cultural Rights, Convention
on Equality for Women

Form of government: Democratic
federal commonwealth
GNP per capita (US$): 12,340
% of GNP spent by state:
On health: 5.1
On military: 2.7
On education: 5.1

FACTORS AFFECTING HUMAN RIGHTS

A relatively prosperous economy, a government answerable to the people, a free press,
though one dominated by a few individuals, and an increasingly cosmopolitan
population ensure general respect for human rights. A national Bill of Rights based on
UN instruments received government support but was eventually shelved. The
improvement of rights for the Aboriginal minority (1.5% of the population) remains
official policy. The major concerns of human rights monitors lie in powers to issue
suppression orders against publishing court evidence, occasional police abuses, and a
degree of corruption at senior government levels.

FREEDOM TO:		**COMMENTS**
1 Travel in own country	YES	Rights respected
2 Travel outside own country	YES	Rights respected
3 Peacefully associate and assemble	YES	Rights respected
4 Teach ideas and receive information	YES	Rights respected
5 Monitor human rights violations	YES	Rights respected
6 Publish and educate in ethnic language	YES	Rights respected

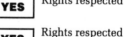

FREEDOM FROM:		**COMMENTS**
7 Serfdom, slavery, forced or child labor Rights respected	YES	Rights respected
8 Extrajudicial killings or "disappearances"	yes	Police accused by a royal commission of possible complicity in deaths of Aboriginals while in custody

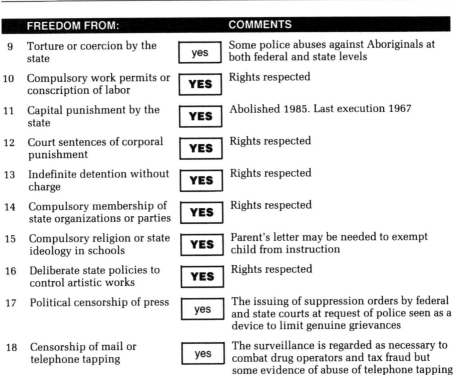

	FREEDOM FROM:		COMMENTS
9	Torture or coercion by the state	yes	Some police abuses against Aboriginals at both federal and state levels
10	Compulsory work permits or conscription of labor	YES	Rights respected
11	Capital punishment by the state	YES	Abolished 1985. Last execution 1967
12	Court sentences of corporal punishment	YES	Rights respected
13	Indefinite detention without charge	YES	Rights respected
14	Compulsory membership of state organizations or parties	YES	Rights respected
15	Compulsory religion or state ideology in schools	YES	Parent's letter may be needed to exempt child from instruction
16	Deliberate state policies to control artistic works	YES	Rights respected
17	Political censorship of press	yes	The issuing of suppression orders by federal and state courts at request of police seen as a device to limit genuine grievances
18	Censorship of mail or telephone tapping	yes	The surveillance is regarded as necessary to combat drug operators and tax fraud but some evidence of abuse of telephone tapping

	FREEDOM FOR OR RIGHTS TO:		COMMENTS
19	Peaceful political opposition	YES	Rights respected
20	Multiparty elections by secret and universal ballot	YES	Voting compulsory from age 18
21	Political and legal equality for women	yes	Of 224 members of federal Senate and House of Representatives, 28 are women
22	Social and economic equality for women	yes	Minor inequalities in pay and employment
23	Social and economic equality for ethnic minorities	yes	Aboriginals, 1.5% of population, suffer inequalities in many areas, including health facilities and employment
24	Independent newspapers	yes	Government limitations to foreign takeovers of newspapers perpetuate monopoly of one corporation with over 50% of national circulation
25	Independent book publishing	YES	Rights respected
26	Independent radio and television networks	yes	A few powerful individual interests dominate the commercial networks. Described as a "government-sanctioned oligopoly"

FREEDOM FOR OR RIGHTS TO: COMMENTS

27 All courts to total   **YES**   Rights respected
    independence

28 Independent trade unions   **YES**   Rights respected. Strong union movement
                                            exercises considerable influence on
                                            government policies

LEGAL RIGHTS: COMMENTS

29 From deprivation of   **YES**   Rights respected
    nationality

30 To be considered innocent   **YES**   Rights respected
    until proved guilty

31 To free legal aid when   yes   Means test
    necessary and counsel of
    own choice

32 From civilian trials in secret **YES** Rights respected

33 To be brought promptly   **YES**   Within 24 hours
    before a judge or court

34 From police searches of   **YES**   Rights respected
    home without a warrant

35 From arbitrary seizure of   **YES**   Rights respected
    personal property

PERSONAL RIGHTS: COMMENTS

36 To interracial, interreligious, **YES** Rights respected. Increasing immigrant
    or civil marriage   population is having a liberalizing effect on
                                              previous forms of social prejudice

37 Equality of sexes during   **YES**   Minimum age 18 for males, 16 for females
    marriage and for divorce
    proceedings

38 To practice any religion   **YES**   Rights respected

39 To use contraceptive pills   **YES**   Rights respected
    and devices

40 To noninterference by state **YES**   Rights respected
    in strictly private affairs

Austria

Human rights rating: 95%

YES 35 **yes** 4 **no** 1 NO 0

Population: 7,600,000
Life expectancy: 74.8
Infant mortality (0–5 years) per 1,000 births: 10
United Nations covenants ratified:
Civil and Political Rights, Economic, Social and Cultural Rights, Convention on Equality for Women

Form of government: Parliamentary democracy
GNP per capita (US$): 15,470
% of GNP spent by state:
On health: 8.4
On military: 1.3
On education: 6.0

FACTORS AFFECTING HUMAN RIGHTS

The individual respected. Government answerable to the people. A prosperous economy, democratic institutions, and a free press. Its geographical position between Eastern and Western Europe, and on the frontier of a disintegrating Yugoslavia, has made the country a haven for refugees and a determined supporter of human rights. European Convention on Human Rights incorporated in the national constitution.

	FREEDOM TO:		COMMENTS
1	Travel in own country	YES	Rights respected
2	Travel outside own country	YES	Rights respected
3	Peacefully associate and assemble	YES	Rights respected
4	Teach ideas and receive information	YES	Rights respected
5	Monitor human rights violations	YES	Rights respected
6	Publish and educate in ethnic language	YES	A small Slovene community protests at minor discriminations

	FREEDOM FROM:		COMMENTS
7	Serfdom, slavery, forced or child labor	YES	Rights respected
8	Extrajudicial killings or "disappearances"	YES	Rights respected
9	Torture or coercion by the state	yes	Well-documented cases of police abuses. In recent years, 33 officers convicted after complaints led to prosecution

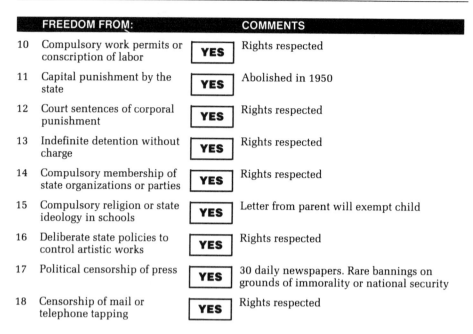

FREEDOM FROM: COMMENTS

10	Compulsory work permits or conscription of labor	YES	Rights respected
11	Capital punishment by the state	YES	Abolished in 1950
12	Court sentences of corporal punishment	YES	Rights respected
13	Indefinite detention without charge	YES	Rights respected
14	Compulsory membership of state organizations or parties	YES	Rights respected
15	Compulsory religion or state ideology in schools	YES	Letter from parent will exempt child
16	Deliberate state policies to control artistic works	YES	Rights respected
17	Political censorship of press	YES	30 daily newspapers. Rare bannings on grounds of immorality or national security
18	Censorship of mail or telephone tapping	YES	Rights respected

FREEDOM FOR OR RIGHTS TO: COMMENTS

19	Peaceful political opposition	YES	Vote from age 19
20	Multiparty elections by secret and universal ballot	YES	Elections every 4 years. National Assembly of 183 members
21	Political and legal equality for women	yes	Despite constitutional equality, women are underrepresented in politics and professions
22	Social and economic equality for women	yes	Inequalities persist in pay and employment
23	Social and economic equality for ethnic minorities	YES	Rights respected
24	Independent newspapers	YES	Rights respected
25	Independent book publishing	YES	Rights respected
26	Independent radio and television networks	YES	Stateowned but independence guaranteed and honored
27	All courts to total independence	YES	Judges appointed for life
28	Independent trade unions	YES	Rights respected

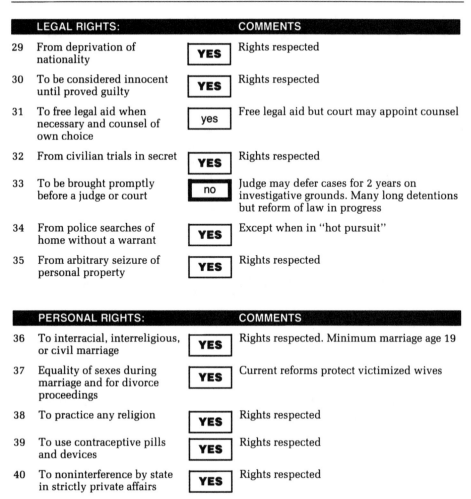

LEGAL RIGHTS:		COMMENTS
29	From deprivation of nationality	**YES**	Rights respected
30	To be considered innocent until proved guilty	**YES**	Rights respected
31	To free legal aid when necessary and counsel of own choice	yes	Free legal aid but court may appoint counsel
32	From civilian trials in secret	**YES**	Rights respected
33	To be brought promptly before a judge or court	no	Judge may defer cases for 2 years on investigative grounds. Many long detentions but reform of law in progress
34	From police searches of home without a warrant	**YES**	Except when in "hot pursuit"
35	From arbitrary seizure of personal property	**YES**	Rights respected

PERSONAL RIGHTS:		COMMENTS
36	To interracial, interreligious, or civil marriage	**YES**	Rights respected. Minimum marriage age 19
37	Equality of sexes during marriage and for divorce proceedings	**YES**	Current reforms protect victimized wives
38	To practice any religion	**YES**	Rights respected
39	To use contraceptive pills and devices	**YES**	Rights respected
40	To noninterference by state in strictly private affairs	**YES**	Rights respected

Bangladesh

Human rights rating: 59%

YES 15 yes 8 no 14 NO 3

Population: 115,600,000
Life expectancy: 51.8
**Infant mortality (0–5 years)
   per 1,000 births**: 184
United Nations covenants ratified:
Convention on Equality for Women

Form of government: Multiparty system
GNP per capita (US$): 170
% of GNP spent by state:
On health: 1.1
On military: 1.5
On education: 1.3

FACTORS AFFECTING HUMAN RIGHTS

The country is one of the poorest in the world and is afflicted by periodic floods over the vast low-lying areas. Since independence from Pakistan in 1971, there have been numerous changes in the leadership, most of the incumbents having come from the military and two having been assassinated. In 1990 violent demonstrations forced the resignation of the executive president, and free elections in 1991 were followed by a multiparty democratic system. Significant improvements are taking place with respect to human rights though these may be restricted by the religious and social traditions of the country.

	FREEDOM TO:		COMMENTS
1	Travel in own country	**YES**	Except to Chittagong Hill Tracts, which is a security risk zone
2	Travel outside own country	**YES**	Rights respected
3	Peacefully associate and assemble	**YES**	Rights respected
4	Teach ideas and receive information	yes	Academic guidelines on sensitive traditional and religious subjects. Criticism of armed forces prejudices career prospects. Prison sentences for students cheating on exams
5	Monitor human rights violations	**YES**	Rights respected
6	Publish and educate in ethnic language	**YES**	Rights respected

	FREEDOM FROM:		COMMENTS
7	Serfdom, slavery, forced or child labor	**NO**	Bonded and child labor, mostly in rural areas

FREEDOM FROM:		COMMENTS
8	Extrajudicial killings or "disappearances"	**no**	Clashes between security forces and demonstrators during 1990 caused nearly 100 deaths, some from excessive violence. The improvement should be regarded as "fragile"
9	Torture or coercion by the state	**no**	Widespread police and prison brutalities at local level. Perpetrators go unpunished. Deaths from prolonged beatings
10	Compulsory work permits or conscription of labor	**YES**	Rights respected
11	Capital punishment by the state	**NO**	By hanging or shooting. For murder, treason, antistate offenses, etc.
12	Court sentences of corporal punishment	**YES**	Rights respected
13	Indefinite detention without charge	**no**	Many detainees under Special Powers Act being released. Position improving
14	Compulsory membership of state organisations or parties	**YES**	Rights respected
15	Compulsory religion or state ideology in schools	**no**	State accord on compulsory religion in Muslim schools
16	Deliberate state policies to control artistic works	yes	Strict obscenity laws and guidelines by Islamic state religion may set restrictive limits. Theater companies, performing arts subject to official scrutiny
17	Political censorship of press	yes	Major improvement since formation of democratic government, which lifted the severe censorship laws of previous administration
18	Censorship of mail or telephone tapping	yes	Occasional surveillance during periods of instability

FREEDOM FOR OR RIGHTS TO:		COMMENTS
19	Peaceful political opposition	yes	The ending of government by executive president in December 1990 is being followed by a multiparty system (described as "Westminster style")
20	Multiparty elections by secret and universal ballot	yes	An amended constitution in mid-1991 enabled multiparty elections to be held later in the year. Proposed parliament reserves 10% of seats for women
21	Political and legal equality for women	**no**	Although the changes to a multiparty system, with women in leading roles, indicate an improvement, traditional inequalities persist, particularly in professions

FREEDOM FOR OR RIGHTS TO:		COMMENTS
22	Social and economic equality for women	no	Oppressive status for women. The general level of poverty, particularly in rural areas, and religious traditions prevent improvement
23	Social and economic equality for ethnic minorities	no	Disadvantages for Hindus, Biharis, and certain smaller tribes; these include land and property rights
24	Independent newspapers	YES	Rights respected
25	Independent book publishing	YES	Rights respected
26	Independent radio and television networks	no	State owned and controlled. The fairer division of airtime for political parties, following the end of the previous regime, may not prove enduring
27	All courts to total independence	yes	The setting aside of the Special Powers Act gives courts greater independence but the reality is an inadequate, underfunded, and corrupt system
28	Independent trade unions	no	Only a minority of workers unionized. Future reforms under the new democratic government are anticipated

LEGAL RIGHTS:		COMMENTS
29	From deprivation of nationality	YES	Rights respected
30	To be considered innocent until proved guilty	no	Many police abuses, particularly in remote areas. Corruption a factor and bribes may prove innocence
31	To free legal aid when necessary and counsel of own choice	NO	The underfunded and inadequate system prevents any consistent countrywide pattern of aid
32	From civilian trials in secret	YES	Rights respected
33	To be brought promptly before a judge or court	no	Despite the dropping of the Special Powers Act, which permitted long detentions, an overburdened system means 100,000 waiting for trial
34	From police searches of home without a warrant	yes	The present change from a repressive regime is too recent to assume that previous police practices cannot recur
35	From arbitrary seizure of personal property	no	Seizures of tribal lands for transfer to Bengalis continued until recently and have not been returned to traditional owners

PERSONAL RIGHTS:		COMMENTS
36	To interracial, interreligious, or civil marriage		No interreligious marriage for practicing Muslims (85% of population). Legal marriage age 21 for males, 18 for females
37	Equality of sexes during marriage and for divorce proceedings		Poverty and traditional inequalities affect status of wives. Religious laws always to the advantage of husbands. Dowry killings (when bride payment not honored) still occasionally occur. Many suicides by wives
38	To practice any religion	YES	Rights respected. Recurring pattern of local religious tensions
39	To use contraceptive pills and devices	YES	State support
40	To noninterference by state in strictly private affairs	YES	Position improved with end of previous regime

Belgium

Human rights rating: 96%

Population: 9,800,000
Life expectancy: 75.2
Infant mortality (0–5 years) per 1,000 births: 13
United Nations covenants ratified: Civil and Political Rights, Economic, Social and Cultural Rights, Convention on Equality for Women

Form of government: Parliamentary monarchy
GNP per capita (US$): 14,490
% of GNP spent by state:
On health: 5.6
On military: 3.1
On education: 5.6

FACTORS AFFECTING HUMAN RIGHTS
The individual respected. Government answerable to the people. Democratic institutions and traditions, a free press, and a high standard of living ensure the honoring of human rights treaties. In 1990 further measures were taken to guarantee respect for human rights following criticism from the European Court of Human Rights. The population is divided between Dutch-speaking Flanders and French-speaking Wallonia (57% and 42% respectively), and an aspect of government policy is to minimize conflicts and rivalries between the two communities.

	FREEDOM TO:		COMMENTS
1	Travel in own country	YES	Rights respected
2	Travel outside own country	YES	Rights respected
3	Peacefully associate and assemble	YES	Rights respected
4	Teach ideas and receive information	YES	Rights respected
5	Monitor human rights violations	YES	Rights respected
6	Publish and educate in ethnic language	YES	Rights respected. Frequent disputes over alleged government priorities between the two major language groups covering 99% of population

	FREEDOM FROM:		COMMENTS
7	Serfdom, slavery, forced or child labor	YES	Rights respected
8	Extrajudicial killings or "disappearances"	YES	Rights respected

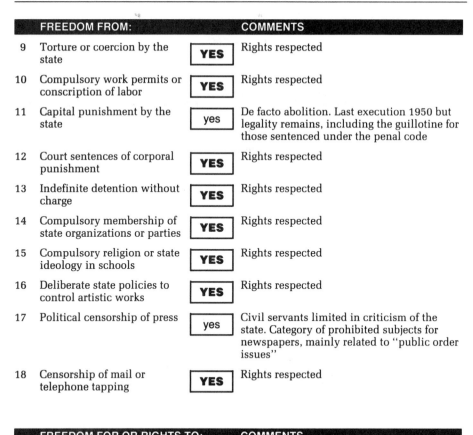

	FREEDOM FROM:		COMMENTS
9	Torture or coercion by the state	**YES**	Rights respected
10	Compulsory work permits or conscription of labor	**YES**	Rights respected
11	Capital punishment by the state	yes	De facto abolition. Last execution 1950 but legality remains, including the guillotine for those sentenced under the penal code
12	Court sentences of corporal punishment	**YES**	Rights respected
13	Indefinite detention without charge	**YES**	Rights respected
14	Compulsory membership of state organizations or parties	**YES**	Rights respected
15	Compulsory religion or state ideology in schools	**YES**	Rights respected
16	Deliberate state policies to control artistic works	**YES**	Rights respected
17	Political censorship of press	yes	Civil servants limited in criticism of the state. Category of prohibited subjects for newspapers, mainly related to "public order issues"
18	Censorship of mail or telephone tapping	**YES**	Rights respected

	FREEDOM FOR OR RIGHTS TO:		COMMENTS
19	Peaceful political opposition	**YES**	Rights respected
20	Multiparty elections by secret and universal ballot	**YES**	Rights respected
21	Political and legal equality for women	yes	Women granted suffrage in 1949. Only limited presence in senior government posts
22	Social and economic equality for women	yes	Pay and employment inequalities. Government body to protect and improve women's rights
23	Social and economic equality for ethnic minorities	yes	Large immigrant groups in main urban areas suffer limited work and pay and social discrimination
24	Independent newspapers	**YES**	Rights respected
25	Independent book publishing	**YES**	Rights respected
26	Independent radio and television networks	**YES**	Main channels are stateowned but guarantees of independence are honored
27	All courts to total independence	**YES**	Judges appointed for life

FREEDOM FOR OR RIGHTS TO:	COMMENTS

28 Independent trade unions   **YES**   Rights respected

LEGAL RIGHTS:	COMMENTS

29 From deprivation of
    nationality   **YES**   Rights respected

30 To be considered innocent
    until proved guilty   yes   Occasional police abuses which may
    prejudice trials

31 To free legal aid when
    necessary and counsel of
    own choice   yes   Means tests. Court usually appoints counsel

32 From civilian trials in secret   **YES**   Rights respected

33 To be brought promptly
    before a judge or court   **YES**   Within 24 hours or released

34 From police searches of
    home without a warrant   **YES**   Rights respected

35 From arbitrary seizure of
    personal property   **YES**   Rights respected

PERSONAL RIGHTS:	COMMENTS

36 To interracial, interreligious,
    or civil marriage   **YES**   Rights respected. Minimum marriage age 18

37 Equality of sexes during
    marriage and for divorce
    proceedings   **YES**   Rights respected

38 To practice any religion   **YES**   Rights respected

39 To use contraceptive pills
    and devices   **YES**   Limited state support

40 To noninterference by state
    in strictly private affairs   **YES**   Rights respected

Benin

Human rights rating: 90%

YES 32 yes 5 no 3 NO 0

Population: 4,600,000
Life expectancy: 47.0
Infant mortality (0–5 years)
per 1,000 births: 150
United Nations covenants ratified:
None

Form of government: Multiparty system
GNP per capita (US$): 390
% of GNP spent by state:
On health: 0.8
On military: 2.3
On education: 3.5

FACTORS AFFECTING HUMAN RIGHTS
The country enjoyed a comprehensive change in 1990 from a Marxist-Leninist single-party government to a democratic system. This culminated in multiparty elections in 1991 and has been followed by much-improved respect for human rights. The future outlook depends on whether the improvements, unprecedented in Africa for their scope, can be maintained and consolidated.

	FREEDOM TO:		COMMENTS
1	Travel in own country	YES	Previous restrictions removed following the 1990 change to a democratic government
2	Travel outside own country	YES	Rights respected
3	Peacefully associate and assemble	YES	Rights respected
4	Teach ideas and receive information	YES	Rights respected
5	Monitor human rights violations	YES	Monitoring freely permitted with the end of the previous military regime
6	Publish and educate in ethnic language	YES	Official language French

	FREEDOM FROM:		COMMENTS
7	Serfdom, slavery, forced or child labor	yes	Traditional customs involve widespread child labor
8	Extrajudicial killings or "disappearances"	YES	Rights respected
9	Torture or coercion by the state	YES	Rights respected
10	Compulsory work permits or conscription of labor	YES	Rights respected

	FREEDOM FROM:		COMMENTS
11	Capital punishment by the state	no	Abolition being considered by the new democratic government
12	Court sentences of corporal punishment	YES	Rights respected
13	Indefinite detention without charge	YES	Rights respected
14	Compulsory membership of state organizations or parties	YES	Rights respected
15	Compulsory religion or state ideology in schools	YES	60% of population follow animist beliefs
16	Deliberate state policies to control artistic works	YES	Rights respected
17	Political censorship of press	YES	Rights respected
18	Censorship of mail or telephone tapping	YES	Rights respected

	FREEDOM FOR OR RIGHTS TO:		COMMENTS
19	Peaceful political opposition	YES	New constitution approved in December 1990 by a national referendum
20	Multiparty elections by secret and universal ballot	YES	1991 election resulted in a multiparty legislature
21	Political and legal equality for women	yes	Traditional inequalities continue. The new constitution and the transformed society have not begun to influence existing practices
22	Social and economic equality for women	no	Inequalities most evident in rural areas or under customary law. Female circumcision still widely practiced
23	Social and economic equality for ethnic minorities	YES	In 1958 three colonial areas were merged into a new state. Little evidence of serious discrimination
24	Independent newspapers	YES	Wide variety appearing following the end of the military regime
25	Independent book publishing	YES	Rights respected
26	Independent radio and television networks	no	State owned and controlled but progress toward guarantees of political neutrality
27	All courts to total independence	YES	Rights respected
28	Independent trade unions	YES	Democratization under new constitution

LEGAL RIGHTS:		COMMENTS
29	From deprivation of nationality	**YES**	Rights respected
30	To be considered innocent until proved guilty	yes	Occasional arbitrary arrests by individual police officers as country adapts to the new democratic reforms
31	To free legal aid when necessary and counsel of own choice	**YES**	Rights respected
32	From civilian trials in secret	**YES**	Rights respected
33	To be brought promptly before a judge or court	yes	Delays caused by inadequate system and local factors
34	From police searches of home without a warrant	**YES**	Rights respected
35	From arbitrary seizure of personal property	**YES**	Rights respected

PERSONAL RIGHTS:		COMMENTS
36	To interracial, interreligious, or civil marriage	**YES**	Rights respected
37	Equality of sexes during marriage and for divorce proceedings	yes	Traditions and customary laws continue, which favor husbands
38	To practice any religion	**YES**	Rights respected
39	To use contraceptive pills and devices	**YES**	Rights respected
40	To noninterference by state in strictly private affairs	**YES**	Rights respected

Bolivia

Human rights rating: 71%

YES 22 yes 11 no 7 NO 0

Population: 7,300,000
Life expectancy: 54.5
Infant mortality (0–5 years) per 1,000 births: 165
United Nations covenants ratified: Civil and Political Rights, Economic, Social and Cultural Rights

Form of government: Multiparty system
GNP per capita (US$): 570
% of GNP spent by state:
On health: 0.4
On military: 2.4
On education: 2.9

FACTORS AFFECTING HUMAN RIGHTS
Democracy was restored in 1982 after many coups and revolutions, with 64 presidents in 175 years. But progress has been affected by the poverty of the country, the failure of government to improve conditions and the standard of living, and the resultant increase in crime, corruption, and excesses by the security forces. Terrorist activities are increasing, as are extrajudicial killings, torture, and arbitrary detentions. Improvements in the human rights situation are linked to an improving economy and efficient government, for which there can be little reason for optimism.

FREEDOM TO: COMMENTS

1	Travel in own country	**YES**	Rights respected
2	Travel outside own country	**YES**	Rights respected
3	Peacefully associate and assemble	**YES**	Rights respected
4	Teach ideas and receive information	**YES**	Rights respected
5	Monitor human rights violations	**YES**	Rights respected
6	Publish and educate in ethnic language	**YES**	Rights respected

---

FREEDOM FROM: COMMENTS

7	Serfdom, slavery, forced or child labor	no	Regional pattern of child labor. Rehabilitation camps for young delinquents sometimes practice forced labor
8	Extrajudicial killings or "disappearances"	no	A number of arbitrary killings by security and police forces in actions against small terrorist groups – some after their capture

FREEDOM FROM:		COMMENTS
9	Torture or coercion by the state	**no**	Electric shocks, beatings, etc., during interrogation and in prisons. Little government action to control practices
10	Compulsory work permits or conscription of labor	**yes**	Some use of prisoners in contravention of international treaties
11	Capital punishment by the state	**no**	Still legal but last execution in 1974
12	Court sentences of corporal punishment	**YES**	Rights respected
13	Indefinite detention without charge	**no**	The situation is worsened by neglect, bad prison conditions, corruption, and official indifference
14	Compulsory membership of state organizations or parties	**YES**	Rights respected
15	Compulsory religion or state ideology in schools	**no**	Compulsory religious instruction in a predominantly Roman Catholic society
16	Deliberate state policies to control artistic works	**YES**	Rights respected
17	Political censorship of press	**YES**	Rights respected
18	Censorship of mail or telephone tapping	**YES**	Rights respected

FREEDOM FOR OR RIGHTS TO:		COMMENTS
19	Peaceful political opposition	**YES**	Vote from age 19
20	Multiparty elections by secret and universal ballot	**YES**	Every 4 years
21	Political and legal equality for women	**yes**	Despite constitution, traditional attitude toward a woman's role in society persists
22	Social and economic equality for women	**yes**	Inequalities in pay, employment, and socially, though some improvement in urban communities
23	Social and economic equality for ethnic minorities	**yes**	Amerindians, although the majority of the population, suffer discrimination in all social and economic areas
24	Independent newspapers	**YES**	Rights respected
25	Independent book publishing	**YES**	Rights respected
26	Independent radio and television networks	**yes**	Both private and state owned though the latter is accused of "manipulation"

FREEDOM FOR OR RIGHTS TO:		COMMENTS
27	All courts to total independence	yes	Documented instances of government attempts to circumscribe Supreme Court. Judges threatened by terrorists. Degrees of official corruption
28	Independent trade unions	yes	Occasional conflicts with government. State of siege declared to break strikes. Force used occasionally

LEGAL RIGHTS:		COMMENTS
29	From deprivation of nationality	YES	Rights respected
30	To be considered innocent until proved guilty	YES	Rights respected
31	To free legal aid when necessary and counsel of own choice	yes	Situation inefficient. Lack of funds and corruption influence court decisions
32	From civilian trials in secret	YES	Rights respected
33	To be brought promptly before a judge or court	yes	Within 48 hours but long delays are caused by the inadequate and incompetent system
34	From police searches of home without a warrant	YES	Rights respected
35	From arbitrary seizure of personal property	YES	Rights respected

PERSONAL RIGHTS:		COMMENTS
36	To interracial, interreligious, or civil marriage	YES	Males may marry at 16 years, females at 14 years
37	Equality of sexes during marriage and for divorce proceedings	no	Many traditional disadvantages affect women's position, including inheritance rights
38	To practice any religion	yes	The predominant Roman Catholic church enjoys many privileges, some to the disadvantage of minority religions
39	To use contraceptive pills and devices	yes	No government support. Roman Catholic church a major influence on birth control methods
40	To noninterference by state in strictly private affairs	YES	Rights respected

Botswana

Human rights rating: 79%

YES 25 yes 10 no 4 NO 1

Population: 1,300,000
Life expectancy: 59.8
Infant mortality (0–5 years)
per 1,000 births: 87
United Nations covenants ratified:
None

Form of government: Multi-party system with executive president
GNP per capita (US$): 1,010
% of GNP spent by state:
On health: 3.8
On military: 2.3
On education: 9.2

FACTORS AFFECTING HUMAN RIGHTS

In 1966 the British crown colony of Bechuanaland became the independent state of Botswana. The economy is diversifying from an essentially pastoral economy. The eight principal tribes adhere to their traditional lives and their chiefs advise the government on relevant matters. Human rights, within the pattern of a customary society, are in general respected.

	FREEDOM TO:		COMMENTS
1	Travel in own country	YES	Rights respected
2	Travel outside own country	YES	Rights respected
3	Peacefully associate and assemble	yes	Police permission usually granted – except when danger of violence
4	Teach ideas and receive information	YES	Rights respected
5	Monitor human rights violations	YES	The country is a rare example of an African state that has not received the attention of international human rights monitors
6	Publish and educate in ethnic language	yes	Minorities complain of restrictions on ethnic languages in schools

	FREEDOM FROM:		COMMENTS
7	Serfdom, slavery, forced or child labor	yes	Regional pattern of child labor
8	Extrajudicial killings or "disappearances"	YES	Rights respected
9	Torture or coercion by the state	yes	Occasional police abuses at local level
10	Compulsory work permits or conscription of labor	YES	Rights respected

FREEDOM FROM:		COMMENTS
11	Capital punishment by the state	**NO**	By hanging. For murder, etc.
12	Court sentences of corporal punishment	no	Flogging for many offenses
13	Indefinite detention without charge	**YES**	Rights respected
14	Compulsory membership of state organizations or parties	**YES**	Rights respected
15	Compulsory religion or state ideology in schools	**YES**	Rights respected
16	Deliberate state policies to control artistic works	**YES**	Rights respected
17	Political censorship of press	yes	Some self-censorship on sensitive issues such as strong criticisms of senior government figures
18	Censorship of mail or telephone tapping	**YES**	Rights respected

FREEDOM FOR OR RIGHTS TO:		COMMENTS
19	Peaceful political opposition	**YES**	Rights respected
20	Multiparty elections by secret and universal ballot	**YES**	Vote from age 18. Executive president responsible to the 38-member National Assembly
21	Political and legal equality for women	yes	Despite legal rights, traditional attitudes perpetuate inequalities at all levels. But some improvement
22	Social and economic equality for women	no	Bride price tradition continues. Women need husband's permission for such commercial transactions as borrowing from a bank or starting a business
23	Social and economic equality for ethnic minorities	yes	Disadvantages relate to communities living in remote areas, following nomadic existence, etc.
24	Independent newspapers	yes	Leading press agency controlled by government. Understood guidelines must be followed
25	Independent book publishing	**YES**	No major publisher. Nearly all books imported
26	Independent radio and television networks	yes	State owned and controlled but safeguards on objectivity of programs. Also a Voice of America relay station on Botswana territory
27	All courts to total independence	**YES**	Rights respected

FREEDOM FOR OR RIGHTS TO:		COMMENTS
28 Independent trade unions	yes	But complex government controls limit freedom to strike

LEGAL RIGHTS:		COMMENTS
29 From deprivation of nationality	YES	Rights respected
30 To be considered innocent until proved guilty	YES	Rights respected
31 To free legal aid when necessary and counsel of own choice	no	Free aid for serious cases only, with court appointing counsel
32 From civilian trials in secret	YES	Rights respected
33 To be brought promptly before a judge or court	YES	Within 48 hours, then detention writs renewable every 14 days
34 From police searches of home without a warrant	YES	Rights respected
35 From arbitrary seizure of personal property	YES	Rights respected

PERSONAL RIGHTS:		COMMENTS
36 To interracial, interreligious, or civil marriage	YES	Marriage may be under common law or customary (tribal) law
37 Equality of sexes during marriage and for divorce proceedings	no	Traditional advantages for husband. One being a right to beat his wife for improper behavior, another to decide on contraceptive methods and practice
38 To practice any religion	YES	Rights respected
39 To use contraceptive pills and devices	YES	Rights respected
40 To noninterference by state in strictly private affairs	YES	Rights respected

Brazil

Human rights rating: 69%

YES 23 yes 9 no 5 NO 3

Population: 150,400,000
Life expectancy: 65.6
Infant mortality (0–5 years) per 1,000 births: 85
United Nations covenants ratified:
Convention on Equality for Women

Form of government: Multiparty federal republic
GNP per capita (US$): 2,160
% of GNP spent by state:
On health: 2.4
On military: 0.9
On education: 3.4

FACTORS AFFECTING HUMAN RIGHTS

In 1990 the country elected its first president in 29 years after a long period of military rule and an interim transitional government. The human rights position has recently deteriorated following the country's economic difficulties and the ensuing social unrest. There has been a significant increase in extrajudicial killings by police and security forces, particularly of Amerindians, trade unionists, and land workers. In numerous land disputes, police and the military have supported estate owners against land workers and peasants. There are also many arbitrary killings and detentions of vagrants and "street children," a phenomenon of impoverished urban areas. The degree of violence throughout the country is the most urgent of human rights problems.

FREEDOM TO: / COMMENTS

#	Freedom	Rating	Comments
1	Travel in own country	YES	Rights respected
2	Travel outside own country	YES	Rights respected
3	Peacefully associate and assemble	YES	Rights respected
4	Teach ideas and receive information	YES	Relatively liberal now that a civilian government is established and military have ceased to threaten some universities
5	Monitor human rights violations	yes	Official cooperation with human rights monitors. But areas of reluctance to investigate, including that of violence of major landowners toward peasant workers
6	Publish and educate in ethnic language	YES	Rights respected

---

FREEDOM FROM: / COMMENTS

#	Freedom	Rating	Comments
7	Serfdom, slavery, forced or child labor	no	Forced labor by large landowners disregarded by authorities. Child labor in rural areas as well as many child nomads

FREEDOM FROM:		COMMENTS
8	Extrajudicial killings or "disappearances" **NO**	Summary executions by police and military of Indians, trade unionists, criminals, and protesters. Hundreds killed each year. Large landowners kill with impunity agitating peasants. Little government intervention
9	Torture or coercion by the state **NO**	Constant pattern of torture. Interrogations also a police device for extracting money and other advantages. Limited official investigation of complaints
10	Compulsory work permits or conscription of labor **YES**	Rights respected
11	Capital punishment by the state **no**	The failure – or refusal – of the government to prevent widespread summary executions by police "death squads" offsets the abolition of the death penalty. Of the hundreds of "street children" killed by these squads, the Pope has said: "Children must not be eliminated on the pretext of preventing crime" (October 1991)
12	Court sentences of corporal punishment **YES**	Rights respected
13	Indefinite detention without charge **yes**	Instances of lengthy detention without a judicial order. Also some condoning of private jails by large landowners
14	Compulsory membership of state organizations or parties **YES**	Rights respected
15	Compulsory religion or state ideology in schools **YES**	Rights respected
16	Deliberate state policies to control artistic works **YES**	Rights respected
17	Political censorship of press **YES**	Rights respected, though the degree of violence in the country has made self-censorship on sensitive issues advisable
18	Censorship of mail or telephone tapping **yes**	Some unlawful surveillance by police

FREEDOM FOR OR RIGHTS TO:		COMMENTS
19	Peaceful political opposition **YES**	Rights respected
20	Multiparty elections by secret and universal ballot **YES**	Compulsory at age 18. Indians recently granted the vote
21	Political and legal equality for women **yes**	Position improving in a country where women have previously been underrepresented in politics and professions
22	Social and economic equality for women **no**	Traditional discrimination in pay, employment, and opportunities

FREEDOM FOR OR RIGHTS TO:		COMMENTS
23	Social and economic equality for ethnic minorities	yes	Discrimination against blacks and the very dark-skinned remains a social reality – with extension into government service. Worst in Bahia state
24	Independent newspapers	**YES**	Rights respected
25	Independent book publishing	**YES**	Rights respected
26	Independent radio and television networks	**YES**	Government licenses to private stations
27	All courts to total independence	**YES**	Rights respected
28	Independent trade unions	yes	But labor leaders frequently murdered by hired assassins of ruthless employers. Union affairs conducted against a background of violence and threats. Limited government action against culprits

LEGAL RIGHTS:		COMMENTS
29	From deprivation of nationality	**YES**	Rights respected
30	To be considered innocent until proved guilty	yes	Harassment of darker-skinned by police. Prejudice encourages suspicion of blacks
31	To free legal aid when necessary and counsel of own choice	**YES**	Rights respected
32	From civilian trials in secret	**YES**	Rights respected
33	To be brought promptly before a judge or court	no	Long delays. Law states within 5 days but preliminary detentions follow an arbitrary pattern
34	From police searches of home without a warrant	**NO**	Many violations by an undisciplined and corrupt military and police force
35	From arbitrary seizure of personal property	no	Little higher supervision of arbitrary seizures (or looting) by the police and military

PERSONAL RIGHTS:		COMMENTS
36	To interracial, interreligious, or civil marriage	**YES**	Legal marriage age 18 for males, 16 for females
37	Equality of sexes during marriage and for divorce proceedings	yes	Husbands favored by long tradition. Many inequalities
38	To practice any religion	**YES**	Rights respected

	PERSONAL RIGHTS:		COMMENTS
39	To use contraceptive pills and devices	**YES**	Some government support
40	To noninterference by state in strictly private affairs	yes	Workers and peasants employed by large landowners suffer intrusion and harassment in their dwellings, occasionally abetted by officials

Bulgaria

Human rights rating: 83%

YES 21 yes 18 no 1 NO 0

Population: 9,000,000
Life expectancy: 72.6
Infant mortality (0–5 years) per 1,000 births: 20
United Nations covenants ratified:
Civil and Political Rights, Economic,
Social and Cultural Rights, Convention
on Equality for Women

Form of government: Multiparty
democracy
GNP per capita (US$): 4,150
% of GNP spent by state:
On health: 3.2
On military: 3.6
On education: 4.4

FACTORS AFFECTING HUMAN RIGHTS
The elections of October 1991 completed the transition, after over 40 years of a Marxist-Leninist regime, to a multiparty democracy. A measure of the fundamental change can be seen in the status of the Turkish minority, 10% of the population. From being a persecuted ethnic group, denied many basic rights and forced to adopt Bulgarian names, they now have their own independent party in the Grand National Assembly. Human rights currently approach the standards of Western Europe, though a matter for concern remains the presence in positions of authority of influential figures from the previous Communist regime.

	FREEDOM TO:		COMMENTS
1	Travel in own country	YES	Rights respected. Previous decrees on registering movements abolished
2	Travel outside own country	YES	Rights respected. New post-Communist laws permit unrestricted issue of passports
3	Peacefully associate and assemble	YES	Rights respected. Situation transformed by the 1989–90 political and social changes
4	Teach ideas and receive information	YES	Rights respected
5	Monitor human rights violations	yes	State security personnel still largely from the previous regime. Minor harassment of monitoring groups
6	Publish and educate in ethnic language	yes	In some instances use of Bulgarian language is compulsory

	FREEDOM FROM:		COMMENTS
7	Serfdom, slavery, forced or child labor	YES	Rights respected
8	Extrajudicial killings or "disappearances"	YES	Rights respected

	FREEDOM FROM:		COMMENTS
9	Torture or coercion by the state	yes	Occasional abuses by security and police, most of which served under the previous Communist regime
10	Compulsory work permits or conscription of labor	yes	Conscripted labor remains a punishment in the penal code
11	Capital punishment by the state	yes	Moratorium. Abolition expected
12	Court sentences of corporal punishment	**YES**	Rights respected
13	Indefinite detention without charge	yes	Most political detainees from the Communist regime have been released. A few ethnic Turks charged with terrorist acts still held
14	Compulsory membership of state organizations or parties	**YES**	Rights respected
15	Compulsory religion or state ideology in schools	**YES**	Rights respected
16	Deliberate state policies to control artistic works	**YES**	Rights respected
17	Political censorship of press	**YES**	Rights respected
18	Censorship of mail or telephone tapping	yes	State security still largely from the previous regime. Less conspicuous surveillance

	FREEDOM FOR OR RIGHTS TO:		COMMENTS
19	Peaceful political opposition	**YES**	Rights respected
20	Multiparty elections by secret and universal ballot	yes	Democratic elections in 1990 to the new National Assembly though the governing party, formerly the Communist party, accused of unfair practices in rural areas during campaign
21	Political and legal equality for women	yes	Despite constitutional equality, underrepresentation at senior levels
22	Social and economic equality for women	yes	Traditional inequality toward women, socially and at work. Worst among Turkish and Gypsy minorities (in total 12% of population)
23	Social and economic equality for ethnic minorities	yes	A correction of the previous "Bulgarianization" of minorities, particularly of Turks (nearly 9% of population) is making progress. Including the old law which made changes of name compulsory
24	Independent newspapers	yes	Newsprint no longer the monopoly of the state. Major newspapers dependent on various political parties – or affiliated

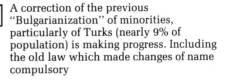

FREEDOM FOR OR RIGHTS TO:		COMMENTS
25	Independent book publishing	yes	Book publishing moving from the previous state monopoly toward independence, with universities among the vanguard
26	Independent radio and television networks	yes	State owned and controlled but supervised by the National Assembly. Politically neutral though some parties allege bias
27	All courts to total independence	YES	Rights respected. System now "depoliticized" but with most judges still former Communists
28	Independent trade unions	YES	The amended constitution has enabled a free union movement to be established

LEGAL RIGHTS:		COMMENTS
29	From deprivation of nationality	YES	Rights respected
30	To be considered innocent until proved guilty	YES	Rights respected
31	To free legal aid when necessary and counsel of own choice	yes	Aid for the needy. Counsel appointed by court
32	From civilian trials in secret	no	Reforms of the law prohibiting secret trials of civilians about to be introduced
33	To be brought promptly before a judge or court	yes	Within 24 hours. Limit occasionally exceeded as country adjusts to a reformed system
34	From police searches of home without a warrant	yes	Warrants not necessary in "urgent" situations – often a pretext for arbitrary violations
35	From arbitrary seizure of personal property	yes	Government action to restore homes and property of Turks who have fled from the previous regime, though operation not yet complete

PERSONAL RIGHTS:		COMMENTS
36	To interracial, interreligious, or civil marriage	YES	Rights respected
37	Equality of sexes during marriage and for divorce proceedings	YES	Rights respected
38	To practice any religion	YES	Rights respected
39	To use contraceptive pills and devices	YES	Rights respected
40	To noninterference by state in strictly private affairs	YES	Rights respected

Burma (Myanmar)

Human rights rating: 17%

YES 3 yes 4 no 9 NO 24

Population: 41,700,000
Life expectancy: 61.3
Infant mortality (0–5 years)
   per 1,000 births: 91
United Nations covenants ratified:
None

Form of government: Military regime
GNP per capita (US$): 200
% of GNP spent by state:
On health: 0.8
On military: 3.1
On education: 2.2

FACTORS AFFECTING HUMAN RIGHTS

The country is dominated by a ruthless regime which, after permitting national elections in 1990, refused to accept the popular result in favor of a multiparty democracy. In its efforts to suppress all forms of dissent, the government's policy is to eliminate opposition. Extrajudicial killings and "disappearances" are numerous, torture is widely practiced, and, under martial law, political opponents are detained indefinitely. A number of separatist and insurgent movements control wide areas of the country, maintaining themselves by a flourishing narcotics trade. The government has chosen to isolate itself from international affairs and any improvement in the human rights situation will depend on extreme political changes.

	FREEDOM TO:		COMMENTS
1	Travel in own country	no	Major areas of country in hands of insurgents, separatists, and others. Frequent night curfew. Internal travel monitored by obligation to report change of address, even when temporary
2	Travel outside own country	no	Exit permits, severe scrutiny of applications. Many restrictions
3	Peacefully associate and assemble	NO	Martial law provisions. No criticism of present military regime. No gatherings of more than five people
4	Teach ideas and receive information	NO	Closure of much of the university system. Academics and students regarded as the most threatening of all opposition groups to the present military ruling council
5	Monitor human rights violations	NO	International monitoring groups denied visits. Local human rights groups arrested. Government denies it practices violations
6	Publish and educate in ethnic language	yes	Restrictions are usually those forced on ethnic groups belonging to underprivileged and neglected communities

FREEDOM FROM:		COMMENTS
7	Serfdom, slavery, forced or child labor	**NO**	Forced labor. Dangerous duties such as being compelled at gunpoint to walk through minefields laid by insurgents. Many casualties
8	Extrajudicial killings or "disappearances"	**NO**	Arbitrary killings and "disappearances". Some under martial law, some from excesses by ill-disciplined security forces
9	Torture or coercion by the state	**NO**	Beatings, electric shocks, semisuffocation, "cutting-the-flesh," and many forms of maltreatment
10	Compulsory work permits or conscription of labor	**NO**	No regular pattern to conscripted labor for diverse tasks
11	Capital punishment by the state	**NO**	By hanging. But not of boys or girls under 7 years of age
12	Court sentences of corporal punishment	no	The unremitting violence on individuals, with official encouragement, must be regarded as a form of judicial corporal punishment
13	Indefinite detention without charge	**NO**	Under martial law. Of political opponents, including Buddhist monks, intellectuals, and political dissidents
14	Compulsory membership of state organizations or parties	**YES**	Rights respected
15	Compulsory religion or state ideology in schools	no	Schools "nationalized" by previous one-party government. A policy of "Burmanization" is being pursued by present one
16	Deliberate state policies to control artistic works	yes	Enforcement of conformism in all areas makes artistic self-expression a challenge to the authorities
17	Political censorship of press	**NO**	Censorship by "press security boards" – of all printed matter
18	Censorship of mail or telephone tapping	**NO**	Total. The military leadership is determined to find and suppress all opposition

FREEDOM FOR OR RIGHTS TO:		COMMENTS
19	Peaceful political opposition	**NO**	Deputy commander in chief has stated: "At present we cannot find any organization that can govern the country in a peaceful and stable manner." Many successful candidates of 1990 election have fled the country
20	Multiparty elections by secret and universal ballot	**NO**	Following seizure of power by the military after the National League for Democracy won 80% of parliament seats in 1990 elections, political parties have been suppressed as "unfit to rule" and "subversive"

FREEDOM FOR OR RIGHTS TO:		COMMENTS
21	Political and legal equality for women	yes	Within the severely circumscribed political life of the country, women enjoy almost equal legal rights but are virtually excluded from the current military government
22	Social and economic equality for women	no	Pay and employment inequalities in a society and an economy suffering from maladministration, oppression, and poverty
23	Social and economic equality for ethnic minorities	no	The many ethnic groups suffer various disadvantages, some because of the "Burmanization" of the civil service, some because they belong to peoples now occupying separatist areas of the country
24	Independent newspapers	**NO**	All state owned. Number of newspapers has dropped from over 30 in 1960 to 6
25	Independent book publishing	no	Severely circumscribed. Also by shortage of paper and printing facilities; subject to government control
26	Independent radio and television networks	**NO**	State owned and controlled. No opposition views on the air
27	All courts to total independence	**NO**	A network of military tribunals, with latest martial law orders, may take over cases from the civil courts
28	Independent trade unions	**NO**	Only unrecognized and unofficial associations of workers. No strikes

LEGAL RIGHTS:		COMMENTS
29	From deprivation of nationality	**NO**	Deprived if leaving country illegally and for a few opponents already residing abroad
30	To be considered innocent until proved guilty	no	Under martial law, situation arbitrary. The constant surveillance can best be described as official paranoia
31	To free legal aid when necessary and counsel of own choice	**NO**	No pretrial rights under current martial law. Counsel will be appointed, if court or military tribunal grants the concession
32	From civilian trials in secret	**NO**	Arbitrary. On the orders of military commanders at local level. These tribunals frequently pass death sentences without public announcements
33	To be brought promptly before a judge or court	**NO**	No legal rights under martial law. Appearance in court can be prevented on the authority of local military commander
34	From police searches of home without a warrant	**NO**	Arbitrary. Police and military may search without a warrant

LEGAL RIGHTS:		COMMENTS
35	From arbitrary seizure of personal property	**NO**	Policy of resettlement for security reasons involved 500,000 people in early 1990. Previous residences seized or demolished without compensation

PERSONAL RIGHTS:		COMMENTS
36	To interracial, interreligious, or civil marriage	**YES**	Rights respected
37	Equality of sexes during marriage and for divorce proceedings	yes	Inequalities vary under different tribal laws, usually to the disadvantage of wives
38	To practice any religion	no	Mosques destroyed and Muslim properties seized in Arakan region. Buddhist monks persecuted and monasteries closed as part of suppression of dissent
39	To use contraceptive pills and devices	**YES**	Rights respected
40	To noninterference by state in strictly private affairs	**NO**	State may interfere in all areas, creating anxiety and terror, which affect private affairs. Overnight guests to be reported, long curfews, etc.

Cambodia

Human rights rating: 33%

YES 6 yes 7 no 10 NO 17

Population: 8,200,000
Life expectancy: 49.7
Infant mortality (0–5 years)
  per 1,000 births: 200
United Nations covenants ratified:
Convention on Equality for Women

Form of government: One-party state
(transitional)
GNP per capita (US$): 130
% of GNP spent by state:
On health: n/a
On military: n/a
On education: n/a

FACTORS AFFECTING HUMAN RIGHTS
The Agreement on a Comprehensive Political Settlement of the Cambodia Conflict was finally concluded in October 1991 by all four factions to the 13-year civil war. A Supreme National Council has been formed to bring about a democratic Cambodia over a transitional period of approximately 18 months. A UN authority has been invited by all factions to monitor and supervise the agreement. As well as assisting with the general military demobilization, it will help to ensure that the articles of the Universal Declaration of Human Rights are honored. The questionnaire below was completed at the time of the signing. The situation remains unpredictable and fears persist of further fighting.

	FREEDOM TO:		COMMENTS
1	Travel in own country	no	Restricted by residential registration laws, permits, security checks, and, in much of the country, occupation by the Khmer Rouge guerrilla movement which perpetrated the genocide of 1975–79 against its own people
2	Travel outside own country	no	Exit permits required but difficult to obtain. However, large bribes can overcome official reluctance
3	Peacefully associate and assemble	NO	Only in support of the Phnom Penh government or the single political party
4	Teach ideas and receive information	NO	Ideas and curricula follow government policy and ideological guidelines
5	Monitor human rights violations	yes	More visits by international monitors permitted as the Phnom Penh government seeks wider recognition
6	Publish and educate in ethnic language	YES	Rights respected

FREEDOM FROM: COMMENTS

7   Serfdom, slavery, forced or child labor — **no** — Apart from regional pattern of child workers, the military of all factions impress labor at local level

8   Extrajudicial killings or "disappearances" — **NO** — Killings by all combatants extend to summary executions of suspected opponents. But the peace negotiations just concluded in Paris have moderated the worst excesses

9   Torture or coercion by the state — **NO** — By all factions. Prisoners in leg irons for long periods

10 Compulsory work permits or conscription of labor — **NO** — By all factions. Conscription adapted to local needs or specific projects

11 Capital punishment by the state — **yes** — Officially abolished by the Phnom Penh government in 1989 but summary executions by its military forces may still continue

12 Court sentences of corporal punishment — **YES** — Rights respected

13 Indefinite detention without charge — **NO** — The 13-year war between large opposing armies has meant the detention of hundreds of political suspects - as well as recognized "prisoners of war"

14 Compulsory membership of state organizations or parties — **YES** — Rights respected

15 Compulsory religion or state ideology in schools — **no** — The teaching by the Phnom Penh government of communist ideology has lessened with the withdrawal of its allies, the Vietnamese army. Buddhist teaching in schools is increasing

16 Deliberate state policies to control artistic works — **no** — Creative artists exercise self-censorship. Little official tolerance of work considered "subversive"

17 Political censorship of press — **NO** — Only clandestine leaflets escape censorship

18 Censorship of mail or telephone tapping — **no** — Within the capacity of the obsolete or war-damaged infrastructure, surveillance of suspected "subversives." These include some Cambodians living abroad

FREEDOM FOR OR RIGHTS TO: COMMENTS

19 Peaceful political opposition — **NO** — The People's Revolutionary Party is the only political organization permitted by the Phnom Penh government

20 Multiparty elections by secret and universal ballot — **NO** — Until the peace treaty and the formation of a Supreme National Council, Cambodia has been a one-party state. Changes will depend on the parties honoring the Paris agreement

	FREEDOM FOR OR RIGHTS TO:		COMMENTS
21	Political and legal equality for women	yes	Traditional and regional factors limit women's participation in politics and professions
22	Social and economic equality for women	no	Traditional inequalities. Women occasionally conscripted by government and the various guerrilla groups
23	Social and economic equality for ethnic minorities	YES	Rights respected. The withdrawal of the Vietnamese army has also meant the departure of many of their nationals resettled in Cambodia and forming a favored minority
24	Independent newspapers	NO	State owned and controlled
25	Independent book publishing	NO	State owned and controlled
26	Independent radio and television networks	NO	State owned and controlled
27	All courts to total independence	NO	Political, ideological, and security considerations prevail. Senior legal officials are usually party members
28	Independent trade unions	NO	Under Communist party control

	LEGAL RIGHTS:		COMMENTS
29	From deprivation of nationality	YES	Rights respected
30	To be considered innocent until proved guilty	yes	Prevailing assumptions of guilt without proof are lessening with government attempts to improve the whole legal system
31	To free legal aid when necessary and counsel of own choice	yes	But counsel chosen from a government panel
32	From civilian trials in secret	NO	In secret, when considered necessary
33	To be brought promptly before a judge or court	no	The system varies, with many dependents sent off to camps for "reeducation." New legal system promised by government
34	From police searches of home without a warrant	NO	Of those suspected of supporting guerrilla movements. Legal formalities not recognized in a bitter civil war.
35	From arbitrary seizure of personal property	no	Looting and banditry by armed factions of all combatants. Situation frequently anarchic

PERSONAL RIGHTS:		COMMENTS
36	To interracial, interreligious, or civil marriage	yes	Permission needed to marry foreigners or nationals living abroad
37	Equality of sexes during marriage and for divorce proceedings	yes	Traditional attitudes toward women – to their disadvantage
38	To practice any religion	no	Buddhism, officially the state religion, under government supervision and sometimes forced to support policies
39	To use contraceptive pills and devices	YES	Rights respected
40	To noninterference by state in strictly private affairs	NO	The exigencies of military and security needs, and civil war conditions, affect all aspects of private and family life. Wide neighborhood surveillance

Cameroon

Human rights rating: 56%

YES 10 **yes** 14 **no** 13 NO 3

Population: 11,800,000
Life expectancy: 53.7
Infant mortality (0–5 years)
 per 1,000 births: 150
United Nations covenants ratified:
Civil and Political Rights, Economic,
Social and Cultural Rights

Form of government: One-party state
GNP per capita (US$): 1,010
% of GNP spent by state:
On health: 0.8
On military: 1.7
On education: 3.0

FACTORS AFFECTING HUMAN RIGHTS
The country continues to be a one-party state despite the government's tentative
declaration of favoring multiparty democracy. Opposition political rallies have been
banned and unlawful ones have provoked violent police reactions, resulting in a
number of deaths. Torture and indefinite detention occur, the extent of these depending
on the perceived threat to the government. Nevertheless, there has recently been an
improvement in the human rights situation. The mixed population consists of nearly
200 different tribes.

	FREEDOM TO:		COMMENTS
1	Travel in own country	yes	Roadblocks in country areas to check identity and "voting" cards may result in arrests – or payment of bribes to corrupt security police
2	Travel outside own country	yes	Passports occasionally denied to political opponents of government
3	Peacefully associate and assemble	no	Frequent bannings though automatic permission given to rallies and events supporting government
4	Teach ideas and receive information	yes	Staff and students under regular surveillance. Some reluctance to discuss provocative political issues
5	Monitor human rights violations	no	Government intention to set up human rights committee still under consideration. Reluctance rather than refusal to cooperate with international monitors
6	Publish and educate in ethnic language	yes	Many of the 230 languages are without a written form. The English-speaking 20% minority protest at minor discrimination

	FREEDOM FROM:		COMMENTS
7	Serfdom, slavery, forced or child labor	yes	Regional pattern of child labor
8	Extrajudicial killings or "disappearances"	no	Usually when undisciplined security forces clash with demonstrators
9	Torture or coercion by the state	**NO**	Torture continues despite government declarations of intention to control violations. Beatings, electric shocks, etc.
10	Compulsory work permits or conscription of labor	**YES**	Rights respected
11	Capital punishment by the state	**NO**	By shooting or hanging but recent sentences not yet carried out
12	Court sentences of corporal punishment	**YES**	Rights respected
13	Indefinite detention without charge	no	Long periods of detention, frequently at local level, for opposition politicians
14	Compulsory membership of state organizations or parties	yes	But membership of government party almost compulsory for those seeking self-advancement
15	Compulsory religion or state ideology in schools	**YES**	25% of population follow animism
16	Deliberate state policies to control artistic works	**YES**	Rights respected
17	Political censorship of press	no	Many instances of censorship. The recent trend toward a freer press appears to be reversing
18	Censorship of mail or telephone tapping	yes	The degree of surveillance depends on the president's assessment of threat to his authority

	FREEDOM FOR OR RIGHTS TO:		COMMENTS
19	Peaceful political opposition	no	In practice opposition restricted. President appoints all senior ministers and administrators
20	Multiparty elections by secret and universal ballot	no	Despite constitutional guarantees, in practice Cameroon is a one-party state
21	Political and legal equality for women	yes	Situation improving but inequalities continue
22	Social and economic equality for women	no	Traditional discrimination varies across an ethnically diversified society. Female circumcision still practiced in Muslim areas
23	Social and economic equality for ethnic minorities	yes	With over 200 tribes, disadvantages are local and not by government design

FREEDOM FOR OR RIGHTS TO:		COMMENTS
24	Independent newspapers	**no**	Public to be protected from "irresponsible journalism." All editions to be checked by censor before distribution
25	Independent book publishing	yes	Understood guidelines have to be followed
26	Independent radio and television networks	no	Government controlled. Some independent programs now permitted
27	All courts to total independence	no	Courts controlled by Ministry of Justice. Political pressures occasionally influence judges' independence. Some corruption
28	Independent trade unions	**NO**	In practice unions come under a single government-controlled organization. Strikes illegal

LEGAL RIGHTS:		COMMENTS
29	From deprivation of nationality	**YES**	Rights respected
30	To be considered innocent until proved guilty	no	Frequently "guilty until proved innocent." French colonial law usually followed
31	To free legal aid when necessary and counsel of own choice	yes	Aid for most needy but court appoints defending counsel
32	From civilian trials in secret	yes	Since end of 1989 state of emergency, secret trials rare
33	To be brought promptly before a judge or court	yes	Within 24 hours but this may be followed by long delays before commencement of trial, particularly in political cases
34	From police searches of home without a warrant	no	Many arbitrary searches when wide interpretation of "subversion" suspected
35	From arbitrary seizure of personal property	**YES**	Rights respected

PERSONAL RIGHTS:		COMMENTS
36	To interracial, interreligious, or civil marriage	**YES**	Rights respected
37	Equality of sexes during marriage and for divorce proceedings	yes	The many marriages under customary (tribal) law always favor husbands
38	To practice any religion	**YES**	Rights respected
39	To use contraceptive pills and devices	**YES**	Rights respected
40	To noninterference by state in strictly private affairs	**YES**	Rights respected though situation may vary under traditional customary law

Canada

Human rights rating: 94%

YES 33 yes 7 no 0 NO 0

Population: 26,500,000
Life expectancy: 77.0
**Infant mortality (0–5 years)
  per 1,000 births**: 8
United Nations covenants ratified:
Civil and Political Rights, Economic,
Social and Cultural Rights, Convention
on Equality for Women

Form of government: Federal
government
GNP per capita (US$): 16,960
% of GNP spent by state:
On health: 6.6
On military: 2.2
On education: 5.6

FACTORS AFFECTING HUMAN RIGHTS
The individual respected. Government answerable to the people. A prosperous country, democratic traditions, and a free press and institutions. The country is absorbing a large influx of immigrants from Eastern Europe and the Far East. Government authority is shared between the federation and the ten provinces and two territories. There are occasional language disputes between the minority French-speaking area and that of the English-speaking majority. Human rights violations are most evident in police abuses against a small black minority and clashes between native Indians and the authorities, which occasionally lead to violence.

	FREEDOM TO:		COMMENTS
1	Travel in own country	**YES**	Rights respected
2	Travel outside own country	**YES**	Rights respected
3	Peacefully associate and assemble	**YES**	Rights respected
4	Teach ideas and receive information	**YES**	Rights respected
5	Monitor human rights violations	**YES**	Rights respected
6	Publish and educate in ethnic language	yes	Action by government of Quebec province, restricting use of English to protect the French language, violates UN covenant. Other area of imposing French language may extend to minority indigenous peoples

	FREEDOM FROM:		COMMENTS
7	Serfdom, slavery, forced or child labor	**YES**	Rights respected

FREEDOM FROM: COMMENTS

#	Item		Comments
8	Extrajudicial killings or "disappearances"	YES	Rights respected
9	Torture or coercion by the state	yes	Occasional abuses by police, usually as violence against non whites
10	Compulsory work permits or conscription of labor	YES	Rights respected
11	Capital punishment by the state	YES	Rights respected. Abolished 1976
12	Court sentences of corporal punishment	YES	Rights respected
13	Indefinite detention without charge	YES	Rights respected
14	Compulsory membership of state organizations or parties	YES	Rights respected
15	Compulsory religion or state ideology in schools	YES	Rights respected
16	Deliberate state policies to control artistic works	YES	Rights respected but a new federal grants scheme to encourage a greater "Canadian content" in the national culture has been criticized
17	Political censorship of press	YES	Rights respected
18	Censorship of mail or telephone tapping	YES	Rights respected

FREEDOM FOR OR RIGHTS TO: COMMENTS

#	Item		Comments
19	Peaceful political opposition	YES	Rights respected
20	Multiparty elections by secret and universal ballot	YES	Elections every 5 years for House of Commons unless dissolved earlier
21	Political and legal equality for women	yes	In practice women underrepresented, particularly in Parliament
22	Social and economic equality for women	yes	Pay and employment inequalities despite new government legislation to counter discrimination
23	Social and economic equality for ethnic minorities	yes	Indian and Eskimo minorities, 1% of population, claim disadvantages, particularly in pay, employment, and land disputes
24	Independent newspapers	YES	Rights respected
25	Independent book publishing	YES	Rights respected

FREEDOM FOR OR RIGHTS TO:		COMMENTS
26	Independent radio and television networks	**YES**	Private and state owned, the latter strictly supervised against political influences or bias
27	All courts to total independence	**YES**	Rights respected
28	Independent trade unions	**YES**	Rights respected

LEGAL RIGHTS:		COMMENTS
29	From deprivation of nationality	**YES**	Rights respected
30	To be considered innocent until proved guilty	**YES**	Province of Quebec law based on Napoleonic Code. Rest of Canada on English common law
31	To free legal aid when necessary and counsel of own choice	yes	Means test. Sometimes only for murder and similar serious crimes
32	From civilian trials in secret	**YES**	National security cases only
33	To be brought promptly before a judge or court	**YES**	Rights respected
34	From police searches of home without a warrant	yes	Local police abuses under guise of drugs or "security" checks
35	From arbitrary seizure of personal property	**YES**	Rights respected

PERSONAL RIGHTS:		COMMENTS
36	To interracial, interreligious, or civil marriage	**YES**	Rights respected
37	Equality of sexes during marriage and for divorce proceedings	**YES**	Rights respected
38	To practice any religion	**YES**	Nearly half the population is Roman Catholic
39	To use contraceptive pills and devices	**YES**	Government support
40	To noninterference by state in strictly private affairs	**YES**	Rights respected

Chile

Human rights rating: 80%

YES 21 yes 18 no 0 NO 1

Population: 13,200,000
Life expectancy: 71.8
Infant mortality (0–5 years)
per 1,000 births: 27
United Nations covenants ratified:
Civil and Political Rights, Economic,
Social and Cultural Rights, Convention
on Equality for Women

Form of government: Multiparty
system
GNP per capita (US$): 1,510
% of GNP spent by state:
On health: 2.1
On military: 3.6
On education: 4.0

FACTORS AFFECTING HUMAN RIGHTS
The end of 17 years of a military dictatorship was followed by presidential and
congressional elections; 95% of the electorate voted in a complex multiparty system.
The previous military regime still enjoys a limited influence and the civilian
government is occasionally in dispute with the armed forces, particularly on the issue
of bringing to trial those guilty of human rights crimes under the previous regime.
Human rights, overall, have improved and many reforms are being introduced.

	FREEDOM TO:		COMMENTS
1	Travel in own country	**YES**	Rights respected
2	Travel outside own country	**YES**	Rights respected
3	Peacefully associate and assemble	**YES**	Rights respected
4	Teach ideas and receive information	**YES**	Rights respected
5	Monitor human rights violations	**YES**	Rights respected
6	Publish and educate in ethnic language	**YES**	Rights respected

	FREEDOM FROM:		COMMENTS
7	Serfdom, slavery, forced or child labor	yes	Limited rural child labor
8	Extrajudicial killings or "disappearances"	yes	The interim period following the end of the military regime has been marked by clashes between security police and left-wing terrorists. Many killings alleged to be the result of the use of excessive force by the police

FREEDOM FROM:		COMMENTS
9	Torture or coercion by the state	yes	Greatly reduced since the change to genuine civilian government
10	Compulsory work permits or conscription of labor	YES	Rights respected
11	Capital punishment by the state	NO	Not yet abolished but a reduction in number of crimes for execution under consideration
12	Court sentences of corporal punishment	YES	Rights respected
13	Indefinite detention without charge	YES	Rights respected
14	Compulsory membership of state organizations or parties	YES	Rights respected
15	Compulsory religion or state ideology in schools	YES	Rights respected
16	Deliberate state policies to control artistic works	YES	Rights respected
17	Political censorship of press	yes	But charges such as "offending" the armed forces or the president, and violent assaults on journalists, have inhibited a free press
18	Censorship of mail or telephone tapping	yes	Some surveillance, particularly against suspected terrorists, though arbitrary practices of previous regime occasionally persist

FREEDOM FOR OR RIGHTS TO:		COMMENTS
19	Peaceful political opposition	YES	Rights respected
20	Multiparty elections by secret and universal ballot	yes	Some aspects of a complex multiparty system adopted in 1989 criticized for being unduly influenced by factions of the previous military regime
21	Political and legal equality for women	yes	Position improving. More participation in government and the professions since end of military regime
22	Social and economic equality for women	yes	Minor discriminations in pay and employment
23	Social and economic equality for ethnic minorities	YES	Rights respected
24	Independent newspapers	yes	A degree of caution persists after 17 years of military rule
25	Independent book publishing	yes	As question 24
26	Independent radio and television networks	yes	Significant progress in change from state control to private or public ownership

FREEDOM FOR OR RIGHTS TO:		**COMMENTS**
27	All courts to total independence	yes	Many judges, despite greater independence, still influenced by practices and service under the previous regime
28	Independent trade unions	yes	Situation being transformed by amendments to liberalize coercive trade union laws

LEGAL RIGHTS:		**COMMENTS**
29	From deprivation of nationality	**YES**	Rights respected
30	To be considered innocent until proved guilty	**YES**	Rights respected
31	To free legal aid when necessary and counsel of own choice	yes	Means test. Counsel may be appointed by court
32	From civilian trials in secret	yes	The influence of a less powerful military may still extend to having cases they regard as "security" tried in secret
33	To be brought promptly before a judge or court	yes	Period of readjustment to a more democratic system. 5 days to appear in court but not always honored
34	From police searches of home without a warrant	yes	New reforms have greatly reduced arbitrary searches and surveillance
35	From arbitrary seizure of personal property	**YES**	Rights respected

PERSONAL RIGHTS:		**COMMENTS**
36	To interracial, interreligious, or civil marriage	**YES**	Males legally from 14 years, females from 12 years
37	Equality of sexes during marriage and for divorce proceedings	yes	Amendments under the new constitution are correcting many inequalities
38	To practice any religion	**YES**	Rights respected
39	To use contraceptive pills and devices	**YES**	Rights respected
40	To noninterference by state in strictly private affairs	**YES**	Rights respected

China

Human rights rating: 21%

YES 2 **yes** 8 **no** 8 NO 22

Population: 1,139,100,000
Life expectancy: 70.1
Infant mortality (0–5 years)
per 1,000 births: 43
United Nations covenants ratified:
Convention on Equality for Women

Form of government: One-party
Communist state
GNP per capita (US$): 330
% of GNP spent by state:
On health: 1.4
On military: 6.0
On education: 2.7

FACTORS AFFECTING HUMAN RIGHTS
The Chinese Communist party has imposed an authoritarian regime on one-fifth of the world's population since 1949. During the last 20 years there have been periods when the government was less repressive, but with the events of June 1989, when a student-led democracy movement in Beijing and elsewhere briefly challenged the regime, the leadership, supported by the security forces and the People's Liberation Army, reverted to harsher policies. During 1991, however, with the democracy movement crushed, about 900 of those involved were released from prison and the government has made a few gestures toward moderating their human rights crimes. The reasons for this may be the vulnerability of the leadership since the virtual end of communism in Europe and, secondly, the need to sustain a program of economic reforms.

FREEDOM TO:		COMMENTS
1 Travel in own country	no	Internal migration subject to permits. These cover work transfers and new places of residence. The recent relaxation of permitting more internal migration of the population, following the Tiananmen Square events, has been reversed
2 Travel outside own country	no	Limited easing of strict controls permitting family reunions abroad; but not for suspected dissidents. Small tourist parties now visiting Soviet Far East and Hong Kong
3 Peacefully associate and assemble	NO	New law ensures no repetition of pretexts for the 1989 demonstrations. Apart from forbidden political issues, no occupying public squares, disturbing city life, or displaying provocative banners
4 Teach ideas and receive information	NO	After Tiananmen Square events, total control imposed on educational curricula, with more emphasis on ideological instruction

FREEDOM TO:		COMMENTS
5 Monitor human rights violations	**NO**	Opposed as intrusion. The idea of Marxist human rights, that is, of collective rights, needing to prevail over individual rights (i.e., capitalist human rights) is the basis of official policy. However, contrary to that policy, there has been a more flexible response to recent human rights initiatives by the US government
6 Publish and educate in ethnic language	yes	Local autonomy in many ethnic minority areas

FREEDOM FROM:		COMMENTS
7 Serfdom, slavery, forced or child labor	no	Labor camps and "reeducation through work" centers. Government figure for new admissions exceeds 200,000 each year
8 Extrajudicial killings or "disappearances"	**NO**	The routine of killings and disappearances continues. Despite reluctance of government to give figures, clandestine reports and estimates for 1990 suggest about 50 victims
9 Torture or coercion by the state	**NO**	The practice increased during the interrogation of detainees after the Tiananmen Square events. Beatings, electric shocks, shackling, and incommunicado detention
10 Compulsory work permits or conscription of labor	**NO**	The "reeducation camps" have been enlarged and boosted following the influx of protestors and students from the 1989 demonstrations
11 Capital punishment by the state	**NO**	More than 40 crimes carry the death penalty. By shooting in the back of the neck. Many hundreds of executions after the 1989 protests
12 Court sentences of corporal punishment	no	Violence toward prison inmates an accepted punishment for "troublemakers"
13 Indefinite detention without charge	**NO**	The many dissidents from earlier periods have been joined by an influx from the student generation of 1989. Described as counterrevolutionaries, the government has remained secretive about their numbers
14 Compulsory membership of state organizations or parties	yes	Although not compulsory, the party membership of nearly 50 million enjoys better social prospects and career opportunities
15 Compulsory religion or state ideology in schools	**NO**	Marxism-Leninism a compulsory subject

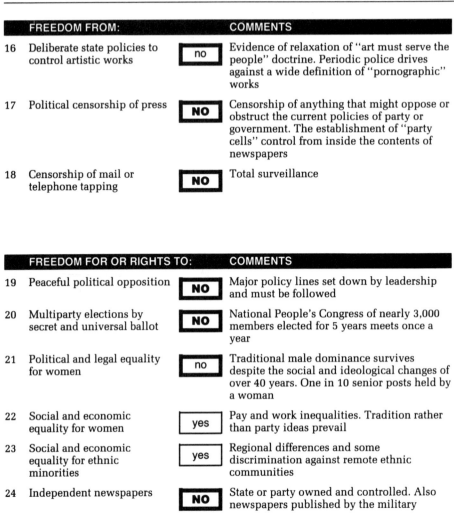

FREEDOM FROM:		COMMENTS
16	Deliberate state policies to control artistic works	no	Evidence of relaxation of "art must serve the people" doctrine. Periodic police drives against a wide definition of "pornographic" works
17	Political censorship of press	NO	Censorship of anything that might oppose or obstruct the current policies of party or government. The establishment of "party cells" control from inside the contents of newspapers
18	Censorship of mail or telephone tapping	NO	Total surveillance

FREEDOM FOR OR RIGHTS TO:		COMMENTS
19	Peaceful political opposition	NO	Major policy lines set down by leadership and must be followed
20	Multiparty elections by secret and universal ballot	NO	National People's Congress of nearly 3,000 members elected for 5 years meets once a year
21	Political and legal equality for women	no	Traditional male dominance survives despite the social and ideological changes of over 40 years. One in 10 senior posts held by a woman
22	Social and economic equality for women	yes	Pay and work inequalities. Tradition rather than party ideas prevail
23	Social and economic equality for ethnic minorities	yes	Regional differences and some discrimination against remote ethnic communities
24	Independent newspapers	NO	State or party owned and controlled. Also newspapers published by the military
25	Independent book publishing	NO	A brief period of relaxation was reversed after the June 1989 events. All books scrutinized for antiparty views. Pulping and burning of books that have become "out of favor"
26	Independent radio and television networks	NO	State owned and controlled by Central Television and Radio
27	All courts to total independence	NO	The Communist party and leadership may influence all levels of courts, even to the point of determining the sentences of those presumed to be guilty before their trial
28	Independent trade unions	NO	Trade unions, in practice, are part of the party. No strikes

LEGAL RIGHTS:		COMMENTS
29	From deprivation of nationality	**YES**	Rights respected
30	To be considered innocent until proved guilty	**NO**	The system permits "sentencing before the actual trial." The notion of "innocence until proved guilty" considered a "bourgeois Western aberration"
31	To free legal aid when necessary and counsel of own choice	no	Counsel a servant of the court. Instances of defending lawyers being imprisoned for advising their clients too well
32	From civilian trials in secret	**NO**	All trials concerned with counterrevolutionary activities may be in secret
33	To be brought promptly before a judge or court	no	24–48 hours for the initial interrogation but many instances of long delays and families of detainees not being informed
34	From police searches of home without a warrant	**NO**	Position arbitrary. Legal safeguards regularly ignored
35	From arbitrary seizure of personal property	yes	The ideological compromises of the government toward a more prosperous society include encouraging greater materialism – and therefore making fewer arbitrary seizures

PERSONAL RIGHTS:		COMMENTS
36	To interracial, interreligious, or civil marriage	yes	Official disapproval of marriage to foreigners. Rare instances of arrests of Chinese partners to prevent such marriages
37	Equality of sexes during marriage and for divorce proceedings	yes	Legal age for marriage: males 22, females 20 years. Illegal arranged marriages and traditional dowry system have not been totally eliminated
38	To practice any religion	yes	Registration of foreign religions depends on restraint, which includes a ban on proselytizing, noninvolvement in any of the political issues of the country, and not being influenced from abroad
39	To use contraceptive pills and devices	**YES**	Active support. Strict policy of family limitation. State-encouraged abortions when limit exceeded
40	To noninterference by state in strictly private affairs	**NO**	The controls on private life extend into the home. Neighborhood watch groups, limitations on family planning, intense social pressures to conform, etc.

Colombia

Human rights rating: 60%

YES 15 yes 11 no 11 NO 3

Population: 33,000,000
Life expectancy: 68.8
Infant mortality (0–5 years) per 1,000 births: 50
United Nations covenants ratified:
Civil and Political Rights, Economic,
Social and Cultural Rights, Convention
on Equality for Women

Form of government: Multiparty system
GNP per capita (US$): 1,180
% of GNP spent by state:
On health: 0.8
On military: 1.0
On education: 2.8

FACTORS AFFECTING HUMAN RIGHTS
The country provides a contrasting picture of a civilian government elected through the democratic process having to contend with a society disrupted by armed violence, corruption, and an enormously rich and powerful narcotics trade. Despite efforts by outside parties, including the US government, to combat the narcotraffickers and reduce the lawlessness, human rights violations of the most brutal kind persist throughout the country. Extrajudicial killings, torture, and disappearances are practiced by the national police, the military, and right-wing assassination squads, while narcotraffickers indulge in the murder of judges, human rights monitors, journalists, and politicians. A third faction, left-wing guerrillas, resorts to more politically motivated violence. Efforts by the government to control the atrocities have had only limited success.

	FREEDOM TO:		COMMENTS
1	Travel in own country	yes	Confrontations between rural guerrillas, armed drug-trafficking gangs, and the military mean that travel in certain dangerous areas is restricted
2	Travel outside own country	YES	Rights respected
3	Peacefully associate and assemble	yes	Police permits dependent on law-and-order situation
4	Teach ideas and receive information	YES	Rights respected
5	Monitor human rights violations	no	The danger to human rights monitors comes from paramilitary police and security forces seeking to obstruct investigations into their excesses. Government has little control. Monitors regularly killed or "disappear"
6	Publish and educate in ethnic language	YES	Rights respected

FREEDOM FROM:		COMMENTS
7	Serfdom, slavery, forced or child labor	**NO**	3 million child workers. Instances of "debt bondage"
8	Extrajudicial killings or "disappearances"	**NO**	Over 300 in 1990 but three times that figure not officially confirmed. Most deaths occur during operations against terrorists and drug gangs, with the police aided by right-wing paramilitary units. Among those killed are many villagers on suspicion of working with the criminal groups
9	Torture or coercion by the state	**NO**	Practiced by the military, the police, the paramilitary, the guerrillas, and the drug traffickers
10	Compulsory work permits or conscription of labor	**YES**	Rights respected
11	Capital punishment by the state	**YES**	Abolished 1910
12	Court sentences of corporal punishment	**YES**	Rights respected
13	Indefinite detention without charge	no	Detentions. Some incommunicado, for long periods. Worst in remote areas of country. Despite legal safeguards, government influence limited
14	Compulsory membership of state organizations or parties	**YES**	Rights respected
15	Compulsory religion or state ideology in schools	**YES**	Rights respected
16	Deliberate state policies to control artistic works	**YES**	Rights respected
17	Political censorship of press	no	As well as state-of-siege restrictions – when imposed – the murder of many journalists has inhibited a free press
18	Censorship of mail or telephone tapping	no	Because of the degree of violence and frequent failure of the government to control its security forces, surveillance is arbitrary and extensive

FREEDOM FOR OR RIGHTS TO:		COMMENTS
19	Peaceful political opposition	yes	The obstacle to peaceful elections is not from the government but from hired assassins and terrorist groups. In the 1990 election, three presidential candidates were murdered
20	Multiparty elections by secret and universal ballot	**YES**	Rights respected

FREEDOM FOR OR RIGHTS TO:		COMMENTS
21	Political and legal equality for women	**yes**	Position improving. Women granted vote in 1954 but limited participation in senior government posts
22	Social and economic equality for women	**yes**	Pay and employment inequalities. Worst in rural areas. The traditional view of woman's role persists
23	Social and economic equality for ethnic minorities	**no**	Colombian Indians and blacks suffer wide discrimination in practice. Uprisings of landless Indians against large estate owners condemned as "leftist" violence
24	Independent newspapers	**yes**	Independence affected by government's failure to control violence, frequently by its own forces. Approximately 20 journalists each year are murdered, kidnapped, or "disappear"
25	Independent book publishing	**yes**	Understood guidelines when state of siege invoked. Murder threats if publications offend some of the powerful armed factions
26	Independent radio and television networks	**yes**	Government controlled but stations are leased to private companies which broadcast freely – though subject to threats from the various violent groups
27	All courts to total independence	**no**	Independence under constitution but the reality is numerous death threats, occasional murder of judges by the many groups of well-armed assassins, and corruption
28	Independent trade unions	**yes**	Less than 15% of workers unionized. Many killed by both right-wing paramilitaries and left-wing guerrillas, for political reasons or during disputes between powerful employers and workers

LEGAL RIGHTS:		COMMENTS
29	From deprivation of nationality	**YES**	Rights respected
30	To be considered innocent until proved guilty	**no**	The overburdened system frequently releases prisoners without trial, the period of being held being a "hit-or-miss" form of punishment
31	To free legal aid when necessary and counsel of own choice	**no**	Inadequate system. Where possible, court appoints counsel for the poor
32	From civilian trials in secret	**yes**	Military tribunals may take over a loose interpretation of cases of sedition. Particularly under state–of–siege decrees

LEGAL RIGHTS:		COMMENTS
33	To be brought promptly before a judge or court		Within 24 hours but many violations of the legal procedures. Arbitrary detentions, sometimes for political reasons, sometimes to extort ransom money from the accused or family
34	From police searches of home without a warrant	no	Overzealous and criminal operations by the military and the security forces result in legal safeguards being consistently ignored
35	From arbitrary seizure of personal property	no	Mostly in remote areas, with the worst culprits being right-wing paramilitary forces

PERSONAL RIGHTS:		COMMENTS
36	To interracial, interreligious, or civil marriage	YES	Rights respected. Marriage at 18 years for both sexes
37	Equality of sexes during marriage and for divorce proceedings		Traditional disadvantages for wives. Worse in rural areas
38	To practice any religion	YES	Rights respected
39	To use contraceptive pills and devices	YES	Government support
40	To noninterference by state in strictly private affairs	YES	Rights respected

Costa Rica

Human rights rating: 90%

YES 30 yes 10 no 0 NO 0

Population: 3,000,000
Life expectancy: 74.9
Infant mortality (0–5 years)
per 1,000 births: 22
United Nations covenants ratified:
Civil and Political Rights, Economic,
Social and Cultural Rights, Convention
on Equality for Women

Form of government: Multiparty
democracy
GNP per capita (US$): 1,690
% of GNP spent by state:
On health: 5.4
On military: nil
On education: 4.5

FACTORS AFFECTING HUMAN RIGHTS
The country enjoys the status of being the most democratic in Central America. In the
1990 presidential election, 80% of the people voted. Human rights are in general
honored; the permanent army was disbanded in 1949. A degree of regional
discrimination against blacks and Indians, though opposed by the government, remains
a social reality, and a number of small indigenous communities live in remote areas and
outside recognized society. The Inter-American Court of Human Rights is based in
Costa Rica.

	FREEDOM TO:		COMMENTS
1	Travel in own country	YES	Rights respected
2	Travel outside own country	YES	Rights respected
3	Peacefully associate and assemble	YES	Rights respected
4	Teach ideas and receive information	YES	Rights respected
5	Monitor human rights violations	YES	Leading supporter of human rights in Latin America
6	Publish and educate in ethnic language	YES	Rights respected

	FREEDOM FROM:		COMMENTS
7	Serfdom, slavery, forced or child labor	yes	Tradition of child labor still continues
8	Extrajudicial killings or "disappearances"	YES	Rights respected
9	Torture or coercion by the state	yes	Occasional instances of arbitrary abuse of prisoners by police

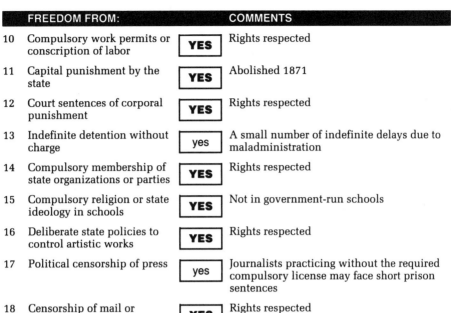

	FREEDOM FROM:		COMMENTS
10	Compulsory work permits or conscription of labor	**YES**	Rights respected
11	Capital punishment by the state	**YES**	Abolished 1871
12	Court sentences of corporal punishment	**YES**	Rights respected
13	Indefinite detention without charge	yes	A small number of indefinite delays due to maladministration
14	Compulsory membership of state organizations or parties	**YES**	Rights respected
15	Compulsory religion or state ideology in schools	**YES**	Not in government-run schools
16	Deliberate state policies to control artistic works	**YES**	Rights respected
17	Political censorship of press	yes	Journalists practicing without the required compulsory license may face short prison sentences
18	Censorship of mail or telephone tapping	**YES**	Rights respected

	FREEDOM FOR OR RIGHTS TO:		COMMENTS
19	Peaceful political opposition	**YES**	Rights respected
20	Multiparty elections by secret and universal ballot	**YES**	Including participation of small Communist parties
21	Political and legal equality for women	yes	Limited political role, particularly in senior government positions
22	Social and economic equality for women	yes	Progress toward equality but traditional discrimination persists
23	Social and economic equality for ethnic minorities	yes	In practice some social and employment discrimination against Amerindians and nonwhites
24	Independent newspapers	**YES**	Rights respected
25	Independent book publishing	**YES**	Rights respected
26	Independent radio and television networks	**YES**	Rights respected
27	All courts to total independence	**YES**	Rights respected
28	Independent trade unions	yes	Prohibitions on some strike actions have brought complaints from the International Labour Organization

LEGAL RIGHTS:		COMMENTS
29	From deprivation of nationality	**YES**	Rights respected
30	To be considered innocent until proved guilty	**YES**	Rights respected
31	To free legal aid when necessary and counsel of own choice	**YES**	Rights respected
32	From civilian trials in secret	**YES**	Rights respected
33	To be brought promptly before a judge or court	yes	Within 24 hours but many long delays, particularly with suspected drug offenders
34	From police searches of home without a warrant	**YES**	Rights respected
35	From arbitrary seizure of personal property	**YES**	Rights respected

PERSONAL RIGHTS:		COMMENTS
36	To interracial, interreligious, or civil marriage	**YES**	Rights respected. Legal age from 15 years
37	Equality of sexes during marriage and for divorce proceedings	yes	Despite constitutional safeguards, many traditional inequalities in favor of husband
38	To practice any religion	**YES**	Rights respected
39	To use contraceptive pills and devices	**YES**	Government support
40	To noninterference by state in strictly private affairs	**YES**	Rights respected

Cuba

Human rights rating: 30%

YES 6 **yes** 6 **no** 7 NO 21

Population: 10,600,000
Life expectancy: 75.4
Infant mortality (0–5 years)
  per 1,000 birth: 14
United Nations covenants ratified:
Convention on Equality for Women

Form of government: One-party
Marxist state
GNP per capita (US$): 2,000
% of GNP spent by state:
On health: 3.2
On military: 7.4
On education: 6.2

FACTORS AFFECTING HUMAN RIGHTS
For over 30 years the country has been a one-party Communist state under the
leadership of President Fidel Castro. All aspects of government, political life, and
society are directed toward the perpetuation of the system and its ideology. Although
there is less overt brutality than in previous years, human rights relating to political
opposition, a free press, the independence of courts, trade unions and seats of learning
are severely circumscribed. With the end of communism in the Soviet Union, the
economic aid and international support given by that country to Cuba are being
terminated, and new austerity measures are being adopted. Human rights
improvements are unlikely until the regime is deposed.

	FREEDOM TO:		COMMENTS
1	Travel in own country		Rights respected
2	Travel outside own country		Exit visas required and difficult to obtain. Women over 40 years and men over 45 receive preference
3	Peacefully associate and assemble		Only in favor of government. A network of informers infiltrates suspected dissident associations
4	Teach ideas and receive information		All seats of learning controlled by government. Curricula planned to consolidate the system and to further communist ideology
5	Monitor human rights violations		Many Cuban monitors continue to be arrested and imprisoned. International investigators such as the United Nations and Red Cross allowed managed visits but their findings repudiated
6	Publish and educate in ethnic language	YES	Rights respected

	FREEDOM FROM:		COMMENTS
7	Serfdom, slavery, forced or child labor	no	Children sent to camps for political indoctrination and to assist in work projects. Degrees of forced labor for political prisoners
8	Extrajudicial killings or "disappearances"	no	A few violations. Usually by use of excessive force by police and while in custody. Government becoming more sensitive to international opinion
9	Torture or coercion by the state	no	Severe maltreatment and psychological coercion practiced rather than a policy of physical violations, etc. Though some beatings do occur
10	Compulsory work permits or conscription of labor	NO	Vagrants and the "work-shy" sent to labor camps for long periods
11	Capital punishment by the state	NO	By shooting. For treason, economic espionage, rape, etc.
12	Court sentences of corporal punishment	YES	Rights respected
13	Indefinite detention without charge	NO	Many. Frequently for demonstrating in favor of human rights and against Cuba's record
14	Compulsory membership of state organizations or parties	yes	Career and social prospects helped by party membership
15	Compulsory religion or state ideology in schools	NO	Marxism-Leninism compulsory in schools
16	Deliberate state policies to control artistic works	no	"Artistic creation is free" but "art must serve the people". Many controls, frequently on grounds of "bourgeois propaganda"
17	Political censorship of press	NO	Journalists are state employees. Total censorship but some criticisms emerging at local level
18	Censorship of mail or telephone tapping	NO	Constant surveillance. The perceived threat from the USA is a major obsession with the regime

	FREEDOM FOR OR RIGHTS TO:		COMMENTS
19	Peaceful political opposition	NO	President Castro has led government uninterruptedly and without recognized opposition since 1959
20	Multiparty elections by secret and universal ballot	NO	Elections limited to Communist party nominees only
21	Political and legal equality for women	yes	But little significant representation at highest levels of government or party
22	Social and economic equality for women	yes	Traditional prejudices survive but women usually have equality in employment

FREEDOM FOR OR RIGHTS TO:		COMMENTS
23	Social and economic equality for ethnic minorities	yes	A communist society but blacks may still suffer minor discrimination because of their skin color
24	Independent newspapers	no	The Roman Catholic newspaper, *Noticias Religioses de Cuba*, a rare independent. Others declared illegal soon after appearance
25	Independent book publishing	**NO**	All publishing houses state controlled. Also imports of foreign books carefully checked
26	Independent radio and television networks	**NO**	State owned and controlled
27	All courts to total independence	**NO**	"Goals of the socialist state must be served by its courts." Lawyers are employed by the state
28	Independent trade unions	**NO**	State controlled. Strikes illegal and "contrary to the Revolution"

LEGAL RIGHTS:		COMMENTS
29	From deprivation of nationality	**NO**	Dissidents have been subjected to forced emigration. Government has used this stratagem to "export" to the USA some of their criminals, psychopaths, etc.
30	To be considered innocent until proved guilty	**NO**	Many cases suffer from unwarrantable assumptions
31	To free legal aid when necessary and counsel of own choice	yes	Free legal service but counsel appointed by court
32	From civilian trials in secret	**NO**	If required by government, secrecy can also cover the identity of the person on trial
33	To be brought promptly before a judge or court	**NO**	Long delays. Timetable decided arbitrarily, depending on whether charge has a political dimension
34	From police searches of home without a warrant	no	Many abuses of constitutional safeguards. Particularly with suspected dissidents
35	From arbitrary seizure of personal property	**NO**	Seizure of property of illegal emigrants and those expelled from country

PERSONAL RIGHTS:		COMMENTS
36	To interracial, interreligious, or civil marriage	**YES**	Rights respected
37	Equality of sexes during marriage and for divorce proceedings	**YES**	Rights respected

PERSONAL RIGHTS:		COMMENTS
38	To practice any religion		Sundry controls on churches and religious activities. Believers may be excluded from state employment and social privileges
39	To use contraceptive pills and devices		State support. Cuba regarded as a model by international population control agency
40	To noninterference by state in strictly private affairs	no	Residential areas have committees of wardens to report on irregularities. Form of harassment on family life. Homosexuals a police target for "bourgeois perversion"

Czechoslovakia   Human rights rating: 97%

YES 35 yes 5 no 0 NO 0

Population: 15,700,000
Life expectancy: 71.8
Infant mortality (0–5 years)
  per 1,000 births: 15
United Nations covenants ratified:
Civil and Political Rights, Economic,
Social and Cultural Rights, Convention
on Equality for Women

Form of government: Multiparty
democracy
GNP per capita (US$): 5,820
% of GNP spent by state:
On health: 4.2
On military: 4.1
On education: 3.6

FACTORS AFFECTING HUMAN RIGHTS
After a period of over 40 years of oppressive Communist rule, the regime yielded to
popular protests and countrywide public demonstrations. Free, multiparty elections
were held in 1991, and the country's political system and human rights record now
compare with the best in Western Europe. Previous political prisoners, approximately
20,000, were freed under a presidential amnesty. Apart from the need to ensure a
successful change to a free market economy from centralized control, a major problem
for the future may be a separatist movement in Slovakia, supported by about 10% of the
population.

	FREEDOM TO:		COMMENTS
1	Travel in own country	**YES**	Rights respected
2	Travel outside own country	**YES**	Rights respected. Situation transformed in 1990. Previous restrictions lifted
3	Peacefully associate and assemble	**YES**	Formality of a police permit for peaceful public rallies frequently disregarded – without consequences
4	Teach ideas and receive information	**YES**	Previous Communist placemen in universities removed in favor of suitably qualified academics. Situation transformed following the end of previous regime
5	Monitor human rights violations	**YES**	The end of Communist rule has been followed by active monitoring by both national and international groups
6	Publish and educate in ethnic language	**YES**	Rights respected though the Hungarian minority considers that their language is not sufficiently supported in education

FREEDOM FROM:		COMMENTS
7	Serfdom, slavery, forced or child labor	**YES**	Rights respected
8	Extrajudicial killings or "disappearances"	**YES**	Rights respected
9	Torture or coercion by the state	**YES**	Rights respected
10	Compulsory work permits or conscription of labour	**YES**	Rights respected
11	Capital punishment by the state	**YES**	Abolished 1990
12	Court sentences of corporal punishment	**YES**	Rights respected
13	Indefinite detention without charge	**YES**	Rights respected. After an amnesty by the new president, 20,000 detainees released
14	Compulsory membership of state organizations or parties	**YES**	Rights respected
15	Compulsory religion or state ideology in schools	**YES**	Rights respected
16	Deliberate state policies to control artistic works	**YES**	Rights respected
17	Political censorship of press	**YES**	Rights respected
18	Censorship of mail or telephone tapping	**YES**	State security abolished. Telephone surveillance only permitted against serious crime, terrorism, and similar. Judicial safeguards

FREEDOM FOR OR RIGHTS TO:		COMMENTS
19	Peaceful political opposition	**YES**	Total democratic political freedom after 42 years of Communist rule
20	Multiparty elections by secret and universal ballot	**YES**	The first parliamentary elections for over 40 years were contested by 22 parties
21	Political and legal equality for women	yes	Despite constitutional equality, underrepresentation at senior levels
22	Social and economic equality for women	yes	Pay and employment inequalities, which may persist or temporarily worsen as the economy changes to free market capitalism
23	Social and economic equality for ethnic minorities	yes	A 1990 law gives the many refugees a protected status. In practice, Gypsies suffer a traditional discrimination, including physical attacks by small hooligan gangs
24	Independent newspapers	**YES**	Rights respected. A free society has produced hundreds of new publications in a single year

FREEDOM FOR OR RIGHTS TO:		COMMENTS
25	Independent book publishing	**YES**	Rights respected
26	Independent radio and television networks	yes	Press law amended in 1990. During the present interim period television and radio remain state owned but with independent management
27	All courts to total independence	yes	Judges previously under the authority of Communist party are being replaced, the situation depending on the availability of suitable appointees
28	Independent trade unions	**YES**	A free union organization has taken over the functions of the previous Communist structure. New union law being drafted

LEGAL RIGHTS:		COMMENTS
29	From deprivation of nationality	**YES**	Rights respected
30	To be considered innocent until proved guilty	**YES**	Rights respected
31	To free legal aid when necessary and counsel of own choice	**YES**	Rights respected
32	From civilian trials in secret	**YES**	Rights respected
33	To be brought promptly before a judge or court	**YES**	Within 24 hours. Limit of 1 year in pretrial custody
34	From police searches of home without a warrant	**YES**	Rights respected
35	From arbitrary seizure of personal property	**YES**	Rights respected

PERSONAL RIGHTS:		COMMENTS
36	To interracial, interreligious, or civil marriage	**YES**	Rights respected
37	Equality of sexes during marriage and for divorce proceedings	**YES**	Rights respected
38	To practice any religion	**YES**	Rights respected
39	To use contraceptive pills and devices	**YES**	Some state support
40	To noninterference by state in strictly private affairs	**YES**	Rights respected

Denmark

Human rights rating: 98%

YES 36 yes 4 no 0 NO 0

Population: 5,100,000
Life expectancy: 75.8
Infant mortality (0–5 years)
  per 1,000 births: 11
United Nations covenants ratified:
Civil and Political Rights, Economic,
Social and Cultural Rights, Convention
on Equality for Women

Form of government: Multiparty
democracy
GNP per capita (US$): 18,450
% of GNP spent by state:
On health: 5.3
On military: 2.1
On education: 7.5

FACTORS AFFECTING HUMAN RIGHTS
The individual respected. The government answerable to the people. A high standard of living, democratic institutions and traditions, a free press,and a cohesive society (over 90% of citizens born in Denmark). The country is among those most concerned about human rights both nationally and internationally.

	FREEDOM TO:		COMMENTS
1	Travel in own country	YES	Rights respected
2	Travel outside own country	YES	Rights respected
3	Peacefully associate and assemble	YES	Permission for public meetings and demonstrations a formality
4	Teach ideas and receive information	YES	Rights respected
5	Monitor human rights violations	YES	Active support by a leading international monitor
6	Publish and educate in ethnic language	YES	Rights respected

	FREEDOM FROM:		COMMENTS
7	Serfdom, slavery, forced or child labor	YES	Rights respected
8	Extrajudicial killings or "disappearances"	YES	Rights respected
9	Torture or coercion by the state	YES	Rights respected
10	Compulsory work permits or conscription of labour	YES	Rights respected

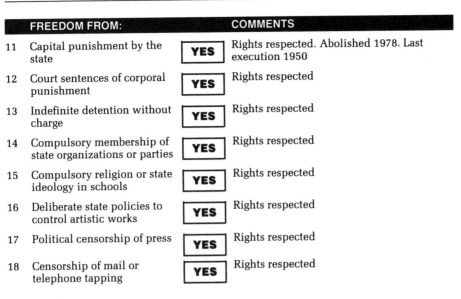

FREEDOM FROM:		COMMENTS
11	Capital punishment by the state	**YES**	Rights respected. Abolished 1978. Last execution 1950
12	Court sentences of corporal punishment	**YES**	Rights respected
13	Indefinite detention without charge	**YES**	Rights respected
14	Compulsory membership of state organizations or parties	**YES**	Rights respected
15	Compulsory religion or state ideology in schools	**YES**	Rights respected
16	Deliberate state policies to control artistic works	**YES**	Rights respected
17	Political censorship of press	**YES**	Rights respected
18	Censorship of mail or telephone tapping	**YES**	Rights respected

FREEDOM FOR OR RIGHTS TO:		COMMENTS
19	Peaceful political opposition	**YES**	Freedom to vote and sit in Parliament at age 18
20	Multiparty elections by secret and universal ballot	**YES**	Complex proportional representation. Legislature has 179 members with two from both Greenland and Faeroe Islands
21	Political and legal equality for women	yes	Underrepresentation despite significant degree of equality
22	Social and economic equality for women	yes	Remaining areas of inequality – pay and work opportunities – being corrected by Equal Rights Council
23	Social and economic equality for ethnic minorities	**YES**	Rights respected
24	Independent newspapers	**YES**	Approximately 50 newspapers
25	Independent book publishing	**YES**	Rights respected
26	Independent radio and television networks	**YES**	Where state owned, a control board ensures independence of management
27	All courts to total independence	**YES**	Rights respected
28	Independent trade unions	**YES**	Rights respected but strikes by public sector workers may occasionally be unlawful

LEGAL RIGHTS: COMMENTS

29	From deprivation of nationality	**YES**	Rights respected
30	To be considered innocent until proved guilty	**YES**	Rights respected
31	To free legal aid when necessary and counsel of own choice	yes	Legally assisted person found guilty must indemnify the public authority
32	From civilian trials in secret	**YES**	Rights respected
33	To be brought promptly before a judge or court	yes	Some instances of long delays but first appearance usually within 24 hours
34	From police searches of home without a warrant	**YES**	Rights respected
35	From arbitrary seizure of personal property	**YES**	Rights respected

PERSONAL RIGHTS: COMMENTS

36	To interracial, interreligious, or civil marriage	**YES**	Rights respected
37	Equality of sexes during marriage and for divorce proceedings	**YES**	Rights respected
38	To practice any religion	**YES**	A growing Muslim refugee minority occasionally conflicts with population (90% Lutheran)
39	To use contraceptive pills and devices	**YES**	Generous state support
40	To noninterference by state in strictly private affairs	**YES**	Rights respected

Dominican Republic Human rights rating: 78%

Population: 7,200,000
Life expectancy: 66.7
Infant mortality (0–5 years)
  per 1,000 births: 80
United Nations covenants ratified:
Civil and Political Rights, Economic,
Social and Cultural Rights, Convention
on Equality for Women

Form of government: Multiparty
system
GNP per capita (US$): 720
% of GNP spent by state:
On health: 1.4
On military: 1.4
On education: 1.6

FACTORS AFFECTING HUMAN RIGHTS
The country is one of the more democratic of the Caribbean region. The multiparty
system has been well established since the adoption of the 1966 constitution though
elections are affected by significant irregularities. There have been large-scale
dismissals of police officers for human rights abuses and the courts and senior officials
are frequently accused of corruption. The sugar industry relies on cheap Haitian labor –
sometimes described as forced labor – and the treatment and living conditions of this
work force, with the complicity of Dominican state agencies, must be regarded as major
human rights violations.

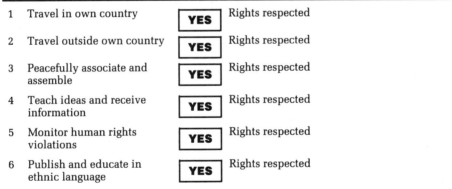

	FREEDOM TO:		COMMENTS
1	Travel in own country	YES	Rights respected
2	Travel outside own country	YES	Rights respected
3	Peacefully associate and assemble	YES	Rights respected
4	Teach ideas and receive information	YES	Rights respected
5	Monitor human rights violations	YES	Rights respected
6	Publish and educate in ethnic language	YES	Rights respected

	FREEDOM FROM:		COMMENTS
7	Serfdom, slavery, forced or child labor	yes	Well-documented evidence of Dominican border guards cooperating with sugar plantation owners in forcing Haitian migrants to work as cheap labor in conditions approaching serfdom

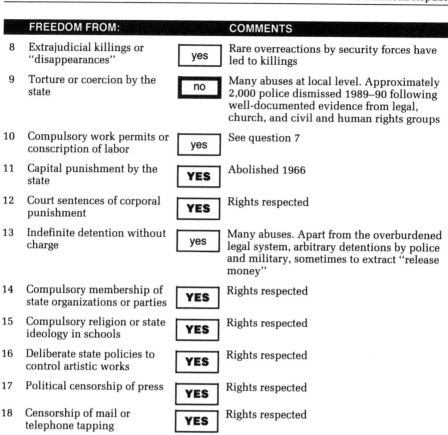

FREEDOM FROM:		COMMENTS
8	Extrajudicial killings or "disappearances"	yes	Rare overreactions by security forces have led to killings
9	Torture or coercion by the state	no	Many abuses at local level. Approximately 2,000 police dismissed 1989–90 following well-documented evidence from legal, church, and civil and human rights groups
10	Compulsory work permits or conscription of labor	yes	See question 7
11	Capital punishment by the state	YES	Abolished 1966
12	Court sentences of corporal punishment	YES	Rights respected
13	Indefinite detention without charge	yes	Many abuses. Apart from the overburdened legal system, arbitrary detentions by police and military, sometimes to extract "release money"
14	Compulsory membership of state organizations or parties	YES	Rights respected
15	Compulsory religion or state ideology in schools	YES	Rights respected
16	Deliberate state policies to control artistic works	YES	Rights respected
17	Political censorship of press	YES	Rights respected
18	Censorship of mail or telephone tapping	YES	Rights respected

FREEDOM FOR OR RIGHTS TO:		COMMENTS
19	Peaceful political opposition	YES	Rights respected
20	Multiparty elections by secret and universal ballot	yes	Many irregularities during 1990 elections. False registrations, government pressures on employees, all in favor of the current administration
21	Political and legal equality for women	yes	Limited representation at senior levels of government and in professions
22	Social and economic equality for women	no	Traditional roles and disadvantages persist, particularly in pay and employment
23	Social and economic equality for ethnic minorities	yes	In practice considerable discrimination against large Haitian work force and the darker-skinned
24	Independent newspapers	YES	Rights respected
25	Independent book publishing	YES	Rights respected

FREEDOM FOR OR RIGHTS TO:		COMMENTS
26	Independent radio and television networks	yes	Occasional government pressures on privately owned stations
27	All courts to total independence	yes	A few instances of corruption of judges. Verdicts may favor the wealthy
28	Independent trade unions	no	New union laws being drafted to protect a labor movement that includes only 15% of work force. Many restrictions on strikes

LEGAL RIGHTS:		COMMENTS
29	From deprivation of nationality	**YES**	Rights respected
30	To be considered innocent until proved guilty	yes	Abuses during government sweeps against black market operators, drug traffickers, etc. An area of regular official corruption
31	To free legal aid when necessary and counsel of own choice	yes	Legal aid for the needy but counsel appointed by court
32	From civilian trials in secret	**YES**	Rights respected
33	To be brought promptly before a judge or court	yes	Some delays beyond the lawful 48 hours. See question 13
34	From police searches of home without a warrant	yes	See question 30
35	From arbitrary seizure of personal property	**YES**	Rights respected

PERSONAL RIGHTS:		COMMENTS
36	To interracial, interreligious, or civil marriage	**YES**	Males may marry at age 16, females at 15
37	Equality of sexes during marriage and for divorce proceedings	yes	Traditional attitudes in society and in the family usually favor husbands. Also apparent in some court judgments
38	To practice any religion	**YES**	Rights respected
39	To use contraceptive pills and devices	**YES**	Government support
40	To noninterference by state in strictly private affairs	**YES**	Rights respected

Ecuador

Human rights rating: 83%

YES 23 yes 14 no 3 NO 0

Population: 10,600,000
Life expectancy: 66.0
Infant mortality (0–5 years) per 1,000 births: 85
United Nations covenants ratified: Civil and Political Rights, Economic, Social and Cultural Rights, Convention on Equality for Women

Form of government: Multiparty system
GNP per capita (US$): 1,120
% of GNP spent by state:
On health: 1.2
On military: 1.6
On education: 4.2

FACTORS AFFECTING HUMAN RIGHTS

A multiparty political system enjoying free elections is nevertheless influenced by the power of the military which, until 1979, was the dominant authority. Torture and similar abuses by the police are widespread and there is considerable discrimination against blacks and Amerindians. Other unsatisfactory areas are the conduct of the courts and prison conditions, both subject to corruption and, in the case of the latter, unrelenting maltreatment of inmates.

	FREEDOM TO:		COMMENTS
1	Travel in own country	YES	Rights respected
2	Travel outside own country	yes	Exit visa dependent on tax clearance and military conscription regulations
3	Peacefully associate and assemble	YES	Only restricted when genuine threat to law and order
4	Teach ideas and receive information	YES	Rights respected
5	Monitor human rights violations	YES	Rights respected
6	Publish and educate in ethnic language	YES	Rights respected

	FREEDOM FROM:		COMMENTS
7	Serfdom, slavery, forced or child labor	yes	Regional pattern of child labor
8	Extrajudicial killings or "disappearances"	YES	Though unproven, rumors of rare "disappearances"
9	Torture or coercion by the state	yes	Local abuses against drug suspects and others under police interrogation

FREEDOM FROM:		COMMENTS
10	Compulsory work permits or conscription of labor	**YES**	Rights respected
11	Capital punishment by the state	**YES**	Abolished 1897
12	Court sentences of corporal punishment	yes	Prison beatings permitted as punishment for misconduct by convicts
13	Indefinite detention without charge	yes	Long delays for trial. Usually because of dilatory and inadequate system
14	Compulsory membership of state organizations or parties	**YES**	Rights respected
15	Compulsory religion or state ideology in schools	**YES**	Rights respected
16	Deliberate state policies to control artistic works	**YES**	Rights respected
17	Political censorship of press	yes	Understood guidelines to avoid censorship. Risks when commenting on the extent of corruption concerning senior government figures
18	Censorship of mail or telephone tapping	**YES**	Rights respected

FREEDOM FOR OR RIGHTS TO:		COMMENTS
19	Peaceful political opposition	**YES**	Wide choice of parties
20	Multiparty elections by secret and universal ballot	**YES**	Voting compulsory
21	Political and legal equality for women	yes	Traditional patterns of female role persist, despite constitutional guarantes. Disadvantages at most levels
22	Social and economic equality for women	no	Pay, employment, and social inequalities
23	Social and economic equality for ethnic minorities	yes	Blacks and Amerindians suffer many disadvantages. Marginal representation in government
24	Independent newspapers	yes	Government may order the publication of information and topics which favor the party in power
25	Independent book publishing	**YES**	Rights respected
26	Independent radio and television networks	yes	Occasional government pressures on the major networks. Rare closures
27	All courts to total independence	yes	But commonly accepted as being vulnerable to manipulation by political and financial influences

FREEDOM FOR OR RIGHTS TO:		COMMENTS
28	Independent trade unions	yes	International Labor Organization has condemned a few restrictions on formation of unions

LEGAL RIGHTS:		COMMENTS
29	From deprivation of nationality	YES	Rights respected
30	To be considered innocent until proved guilty	YES	Rights respected
31	To free legal aid when necessary and counsel of own choice	yes	Legal aid is provided by private associations of lawyers
32	From civilian trials in secret	YES	Rights respected
33	To be brought promptly before a judge or court	no	The 24-hour legal requirement is frequently ignored. Long periods under indictment sometimes ended by bribing officials
34	From police searches of home without a warrant	YES	Rights respected
35	From arbitrary seizure of personal property	YES	Rights respected

PERSONAL RIGHTS:		COMMENTS
36	To interracial, interreligious, or civil marriage	YES	Legal marriage at 14 for males and 12 for females
37	Equality of sexes during marriage and for divorce proceedings	yes	Social customs favor husbands, including divorce settlements and property rights
38	To practice any religion	YES	Rights respected
39	To use contraceptive pills and devices	YES	Government support
40	To noninterference by state in strictly private affairs	no	Prison sentences for homosexual acts between consenting adults

Egypt	Human rights rating:	50%

Population: 52,400,000
Life expectancy: 60.3
Infant mortality (0–5 years)
per 1,000 births: 94
United Nations covenants ratified:
Civil and Political Rights, Economic,
Social and Cultural Rights, Convention
on Equality for Women

Form of government: Executive
president
GNP per capita (US$): 660
% of GNP spent by state:
On health: 1.1
On military: 8.9
On education: 5.5

FACTORS AFFECTING HUMAN RIGHTS

The National Democratic party, with an executive president, has governed the country since 1978, reinforcing its powers for most of that period with the Emergency Law. This has given a legal basis to oppressive measures against the two major opponents of the government, Muslim extremists and left-wing activists. Human rights violations occasionally include extrajudicial killings and, more frequently, torture and indefinite detentions. The nature of the political structure of the country, while permitting a degree of opposition, ensures the continuation in power of the present government.

FREEDOM TO: | | COMMENTS

#	Freedom to		Comments
1	Travel in own country	yes	Except in military zones, which may be extended under Emergency Law
2	Travel outside own country	no	Women particularly affected. Wives need husband's permission to obtain passport, daughters under 21 their father's permission
3	Peacefully associate and assemble	no	Forceful measures against demonstrations by Islamic fundamentalists. All public rallies require Ministry of Interior permission
4	Teach ideas and receive information	no	No atheism or communism to be encouraged. Under Emergency Law, almost continuous since 1981, academics can be arrested for "nonconformism"
5	Monitor human rights violations	yes	Obstruction rather than direct suppression of monitoring
6	Publish and educate in ethnic language	YES	Rights respected

FREEDOM FROM: | | COMMENTS

#	Freedom from		Comments
7	Serfdom, slavery, forced or child labor	yes	Widespread regional child labor.

FREEDOM FROM:		COMMENTS
8	Extrajudicial killings or "disappearances"	**NO**	The upsurge of violent Muslim fundamentalism has resulted in many deaths, some during demonstrations, others while in police custody
9	Torture or coercion by the state	no	Torture commonplace. Beatings, scorchings, electric shocks, etc. Some resulting in deaths. Islamic fundamentalists, left-wing dissidents, and journalists most affected
10	Compulsory work permits or conscription of labor	**YES**	Rights respected
11	Capital punishment by the state	**NO**	Military offenders shot, civilians hanged. Rape may be a capital offense
12	Court sentences of corporal punishment	**YES**	Rights respected
13	Indefinite detention without charge	**NO**	Approximately 1,000 detained for political and religious reasons. Many under the comprehensive Emergency Law
14	Compulsory membership of state organizations or parties	**YES**	Rights respected
15	Compulsory religion or state ideology in schools	no	State support for schools which insist on compulsory religious instruction
16	Deliberate state policies to control artistic works	no	State considers art forms capable of subversion. Controls include censorship of plays and film scripts before production
17	Political censorship of press	yes	Despite a watchful press council and self-censorship, the opposition press is vigorous in its protests of human rights violations
18	Censorship of mail or telephone tapping	**NO**	Constant surveillance under Emergency Law. Correspondence intercepted

FREEDOM FOR OR RIGHTS TO:		COMMENTS
19	Peaceful political opposition	yes	Opposition complains of electoral fraud, particularly at president's election by an allegedly compliant People's Assembly
20	Multiparty elections by secret and universal ballot	yes	But wide presidential powers and a dominant governing party control both the country and the People's Assembly
21	Political and legal equality for women	no	Women hold fewer than one in 10 seats in the assembly. Underrepresented in male-dominated professions
22	Social and economic equality for women	no	Traditional discrimination socially, in pay, and in employment. Modern women's groups campaigning against female circumcision. Alleged to be as high as 80% in rural areas

FREEDOM FOR OR RIGHTS TO:		COMMENTS
23	Social and economic equality for ethnic minorities	**YES**	Rights respected
24	Independent newspapers	yes	Major newspapers are government owned. Editors appointed for their loyalty to the president. Other dailies conform to understood guidelines
25	Independent book publishing	no	Seizures and banning of controversial books. Nobel Prize winner's novel banned. Other novelists face imprisonment, usually for affronting the religious censors. Also nothing which is "sexually stimulating"
26	Independent radio and television networks	no	State owned and controlled but some independent programs
27	All courts to total independence	yes	Only the president, under the Emergency Law, may overturn convictions
28	Independent trade unions	yes	Degree of government control has brought complaints from the International Labour Organization

LEGAL RIGHTS:		COMMENTS
29	From deprivation of nationality	**YES**	Rights respected
30	To be considered innocent until proved guilty	yes	But Emergency Law has resulted in many arbitrary arrests
31	To free legal aid when necessary and counsel of own choice	no	Serious crimes only. Court appoints counsel
32	From civilian trials in secret	yes	State Security Court may vary procedures. Some trials partly in secret under wide definition of "national interest"
33	To be brought promptly before a judge or court	no	Within 48 hours but indefinitely if cases come under Emergency Law provisions
34	From police searches of home without a warrant	no	Many violations by security police in actions against Muslim fundamentalists; also against wide interpretation of security "risks"
35	From arbitrary seizure of personal property	no	Confiscation of property. Punishment permitted under Emergency Law

PERSONAL RIGHTS:		COMMENTS
36	To inter-racial, interreligious civil marriage	no	The religious ministry represents Islamic law, the religion of 94% of population, which forbids all forms of intermarriage

PERSONAL RIGHTS:		COMMENTS
37	Equality of sexes during marriage and for divorce proceedings	yes	But husband may have more than one wife and enjoys a higher traditional status
38	To practice any religion	no	Converts to Christianity from Islam face imprisonment. Minority religions need special permits to construct places of worship. Coptic Christians allege discrimination
39	To use contraceptive pills and devices	**YES**	State support
40	To noninterference by state in strictly private affairs	yes	Except for intrusions, searches, etc., under Emergency Law

El Salvador

Human rights rating: 53%

YES 12 yes 14 no 6 NO 8

Population: 5,300,000
Life expectancy: 64.4
Infant mortality (0–5 years)
per 1,000 births: 90
United Nations covenants ratified:
Civil and Political Rights, Economic,
Social and Cultural Rights, Convention
on Equality for Women

Form of government: Multiparty
system
GNP per capita (US$): 940
% of GNP spent by state:
On health: 0.8
On military: 3.7
On education: 1.9

FACTORS AFFECTING HUMAN RIGHTS
The government has been involved since 1979 in a major military operation against the
National Liberation Front. Despite the heavy losses on both sides, the greatest casualties
have been suffered by civilians. Further killings are committed by "death squads",
these usually being plainclothes soldiers or police. At the end of 1991, after many
attempts, the UN appeared finally to have arranged a peace accord but the environment
of violence, lawlessness, cruelty, and corruption may continue to inhibit human rights
improvements.

	FREEDOM TO:		COMMENTS
1	Travel in own country	yes	The existence of war zones has meant arrests of those who have failed to produce satisfactory identity documents
2	Travel outside own country	YES	Rights respected
3	Peacefully associate and assemble	yes	Despite constitutional rights, the reality of violence by opposing groups, including those of the state, is an inhibiting factor
4	Teach ideas and receive information	YES	Rights respected
5	Monitor human rights violations	YES	By national and international bodies. Monitoring, in view of the widespread violence, may be a high-risk occupation
6	Publish and educate in ethnic language	YES	Rights respected

---

	FREEDOM FROM:		COMMENTS
7	Serfdom, slavery, forced or child labor	yes	Regional pattern of child labor

FREEDOM FROM:		COMMENTS
8	Extrajudicial killings or "disappearances"	**NO**	Action against a major guerrilla movement has encouraged unlawful killings of civilians and peasants by the military and paramilitary "death squads"
9	Torture or coercion by the state	**NO**	Widespread. Beatings, electric shocks, semisuffocation, rape, sleep deprivation, etc.
10	Compulsory work permits or conscription of labor	yes	Abuses include the use of prisoners as laborers for public projects
11	Capital punishment by the state	yes	Abolished for ordinary crimes in 1983 but the murders committed by the military and security forces are frequently justified by the government
12	Court sentences of corporal punishment	yes	The extent of torture must be regarded as condoned corporal punishment by the government
13	Indefinite detention without charge	**NO**	Disregard for lawful procedures has resulted in long detentions, a risk of death for some detainees, and the failure of the security forces to obey orders to inform the Office of Information on Detainees of arrests
14	Compulsory membership of state organizations or parties	**YES**	Rights respected
15	Compulsory religion or state ideology in schools	**YES**	Rights respected
16	Deliberate state policies to control artistic works	**YES**	Rights respected
17	Political censorship of press	yes	Mostly self-censorship, particularly because of murders of many journalists and editors by both terrorist factions and right-wing "death squads"
18	Censorship of mail or telephone tapping	no	Wide surveillance by security forces, treasury police, etc., particularly during periods of state of emergency

FREEDOM FOR OR RIGHTS TO:		COMMENTS
19	Peaceful political opposition	yes	New electoral reforms, if successful, will improve the prospects of peaceful opposition, constantly affected by the violence of opposing factions
20	Multiparty elections by secret and universal ballot	yes	If the guerrilla front is brought into the electoral process, less violence during and after elections may be anticipated
21	Political and legal equality for women	no	Despite the constitution, women remain victims of the social and political violence and traditional attitudes

FREEDOM FOR OR RIGHTS TO:	COMMENTS
22 Social and economic equality for women	**no** — Wide differences in pay and employment opportunities. Regional factors as well as a depressed economy
23 Social and economic equality for ethnic minorities	**YES** — Rights respected
24 Independent newspapers	**yes** — Independence affected by threats from both "death squads" and pro-guerrilla elements. Political prudence added to commercial considerations
25 Independent book publishing	**yes** — As question 24
26 Independent radio and television networks	**yes** — Both state controlled and private. Risk of violence may affect operations
27 All courts to total independence	**yes** — Judges subject to death threats. May also be influenced by their political affiliations
28 Independent trade unions	**no** — Many limitations on organizing unions and strike actions. A newly formed commission is considering reforms

LEGAL RIGHTS:	COMMENTS
29 From deprivation of nationality	**YES** — Rights respected
30 To be considered innocent until proved guilty	**NO** — Deaths and torture are often inflicted by security forces in their efforts to extract confessions
31 To free legal aid when necessary and counsel of own choice	**no** — Resources not usually available for free legal aid
32 From civilian trials in secret	**NO** — Dependent on the insistence of the military or security forces to hold a civilian trial. Many trials in secret – and secret from the government
33 To be brought promptly before a judge or court	**NO** — Arbitrary situation. Within 72 hours but long delays or even "disappearances" of the accused
34 From police searches of home without a warrant	**NO** — Arbitrary situation. Legal procedures constantly ignored with only limited efforts by the government to correct this
35 From arbitrary seizure of personal property	**NO** — Particularly in the war zones. Peasants and small communes suffer most. Significant corruption by the military; looting, etc.

PERSONAL RIGHTS:		COMMENTS
36	To interracial, interreligious, or civil marriage	**YES**	Males may marry at 16 years, females at 14
37	Equality of sexes during marriage and for divorce proceedings	yes	Traditional inequalities of the region usually to disadvantage of wives
38	To practice any religion	**YES**	Rights respected
39	To use contraceptive pills and devices	**YES**	State support
40	To noninterference by state in strictly private affairs	no	The extent of violence, lawlessness, and harassment of communities and individuals creates constant fear in private lives

Finland

Human rights rating: 99%

YES 38 **yes** 2 **no** 0 NO 0

Population: 5,000,000
Life expectancy: 75.5
**Infant mortality (0–5 years)
  per 1,000 births**: 7
United Nations covenants ratified:
Civil and Political Rights, Economic,
Social and Cultural Rights, Convention
on Equality for Women

Form of government: Parliamentary
democracy
GNP per capita (US$): 18,590
% of GNP spent by state:
On health: 6.0
On military: 1.7
On education: 5.9

FACTORS AFFECTING HUMAN RIGHTS
The individual respected. Government answerable to the people. A prosperous
economy, democratic traditions, and a free press. In the past the long frontier with its
powerful neighbor, the USSR, has inhibited the expression of views or actions which
that country might have found unwelcome, but with the end of communism this
restraint no longer applies. Finland is influencing the democratic process in the newly
independent Baltic states.

	FREEDOM TO:		COMMENTS
1	Travel in own country	YES	Rights respected
2	Travel outside own country	YES	Rights respected
3	Peacefully associate and assemble	YES	Formality of police permit for public rallies and demonstrations
4	Teach ideas and receive information	YES	Rights respected
5	Monitor human rights violations	YES	Encouraged by government
6	Publish and educate in ethnic language	YES	Rights respected

	FREEDOM FROM:		COMMENTS
7	Serfdom, slavery, forced or child labor	YES	Rights respected
8	Extrajudicial killings or "disappearances"	YES	Rights respected
9	Torture or coercion by the state	YES	Rights respected

FREEDOM FROM:		COMMENTS
10	Compulsory work permits or conscription of labor	**YES**	Rights respected
11	Capital punishment by the state	**YES**	Abolished in 1972 for all crimes
12	Court sentences of corporal punishment	**YES**	Rights respected
13	Indefinite detention without charge	**YES**	Rights respected
14	Compulsory membership of state organizations or parties	**YES**	Rights respected
15	Compulsory religion or state ideology in schools	**YES**	Rights respected
16	Deliberate state policies to control artistic works	**YES**	Rights respected
17	Political censorship of press	**YES**	With the changed situation in the USSR, previous restraints on issues likely to be provocative no longer necessary
18	Censorship of mail or telephone tapping	**YES**	Rights respected

FREEDOM FOR OR RIGHTS TO:		COMMENTS
19	Peaceful political opposition	**YES**	Rights respected
20	Multiparty elections by secret and universal ballot	**YES**	200-member Parliament by proportional representation. Elections every 4 years. Vote from age 18
21	Political and legal equality for women	**YES**	Women members are approximately one third of parliament
22	Social and economic equality for women	yes	Minor inequalities in pay, employment, and socially, which the Equality Council is seeking to correct
23	Social and economic equality for ethnic minorities	**YES**	Rights respected
24	Independent newspapers	**YES**	Rights respected
25	Independent book publishing	**YES**	Rights respected
26	Independent radio and television networks	**YES**	Independent bodies ensure freedom for state controlled stations
27	All courts to total independence	**YES**	Rights respected
28	Independent trade unions	**YES**	Rights respected

LEGAL RIGHTS:		COMMENTS
29	From deprivation of nationality	**YES**	Rights respected
30	To be considered innocent until proved guilty	**YES**	Rights respected
31	To free legal aid when necessary and counsel of own choice	**YES**	Rights respected
32	From civilian trials in secret	**YES**	Rights respected
33	To be brought promptly before a judge or court	yes	Up to 7 days' detention in towns and 30 days in country areas
34	From police searches of home without a warrant	**YES**	Rights respected
35	From arbitrary seizure of personal property	**YES**	Rights respected

PERSONAL RIGHTS:		COMMENTS
36	To interracial, interreligious, or civil marriage	**YES**	Rights respected
37	Equality of sexes during marriage and for divorce proceedings	**YES**	Rights respected
38	To practice any religion	**YES**	Nearly 90% follow Lutheran faith
39	To use contraceptive pills and devices	**YES**	State support
40	To noninterference by state in strictly private affairs	**YES**	Rights respected

France

Human rights rating: 94%

Population: 56,100,000
Life expectancy: 76.4
Infant mortality (0–5 years) per 1,000 births: 10
United Nations covenants ratified:
Civil and Political Rights, Economic, Social and Cultural Rights, Convention on Equality for Women

Form of government: Democratic republic
GNP per capita (US$): 16,090
% of GNP spent by state:
On health: 6.6
On military: 3.9
On education: 5.9

FACTORS AFFECTING HUMAN RIGHTS

The individual respected. Government answerable to the people. A high standard of living, a dedication to human rights, and a free and cultured press. But the large influx of immigrants from the ex-French colonies, particularly from North Africa, is causing social tensions and sometimes outbreaks of violence. The relatively high rate of unemployment is aggravating these developments, a situation being exploited by extreme right-wing political factions and others calling for a "France for the French." The police and security forces are frequently accused of giving covert support to such views.

	FREEDOM TO:		COMMENTS
1	Travel in own country	YES	Rights respected
2	Travel outside own country	YES	Rights respected
3	Peacefully associate and assemble	yes	Recent unrest among ethnic groups has resulted in some limitations on public meetings normally considered permissible
4	Teach ideas and receive information	YES	Rights respected
5	Monitor human rights violations	YES	Rights respected
6	Publish and educate in ethnic language	YES	Rights respected

	FREEDOM FROM:		COMMENTS
7	Serfdom, slavery, forced or child labor	YES	Rights respected
8	Extrajudicial killings or "disappearances"	YES	Rights respected

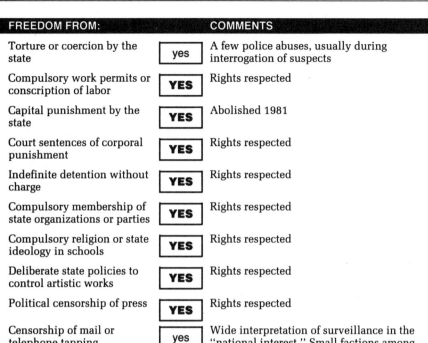

FREEDOM FROM:		COMMENTS
9	Torture or coercion by the state	yes	A few police abuses, usually during interrogation of suspects
10	Compulsory work permits or conscription of labor	YES	Rights respected
11	Capital punishment by the state	YES	Abolished 1981
12	Court sentences of corporal punishment	YES	Rights respected
13	Indefinite detention without charge	YES	Rights respected
14	Compulsory membership of state organizations or parties	YES	Rights respected
15	Compulsory religion or state ideology in schools	YES	Rights respected
16	Deliberate state policies to control artistic works	YES	Rights respected
17	Political censorship of press	YES	Rights respected
18	Censorship of mail or telephone tapping	yes	Wide interpretation of surveillance in the "national interest." Small factions among ethnic minorities regarded as security "risks"

FREEDOM FOR OR RIGHTS TO:		COMMENTS
19	Peaceful political opposition	YES	Rights respected
20	Multiparty elections by secret and universal ballot	YES	Rights respected
21	Political and legal equality for women	yes	Women underrepresented in the National Assembly and Senate. Approximately 5%
22	Social and economic equality for women	yes	Minor inequalities in pay and employment
23	Social and economic equality for ethnic minorities	yes	Many areas of discrimination against blacks and North African Muslims in pay and employment. Position aggravated by right-wing groups agitating for a "France for the French"
24	Independent newspapers	YES	Rights respected
25	Independent book publishing	YES	Rights respected
26	Independent radio and television networks	YES	Rights respected but allegations of government pressures on sensitive issues and pro-government appointees to state owned stations

FREEDOM FOR OR RIGHTS TO:		COMMENTS
27	All courts to total independence	**YES**	Rights respected
28	Independent trade unions	**YES**	Rights respected

LEGAL RIGHTS:		COMMENTS
29	From deprivation of nationality	**YES**	Rights respected
30	To be considered innocent until proved guilty	**YES**	Rights respected
31	To free legal aid when necessary and counsel of own choice	**YES**	Rights respected
32	From civilian trials in secret	**YES**	Rights respected
33	To be brought promptly before a judge or court	yes	4 days maximum but pretrial detention, particularly in drug cases, can extend to a year or longer. Occasional arbitrary abuses by police
34	From police searches of home without a warrant	**YES**	Rights respected
35	From arbitrary seizure of personal property	**YES**	Rights respected

PERSONAL RIGHTS:		COMMENTS
36	To interracial, interreligious, or civil marriage	**YES**	Rights respected
37	Equality of sexes during marriage and for divorce proceedings	**YES**	Rights respected
38	To practice any religion	**YES**	Rights respected
39	To use contraceptive pills and devices	**YES**	State support
40	To noninterference by state in strictly private affairs	**YES**	Rights respected

Germany

Human rights rating: 98%

YES 36 **yes** 4 **no** 0 NO 0

Population: 77,600,000
Life expectancy: 75.2
Infant mortality (0–5 years) per 1,000 births: 10
United Nations covenants ratified: Civil and Political Rights, Economic, Social and Cultural Rights, Convention on Equality for Women

Form of government: Multiparty democracy
GNP per capita (US$): 16,570
% of GNP spent by state:
On health: 6.3
On military: 3.1
On education: 4.5

FACTORS AFFECTING HUMAN RIGHTS

With the unification of East Germany and the Federal Republic in 1990, citizens of the former Communist state now enjoy a multiparty democratic society and the protection of the Basic Law, which enshrines human rights. These are usually respected in practice, but the immediate problem for the country is a population increase of nearly 20% in the prosperous Federal Republic. The burden of a backward state economy and the need for its adaptation to a free market system have meant an increase in unemployment, particularly in the former East Germany. This has provoked hostility and violence toward the many immigrant workers, many from Asia and Africa, a situation exploited by right-wing xenophobic groups.

	FREEDOM TO:		COMMENTS
1	Travel in own country	YES	Rights respected
2	Travel outside own country	YES	Rights respected
3	Peacefully associate and assemble	YES	Police permits only refused when genuine threat to law and order
4	Teach ideas and receive information	YES	Rights respected
5	Monitor human rights violations	YES	Rights respected
6	Publish and educate in ethnic language	YES	Rights respected

	FREEDOM FROM:		COMMENTS
7	Serfdom, slavery, forced or child labor	YES	Rights respected
8	Extrajudicial killings or "disappearances"	YES	Rights respected

FREEDOM FROM:		COMMENTS
9	Torture or coercion by the state	**YES**	Rights respected
10	Compulsory work permits or conscription of labor	**YES**	Rights respected
11	Capital punishment by the state	**YES**	Abolished 1949
12	Court sentences of corporal punishment	**YES**	Rights respected
13	Indefinite detention without charge	**YES**	Rights respected
14	Compulsory membership of state organizations or parties	**YES**	Rights respected
15	Compulsory religion or state ideology in schools	**YES**	Rights respected
16	Deliberate state policies to control artistic works	**YES**	Rights respected
17	Political censorship of press	**YES**	Rights respected
18	Censorship of mail or telephone tapping	**YES**	Rights respected

FREEDOM FOR OR RIGHTS TO:		COMMENTS
19	Peaceful political opposition	**YES**	The Basic Law bans antidemocratic and neo-Nazi parties
20	Multiparty elections by secret and universal ballot	**YES**	Rights respected
21	Political and legal equality for women	yes	Women are increasingly participating in politics at senior levels and in the professions
22	Social and economic equality for women	yes	Inequalities persist in pay and employment, a situation worsened by the union of the two parts of the country and the consequent higher unemployment
23	Social and economic equality for ethnic minorities	yes	Unification of the two parts of Germany has increased unemployment in the former East Germany. This has worsened discrimination against blacks and foreign workers and recent immigrants
24	Independent newspapers	**YES**	Rights respected
25	Independent book publishing	**YES**	Rights respected
26	Independent radio and television networks	**YES**	Rights respected

#	FREEDOM FOR OR RIGHTS TO:		COMMENTS
27	All courts to total independence	YES	Rights respected
28	Independent trade unions	YES	Rights respected

#	LEGAL RIGHTS:		COMMENTS
29	From deprivation of nationality	YES	Rights respected
30	To be considered innocent until proved guilty	YES	Rights respected
31	To free legal aid when necessary and counsel of own choice	yes	Means test. Court appoints counsel
32	From civilian trials in secret	YES	Rights respected
33	To be brought promptly before a judge or court	YES	Within 24 hours. Pre-trial custody may be extended in some cases
34	From police searches of home without a warrant	YES	Rights respected
35	From arbitrary seizure of personal property	YES	Rights respected

#	PERSONAL RIGHTS:		COMMENTS
36	To interracial, interreligious, or civil marriage	YES	Marriage legal at age 18 for both sexes
37	Equality of sexes during marriage and for divorce proceedings	YES	Rights respected
38	To practice any religion	YES	Rights respected
39	To use contraceptive pills and devices	YES	Government support
40	To noninterference by state in strictly private affairs	YES	Rights respected

Ghana

Human rights rating: 53%

YES 10 yes 12 **no** 10 NO 8

Population: 15,000,000
Life expectancy: 55.0
Infant mortality (0–5 years) per 1,000 births: 143
United Nations covenants ratified:
Convention on Equality for Women

Form of government: National council
GNP per capita (US$): 400
% of GNP spent by state:
On health: 1.2
On military: 0.9
On education: 3.6

FACTORS AFFECTING HUMAN RIGHTS

The Provisional National Defence Council under a former military officer has enjoyed absolute power since 1981. Its oppressive policies and its disregard for human rights are explained as being necessary to restore the country's prosperity. There has been some public discussion of more democratic freedoms and tolerance of limited criticism within understood guidelines but anything that seriously challenges the government's authority is quickly suppressed. About 200 political opponents were understood to be indefinitely detained at mid-1991.

FREEDOM TO:		COMMENTS
1 Travel in own country	yes	Minor road checks and searches, usually against smuggling, in both countryside and towns
2 Travel outside own country	no	Opponents of regime frequently refused passports
3 Peacefully associate and assemble	no	Only for rallies and associations in favor of the government
4 Teach ideas and receive information	yes	Understood guidelines for academics and students. Some tolerance of open criticism, provided it is constructive
5 Monitor human rights violations	no	No government cooperation with international monitors. Policy of obstruction rather than banning of international bodies
6 Publish and educate in ethnic language	YES	Rights respected

FREEDOM FROM:		COMMENTS
7 Serfdom, slavery, forced or child labor	yes	Traditional child labor, particularly in rural areas
8 Extrajudicial killings or "disappearances"	yes	There have been no recent coup attempts therefore no mass killings of opponents

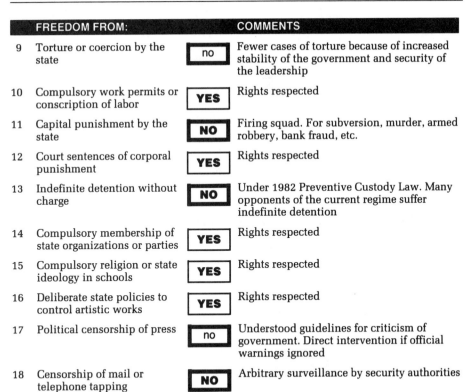

FREEDOM FROM:		COMMENTS
9	Torture or coercion by the state	no	Fewer cases of torture because of increased stability of the government and security of the leadership
10	Compulsory work permits or conscription of labor	YES	Rights respected
11	Capital punishment by the state	NO	Firing squad. For subversion, murder, armed robbery, bank fraud, etc.
12	Court sentences of corporal punishment	YES	Rights respected
13	Indefinite detention without charge	NO	Under 1982 Preventive Custody Law. Many opponents of the current regime suffer indefinite detention
14	Compulsory membership of state organizations or parties	YES	Rights respected
15	Compulsory religion or state ideology in schools	YES	Rights respected
16	Deliberate state policies to control artistic works	YES	Rights respected
17	Political censorship of press	no	Understood guidelines for criticism of government. Direct intervention if official warnings ignored
18	Censorship of mail or telephone tapping	NO	Arbitrary surveillance by security authorities

FREEDOM FOR OR RIGHTS TO:		COMMENTS
19	Peaceful political opposition	NO	All political activities suspended. Government claims its policy is "to encourage the democratic process"
20	Multiparty elections by secret and universal ballot	NO	All legislative power is with the ruling council though nonparty local elections permitted from 1989
21	Political and legal equality for women	yes	Male domination of senior government posts but women well represented at lower levels
22	Social and economic equality for women	no	Pay and employment inequalities with worst extremes in rural areas. Forms of traditional near-bondage of women persist
23	Social and economic equality for ethnic minorities	YES	Rights respected
24	Independent newspapers	no	The two major newspapers government owned. Others conform to accepted guidelines
25	Independent book publishing	yes	No publications hostile to the regime or insulting to the leadership

FREEDOM FOR OR RIGHTS TO:		COMMENTS
26	Independent radio and television networks	**NO**	State owned and controlled. Voice of the government
27	All courts to total independence	no	Government retains power to dismiss judges arbitrarily. Public tribunals sometimes try political and security cases. Their legitimacy not recognized by the Ghana Bar Association
28	Independent trade unions	yes	Unions enjoy limited freedom to organize, strike, and criticize government in a constructive manner. But no major challenges permitted

LEGAL RIGHTS:		COMMENTS
29	From deprivation of nationality	**YES**	Rights respected
30	To be considered innocent until proved guilty	no	When government feels threatened, many arrests "on suspicion." Position currently stable
31	To free legal aid when necessary and counsel of own choice	yes	Free legal aid but counsel appointed from a selected panel
32	From civilian trials in secret	**NO**	Secret trials of political opponents and of military charged with "plotting"
33	To be brought promptly before a judge or court	no	Technically within 48 hours but long delays frequent
34	From police searches of home without a warrant	**NO**	Periodic searches in pursuit of "security risks and plotters." Neighborhood defense committees act as informers
35	From arbitrary seizure of personal property	yes	Political opponents occasionally suffer seizure of property

PERSONAL RIGHTS:		COMMENTS
36	To interracial, interreligious, or civil marriage	**YES**	Marriage age 21 for both sexes though marriages under tribal law may vary
37	Equality of sexes during marriage and for divorce proceedings	yes	Government encouraging equality but, in practice, tribal customs still strong – and to disadvantage of wives
38	To practice any religion	yes	Consequences serious if churches suspected of interference in politics
39	To use contraceptive pills and devices	**YES**	Government support
40	To noninterference by state in strictly private affairs	yes	Except when perceived threat to government starts security alert and intrusions

Greece

Human rights rating: 87%

YES 25 yes 15 no 0 NO 0

Population: 10,000,000
Life expectancy: 76.1
Infant mortality (0–5 years)
  per 1,000 births: 18
United Nations covenants ratified:
Economic, Social and Cultural Rights,
Convention on Equality for Women

Form of government: Multiparty
democracy
GNP per capita (US$): 4,800
% of GNP spent by state:
On health: 3.5
On military: 5.7
On education: 2.5

FACTORS AFFECTING HUMAN RIGHTS

With membership in the European Community, Greece has become more closely
identified with Western Europe. Democratic institutions, a government that is
answerable to the people, an improving standard of living, and a free press are
contributing to the consolidation of the system after the overthrow of the military
dictatorship in 1974. The treatment of the Turkish minority reflects traditional Greek
hostility toward its neighbor and there are complaints about discrimination against
Muslims, particularly those living near the frontier.

	FREEDOM TO:		COMMENTS
1	Travel in own country	YES	Rights respected
2	Travel outside own country	YES	Rights respected
3	Peacefully associate and assemble	YES	Rights respected
4	Teach ideas and receive information	YES	Rights respected
5	Monitor human rights violations	yes	Some reluctance to cooperate with foreign investigators of rights of Muslim minority
6	Publish and educate in ethnic language	yes	Turkish minority frequently protests discrimination. Complaints also from Macedonian minority

	FREEDOM FROM:		COMMENTS
7	Serfdom, slavery, forced or child labor	yes	Some child labor. In some instances legal minimum age is 12 years. Little government supervision
8	Extrajudicial killings or "disappearances"	YES	Rights respected
9	Torture or coercion by the state	yes	Occasional cases of police brutality and overzealous interrogation

FREEDOM FROM:		COMMENTS
10	Compulsory work permits or conscription of labor	**YES**	Rights respected
11	Capital punishment by the state	yes	De facto abolition. Last execution 1972 but legality remaining. Last commuted death sentence 1988
12	Court sentences of corporal punishment	**YES**	Rights respected
13	Indefinite detention without charge	**YES**	Rights respected
14	Compulsory membership of state organizations or parties	**YES**	Rights respected
15	Compulsory religion or state ideology in schools	yes	Compulsory religion in the many Greek Orthodox schools
16	Deliberate state policies to control artistic works	yes	Rare bannings of cultural events held by Turkish minority
17	Political censorship of press	yes	A seldom-exercised law permits newspapers to be seized for insulting the president, offending religion, printing proclamation by terrorist groups, etc.
18	Censorship of mail or telephone tapping	yes	Occasional infringements

FREEDOM FOR OR RIGHTS TO:		COMMENTS
19	Peaceful political opposition	**YES**	Rights respected
20	Multiparty elections by secret and universal ballot	**YES**	Rights respected
21	Political and legal equality for women	yes	Despite constitutional equality, few women in senior government posts
22	Social and economic equality for women	yes	Discrimination persists in pay and employment
23	Social and economic equality for ethnic minorities	yes	Traditional disadvantages for Muslim Turkish minority, which forms over 1% of population
24	Independent newspapers	**YES**	Rights respected
25	Independent book publishing	**YES**	Rights respected
26	Independent radio and television networks	yes	Although private networks have recently been permitted, most stations are state owned or controlled. Guarantees of impartiality sometimes disputed
27	All courts to total independence	**YES**	Rights respected

FREEDOM FOR OR RIGHTS TO:		COMMENTS
28 Independent trade unions	YES	Rights respected though antistrike legislation causes frequent government confrontation with unions

LEGAL RIGHTS:		COMMENTS
29 From deprivation of nationality	yes	Instances of nonethnic Greek Muslim citizens being deprived of nationality on settling in Turkey
30 To be considered innocent until proved guilty	YES	Rights respected
31 To free legal aid when necessary and counsel of own choice	YES	Means test. Choice of lawyer from a court-appointed panel
32 From civilian trials in secret	YES	Rights respected
33 To be brought promptly before a judge or court	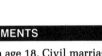 YES	24 hours. Release within 3 days if warrant not issued
34 From police searches of home without a warrant	YES	Rights respected
35 From arbitrary seizure of personal property	YES	Rights respected though Muslims complain of land expropiation in sensitive areas on frontier with Turkey

PERSONAL RIGHTS:		COMMENTS
36 To interracial, interreligious, or civil marriage	YES	From age 18. Civil marriage first introduced in 1984
37 Equality of sexes during marriage and for divorce proceedings	YES	Greater liberalization of marriage and divorce laws is transforming a previously male-dominated society
38 To practice any religion	yes	Arrests of Jehovah's Witnesses for proselytizing. 98% of population belong nominally to Christian Eastern Orthodox faith
39 To use contraceptive pills and devices	YES	Limited government support
40 To noninterference by state in strictly private affairs	YES	Rights respected

Guatemala

Human rights rating: 62%

Population: 9,200,000
Life expectancy: 63.4
Infant mortality (0–5 years)
per 1,000 births: 97
United Nations covenants ratified:
Economic, Social and Cultural Rights,
Convention on Equality for Women

Form of government: Multiparty system
GNP per capita (US$): 900
% of GNP spent by state:
On health: 0.7
On military: 1.3
On education: 1.8

FACTORS AFFECTING HUMAN RIGHTS

The holding of multiparty democratic elections, regarded by foreign monitors as genuine and equitable, contrasts with the reality of one of the most violent societies in the world. As well as conducting operations against guerrilla groups, the army and security forces follow a policy of arbitrary killings of many of those opposed to their right-wing dogmas. Their victims therefore include human rights monitors, academics, students, trade unionists, liberal politicians, and journalists who risk criticizing them. The death toll exceeded 5,000 in 1990, a proportion being victims of the guerrillas. Attempts by the government to control its own forces are usually ineffective, partly from fear of the consequences and partly because the military and the police seem to be above the law.

	FREEDOM TO:		COMMENTS
1	Travel in own country	yes	Travel usually restricted by operations against guerrilla groups or because areas of the country are occupied by them. But ill-disciplined government forces also pose a threat to the traveler
2	Travel outside own country	YES	Rights respected
3	Peacefully associate and assemble	yes	Rallies called to protest against the government's failure to protect human rights or left-wing causes frequently attacked by the government's own security forces
4	Teach ideas and receive information	yes	Academics and students frequently killed or abducted, inhibiting university life. The government lacks resolution in pursuing the perpetrators
5	Monitor human rights violations	yes	Although there is freedom to monitor, human rights activists are frequently the target for unlawful killings by the military and security forces. Limited investigations by the human rights ombudsman

FREEDOM TO:		COMMENTS
6 Publish and educate in ethnic language	yes	Indigenous peoples usually lack the resources either to publish or educate in their own language

FREEDOM FROM:		COMMENTS
7 Serfdom, slavery, forced or child labor	yes	Regional pattern of child labor
8 Extrajudicial killings or "disappearances"	NO	Murders during 1990 exceeded 5,000, a good proportion being extrajudicial killings by the security forces, either after the capture of guerrillas, in terrorist acts against human rights activists or political opponents, in random shootings of criminals. Many killings of "street" children by police in inner cities
9 Torture or coercion by the state	NO	Widespread torture by all sections of the security forces. Prisoners or the bodies of the deceased show evidence of savage beatings and mutilation
10 Compulsory work permits or conscription of labor	yes	The large security and military forces needed to combat terrorist groups are augmented by the conscription of males from the indigenous peoples. Also used as laborers
11 Capital punishment by the state	NO	Shooting by firing squad. Cannot be imposed on women or those over 70 years. Last official executions 1983
12 Court sentences of corporal punishment	YES	Rights respected
13 Indefinite detention without charge	yes	Usually by arbitrary and unlawful actions by security forces. These have a network of "safe houses," functioning outside the government's own judicial system, where prisoners are held and interrogated
14 Compulsory membership of state organizations or parties	YES	Rights respected
15 Compulsory religion or state ideology in schools	YES	Rights respected
16 Deliberate state policies to control artistic works	YES	Rights respected
17 Political censorship of press	yes	The climate of fear, caused by deaths of reporters and bombings of premises, which the government cannot control, inhibits press freedom
18 Censorship of mail or telephone tapping	yes	Some surveillance by security forces of human rights groups and "Communists"

FREEDOM FOR OR RIGHTS TO: COMMENTS

19 Peaceful political opposition   **yes**   Violent attempts by guerrilla groups to prevent peaceful elections. Deaths and intimidation, particularly in communities under their control

20 Multiparty elections by secret and universal ballot   **YES**   Rights respected but strict security measures necessary to ensure fair elections. Periodic attempts to bring guerrillas into the electoral process

21 Political and legal equality for women   **yes**   Constitutional guarantees of equality do not correct women's limited role in politics and in the professions

22 Social and economic equality for women   **no**   Regional inequalities. Particularly in pay and employment

23 Social and economic equality for ethnic minorities   **yes**   Discrimination socially and economically against indigenous peoples. Amerindians, 40% of the population, usually a source of cheap labor

24 Independent newspapers   **yes**   These are mostly right-wing in their politics and less likely to provoke the military or security forces

25 Independent book publishing   **yes**   Commercial and other considerations ensure that publications do not offend those parties who might resort to violence

26 Independent radio and television networks   **yes**   Both state owned and private. The climate of terror is an influence on the independence of programs

27 All courts to total independence   **yes**   In practice independence is affected by threats to the judiciary, corruption, and "low morale"

28 Independent trade unions   **yes**   Little organized unionism. Less than 10% of work force in unions. Strike actions limited

LEGAL RIGHTS: COMMENTS

29 From deprivation of nationality   **YES**   Rights respected

30 To be considered innocent until proved guilty   **yes**   Situation complicated by arbitrary actions of security and paramilitary "death squads," which might result in assumption of guilt without trial

31 To free legal aid when necessary and counsel of own choice   Means test. Counsel appointed by court

32 From civilian trials in secret   **YES**   Rights respected

LEGAL RIGHTS: COMMENTS

33 To be brought promptly | yes | Within 6 hours but many violations,
    before a judge or court   particularly by police and security forces
                                            apparently outside the control of the
                                            government

34 From police searches of | yes | Despite constitutional guarantees, arbitrary
    home without a warrant   pattern of intrusions

35 From arbitrary seizure of | yes | Many unlawful seizures by corrupt military
    personal property   and paramilitary groups in a violent society
                                            and where the government exercises little
                                            control over its own forces

PERSONAL RIGHTS: COMMENTS

36 To interracial, interreligious, **YES**   Rights respected
    or civil marriage

37 Equality of sexes during | yes | Regional and traditional patterns of marriage
    marriage and for divorce   favor the husband. Particularly in rural areas
    proceedings

38 To practice any religion   **YES**   Rights respected

39 To use contraceptive pills   **YES**   Rights respected
    and devices

40 To noninterference by state   **NO**   Forced relocation into "model villages" of
    in strictly private affairs   indigenous communities living in areas
                                            exposed to guerrilla actions. Assumed that
                                            such communities, possibly in sympathy
                                            with guerrillas, can be more easily watched

Honduras

Human rights rating: 65%

YES 16 yes 17 no 5 NO 2

Population: 5,100,000
Life expectancy: 64.9
Infant mortality (0–5 years)
per 1,000 births: 84
United Nations covenants ratified:
Economic, Social and Cultural Rights,
Convention on Equality for Women

Form of government: Multiparty
system
GNP per capita (US$): 860
% of GNP spent by state:
On health: 2.6
On military: 5.9
On education: 5.0

FACTORS AFFECTING HUMAN RIGHTS

As with other countries of Central America, the holding of democratic multiparty elections appears to conflict with the reality of a society in which the military and other security forces are usually immune from their legal and constitutional obligations. Among the victims of extrajudicial killings, torture, and incommunicado detention are human rights activists, students, intellectuals, journalists, and peasant leaders. Promises by the government to punish the unlawful actions of many of its own forces are only partially honored, and the sporadic killings by the terrorist People's Liberation Movement provide justification for the failure to assert its authority.

FREEDOM TO: / COMMENTS

1	Travel in own country	**YES**	Rights respected
2	Travel outside own country	yes	Exit visas required and freely issued to political dissidents
3	Peacefully associate and assemble	**YES**	Permits refused only when threat of violence
4	Teach ideas and receive information	**YES**	Rights respected
5	Monitor human rights violations	**YES**	Rights respected. Greater cooperation from government during the last year
6	Publish and educate in ethnic language	**YES**	Rights respected

---

FREEDOM FROM: / COMMENTS

7	Serfdom, slavery, forced or child labor	yes	Regional pattern of child labor
8	Extrajudicial killings or "disappearances"	**NO**	By security forces and the military. These arbitrary killings and "disappearances" include human rights activists and are seldom investigated by superior authority

FREEDOM FROM:		COMMENTS
9	Torture or coercion by the state	**NO**	Beatings, electric shocks, semisuffocation, maltreatment, etc.
10	Compulsory work permits or conscription of labor	yes	Some forced conscription into the army – usually of the "lower classes"
11	Capital punishment by the state	**YES**	Abolished 1956
12	Court sentences of corporal punishment	**YES**	Rights respected
13	Indefinite detention without charge	no	Incommunicado detention. The law frequently ignored by security forces
14	Compulsory membership of state organizations or parties	**YES**	Rights respected
15	Compulsory religion or state ideology in schools	**YES**	Rights respected
16	Deliberate state policies to control artistic works	**YES**	Rights respected
17	Political censorship of press	yes	A degree of self-censorship practiced. Journalists are compelled to be members of a professional guild, which eliminates many foreign reporters, and occasionally receive death threats
18	Censorship of mail or telephone tapping	yes	Some surveillance by security agents, principally of suspected subversives and the left-wing terrorist movement

FREEDOM FOR OR RIGHTS TO:		COMMENTS
19	Peaceful political opposition	**YES**	Rights respected
20	Multiparty elections by secret and universal ballot	yes	The 1989 election was followed by complaints of vote rigging. Reforms of the electoral process being considered by Congress
21	Political and legal equality for women	no	Despite constitutional rights, underrepresentation in politics and professions
22	Social and economic equality for women	no	Regional discrimination against women, particularly in pay and employment
23	Social and economic equality for ethnic minorities	yes	Social discrimination against Amerindians and other small indigenous groups
24	Independent newspapers	yes	Death threats and harassment, particularly of left-wing publishers, may influence independence. Reporters prepared to "misreport" for bribes

FREEDOM FOR OR RIGHTS TO: COMMENTS

25	Independent book publishing	yes	Prudence needed. A provocative publication can lead to violence in a politically unstable society
26	Independent radio and television networks	yes	Mostly private but licenses controlled by government. This enables it to exercise discreet pressures
27	All courts to total independence	no	Constant conflicts between the differing rights of civil and military courts. Most judges appointed by justice minister. Independence affected by low pay, bribes, political pressures, threats, etc.
28	Independent trade unions	yes	The main threat to unions comes from right-wing antiunion employers supported unlawfully by henchmen in security forces. Union leaders occasionally murdered

LEGAL RIGHTS: COMMENTS

29	From deprivation of nationality	YES	Rights respected
30	To be considered innocent until proved guilty	yes	Violations by security forces, particularly by the National Department of Investigation
31	To free legal aid when necessary and counsel of own choice	yes	Underfunded and inadequate judicial system means limited "public defender" service for the poor. Some aid from a US legal agency
32	From civilian trials in secret	YES	Rights respected
33	To be brought promptly before a judge or court	no	Within 24 hours but the law is regularly ignored. Security forces and police seldom disciplined for violations
34	From police searches of home without a warrant	yes	Occasional arbitrary entry, particularly by right-wing security elements. Searches may be a pretext for looting. Corruption widespread
35	From arbitrary seizure of personal property	yes	As question 34

PERSONAL RIGHTS: COMMENTS

36	To interracial, interreligious, or civil marriage	YES	Rights respected. Males may marry at 18 years, females at 16
37	Equality of sexes during marriage and for divorce proceedings	yes	Traditional and ethnic influences favor husbands

PERSONAL RIGHTS: COMMENTS

38	To practice any religion	yes	Harassment and death threats by unidentified security agents against Roman Catholic priests. The Catholic church has been a leading opponent of state oppression
39	To use contraceptive pills and devices	YES	Rights respected
40	To noninterference by state in strictly private affairs	YES	Rights respected

Hong Kong

Human rights rating: 79%

YES 19 yes 18 no 3 NO 0

Population: 5,900,000
Life expectancy: 77.3
Infant mortality (0–5 years)
  per 1,000 births: 8
United Nations covenants ratified: Not a member of the United Nations

Form of government: British crown colony
GNP per capita (US$): 9,220
% of GNP spent by state:
On health: n/a
On military: n/a
On education: 2.8

FACTORS AFFECTING HUMAN RIGHTS

The British crown colony is passing through a period which may be described as the end of an era. It reverts to Chinese sovereignty in 1997, and British policy is currently concerned with ensuring a smooth transition and the honoring by China of the Sino-British Joint Declaration. This guarantees human rights as set out in the UN covenants. A skeptical population remains in a state of considerable anxiety, many emigrating and others seeking passports "just in case". In its policy of prudence, the Hong Kong government is following "understood guidelines" in moderating media comment, films, books, etc., which might "offend a neighboring country." Similarly with the handing back to China of "illegal immigrants." The human rights situation is also complicated by the presence in camps of 64,000 Vietnamese refugees, with forcible repatriation agreed with Vietnam.

	FREEDOM TO:		COMMENTS
1	Travel in own country	YES	Rights respected
2	Travel outside own country	YES	Rights respected
3	Peacefully associate and assemble	yes	Limitations on rallies and meetings that may provoke protests from China, scheduled to have the colony restored to it in 1997. Under the Summary Offences Ordinance, convictions for addressing crowds with a megaphone
4	Teach ideas and receive information	yes	The growing pressures on the Hong Kong government and on society from China, which seeks to exercise influence before the takeover date of 1997, is creating unease in universities, accelerating the emigration of many academics and students. Need for prudence increasing

FREEDOM TO:		COMMENTS
5	Monitor human rights violations	yes	Although there are no limits on monitoring, some groups of human rights activists clash with the police, and foreigners visiting Hong Kong with valid visas have been refused entry on grounds of possible provocation of China
6	Publish and educate in ethnic language	**YES**	Rights respected

FREEDOM FROM:		COMMENTS
7	Serfdom, slavery, forced or child labor	**YES**	Rights respected
8	Extrajudicial killings or "disappearances"	**YES**	Rights respected
9	Torture or coercion by the state	yes	Minor abuses increasing, particularly toward illegal immigrants. Crime rate rising with cross-border operations by Hong Kong and China crime syndicates
10	Compulsory work permits or conscription of labor	no	Work permits introduced because of illegal immigrants from China and the increasingly numerous Vietnamese refugees
11	Capital punishment by the state	yes	To be abolished from 1992. Sentences commuted since 1966
12	Court sentences of corporal punishment	yes	Rare instances of caning sentences – 18 strokes maximum
13	Indefinite detention without charge	**YES**	48 hours maximum
14	Compulsory membership of state organizations or parties	**YES**	Rights respected
15	Compulsory religion or state ideology in schools	**YES**	Rights respected
16	Deliberate state policies to control artistic works	yes	Hong Kong is a major Far East film producer but government is increasing its surveillance to avoid offending "a neighboring country"
17	Political censorship of press	yes	Gradual pressures of self-censorship as the colony approaches the takeover date of 1997, despite guarantees in the 1991 Bill of Rights
18	Censorship of mail or telephone tapping	yes	Allegations of arbitrary surveillance as government attempts to avoid tension with the People's Republic

FREEDOM FOR OR RIGHTS TO:		COMMENTS
19	Peaceful political opposition	yes	Careful official encouragement of the process of democratization. Needed to strengthen the political structure of the colony before 1997

FREEDOM FOR OR RIGHTS TO:		COMMENTS
20	Multiparty elections by secret and universal ballot	no	The colony's law-making body is the Legislative Council, which is elected on a limited suffrage. It is not a parliament for dealing with international issues, which are the concern of the British government and the governor
21	Political and legal equality for women	yes	Women are increasing their participation in male-dominated politics and professions
22	Social and economic equality for women	no	Confucian traditions still prevail in "old-style" Chinese families. Pay and employment inequalities
23	Social and economic equality for ethnic minorities	yes	Illegal immigrants are now a source of cheap labor, prevented by their situation from benefiting from any social or official protection
24	Independent newspapers	yes	67 registered newspapers but some apprehension that independence will end in 1997. The proximity of the Hong Kong branch of Xinhua, the Chinese press agency, is becoming more intimidating
25	Independent book publishing	yes	Understood guidelines on books liable to offend the People's Republic of China even though some publishers favor the Taiwan regime
26	Independent radio and television networks	YES	Both private and government owned
27	All courts to total independence	YES	Rights respected
28	Independent trade unions	YES	Rights respected

LEGAL RIGHTS:		COMMENTS
29	From deprivation of nationality	yes	With the approach of 1997, a growing number of Hong Kong Chinese wish to emigrate. The failure of the British government to offer British passports "as of right" has created a complex and embittered situation
30	To be considered innocent until proved guilty	YES	The new Bill of Rights, enacted in June 1991, is intended to strengthen the existing law after 1997. Population skeptical
31	To free legal aid when necessary and counsel of own choice	YES	Counsel of own choice through the Legal Aid Department
32	From civilian trials in secret	YES	Rights respected
33	To be brought promptly before a judge or court	YES	Rights respected

LEGAL RIGHTS:		COMMENTS
34	From police searches of home without a warrant	yes	Arbitrary searches because of the big increase in crime, the drive against drug traffickers, triads, syndicates, illegal immigrants, etc.
35	From arbitrary seizure of personal property	yes	Frequent accusations of corruption, bribery, forced entry of business premises, etc. against the police

PERSONAL RIGHTS:		COMMENTS
36	To interracial, interreligious, or civil marriage	**YES**	Rights respected
37	Equality of sexes during marriage and for divorce proceedings	yes	Confucian traditions based on male superiority still followed by the older generation
38	To practice any religion	**YES**	Rights respected
39	To use contraceptive pills and devices	**YES**	Rights respected
40	To noninterference by state in strictly private affairs	**YES**	Rights respected

Hungary

Human rights rating: 97%

Population: 10,600,000
Life expectancy: 70.9
Infant mortality (0–5 years)
per 1,000 births: 18
United Nations covenants ratified:
Civil and Political Rights, Economic,
Social and Cultural Rights, Convention
on Equality for Women

Form of government: Multiparty
democracy
GNP per capita (US$): 2,460
% of GNP spent by state:
On health: 3.2
On military: 2.4
On education: 3.8

FACTORS AFFECTING HUMAN RIGHTS

After 40 years of almost unbroken Communist rule, Hungary held its first multiparty
democratic elections in 1990. The country has embraced its new freedom in every area
and must be considered among the freest in the world. The exception to this may be the
status of the Gypsy minority, 5% of the population, but the discrimination, where it is
practiced, is traditional and not encouraged by the government. The formation of a
national council seeks to improve the position of Gypsies.

	FREEDOM TO:		COMMENTS
1	Travel in own country	YES	Rights respected
2	Travel outside own country	YES	Rights respected
3	Peacefully associate and assemble	YES	Rights respected
4	Teach ideas and receive information	YES	Rights respected
5	Monitor human rights violations	YES	Rights respected
6	Publish and educate in ethnic language	YES	Rights respected

	FREEDOM FROM:		COMMENTS
7	Serfdom, slavery, forced or child labor	YES	Rights respected
8	Extrajudicial killings or "disappearances"	YES	Rights respected
9	Torture or coercion by the state	YES	Rights respected

	FREEDOM FROM:		COMMENTS
10	Compulsory work permits or conscription of labor	**YES**	Rights respected
11	Capital punishment by the state	**YES**	Abolished 1990
12	Court sentences of corporal punishment	**YES**	Rights respected
13	Indefinite detention without charge	**YES**	Rights respected
14	Compulsory membership of state organizations or parties	**YES**	Rights respected
15	Compulsory religion or state ideology in schools	**YES**	Rights respected
16	Deliberate state policies to control artistic works	**YES**	Rights respected
17	Political censorship of press	**YES**	Rights respected
18	Censorship of mail or telephone tapping	yes	Permitted under exceptional circumstances. Last of "old guard" practices at Interior Ministry being abolished

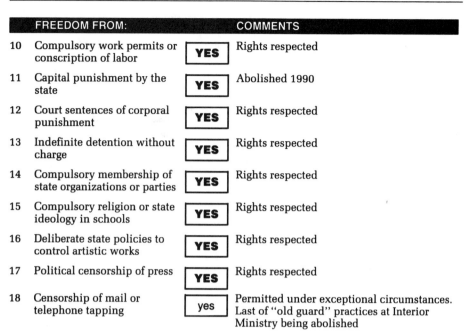

	FREEDOM FOR OR RIGHTS TO:		COMMENTS
19	Peaceful political opposition	**YES**	Rights respected
20	Multiparty elections by secret and universal ballot	**YES**	First free elections since 1947 in 1990
21	Political and legal equality for women	yes	Limited participation of women in politics and professions but situation improving
22	Social and economic equality for women	yes	Discrimination in pay and employment. The economic switch from a state-dominated system to a free market economy bringing temporary but limited disadvantages
23	Social and economic equality for ethnic minorities	yes	Long tradition of discrimination against Romanian and Gypsy minorities
24	Independent newspapers	**YES**	Rights respected. The changes following the end of state control have meant the launching of many new journals
25	Independent book publishing	**YES**	Rights respected
26	Independent radio and television networks	yes	Although most stations are state owned, objectivity and a wide variety of views are usual
27	All courts to total independence	**YES**	Senior judges must not have political affiliations

FREEDOM FOR OR RIGHTS TO:	COMMENTS

28 Independent trade unions **YES** Unions are in the process of adjusting to a free market economy, with government cooperation

LEGAL RIGHTS:	COMMENTS

29 From deprivation of nationality **YES** Rights respected

30 To be considered innocent until proved guilty **YES** Rights respected

31 To free legal aid when necessary and counsel of own choice **YES** Rights respected

32 From civilian trials in secret **YES** Rights respected

33 To be brought promptly before a judge or court **YES** Within 72 hours

34 From police searches of home without a warrant **YES** Rights respected

35 From arbitrary seizure of personal property **YES** Rights respected

PERSONAL RIGHTS:	COMMENTS

36 To interracial, interreligious, or civil marriage **YES** Males may marry at 18 years, females at 16 years

37 Equality of sexes during marriage and for divorce proceedings **YES** Rights respected

38 To practice any religion **YES** Rights respected. Property previously confiscated by the state being restored to owners

39 To use contraceptive pills and devices **YES** State support

40 To noninterference by state in strictly private affairs **YES** Rights respected

India

Human rights rating: 54%

YES 9 **yes** 16 **no** 12 NO 3

Population: 853,000,000
Life expectancy: 59.1
Infant mortality (0–5 years) per 1,000 births: 145
United Nations covenants ratified: Civil and Political Rights, Economic, Social and Cultural Rights

Form of government: Federal democracy
GNP per capita (US$): 340
% of GNP spent by state:
On health: 0.9
On military: 3.5
On education: 3.4

FACTORS AFFECTING HUMAN RIGHTS

Despite a reputation for being the "largest democracy in the world," and the enshrining in the constitution of fundamental and human rights, the nature of Indian society prevents the government from implementing its legal obligations. The country is divided by religion, ethnic groups, regional and communal disputes, and the unresolved problem of a divided Kashmir. These result in frequent clashes, violent demonstrations, and militant separatist movements, all calling for vigorous responses from the army and security forces. These responses have frequently resulted in extrajudicial killings and widespread torture and, under the National Security Act, a figure of 15,000 "political prisoners" are claimed to be under detention. In the remoter areas of the country, the police and security forces have rounded up village women for gang rapes. The prospects for human rights improvements are limited.

	FREEDOM TO:		COMMENTS
1	Travel in own country	**YES**	But periodic controls in areas of racial, communal, religious, or separatist conflicts
2	Travel outside own country	**YES**	Currency exchange regulations limit foreign travel
3	Peacefully associate and assemble	yes	Rallies frequently turn to riots, provoking swift police action. Under the National Security Act (1980) all meetings may be banned and curfews imposed
4	Teach ideas and receive information	**YES**	Rights respected
5	Monitor human rights violations	yes	Limits on visits by international monitors to areas of greatest communal conflicts – and therefore areas of possible excesses of security forces
6	Publish and educate in ethnic language	yes	Minor limitations on Tamil and other languages

FREEDOM FROM:		COMMENTS
7	Serfdom, slavery, forced or child labor	**no**	Bonded labor continues in practice. Proportion of child labor largest in world
8	Extrajudicial killings or "disappearances"	**NO**	Many deaths by ill-disciplined police in communal rioting. Security forces stage "fake encounters" to permit killings. Worst in areas of separatist movements
9	Torture or coercion by the state	**NO**	Beatings, electric shocks, rape, burns, etc. The size of the country and the extent of social and communal violence limit the government's ability to prevent widespread torture
10	Compulsory work permits or conscription of labor	**YES**	Rights respected
11	Capital punishment by the state	**NO**	Hanging or shooting. Death penalty for "bride-burning," an unlawful ritualistic practice
12	Court sentences of corporal punishment	**YES**	Rights respected
13	Indefinite detention without charge	**no**	The extent of communal violence has led to mass detentions under the National Security Act and other acts. Figures range up to 15,000 "political prisoners." An inadequate judicial system does not permit early trials
14	Compulsory membership of state organizations or parties	**YES**	Rights respected
15	Compulsory religion or state ideology in schools	**YES**	Rights respected
16	Deliberate state policies to control artistic works	**yes**	The Censor Board imposes restrictions on art and drama. Periodic morality campaigns and publicizing of its own "propaganda"
17	Political censorship of press	**no**	The government uses its censorship powers on security matters to suppress criticism. Some harassment of journalists
18	Censorship of mail or telephone tapping	**no**	Surveillance intensified during periods of communal unrest. Permitted under the National Security Act

FREEDOM FOR OR RIGHTS TO:		COMMENTS
19	Peaceful political opposition	**yes**	Occasional violence between Hindus and Muslims. Also by Sikhs and other minorities. Political leaders periodically murdered. Corruption and harassment
20	Multiparty elections by secret and universal ballot	**yes**	Regular elections are marked by assassinations of candidates, "boothcapturing" and corruption

	FREEDOM FOR OR RIGHTS TO:		COMMENTS
21	Political and legal equality for women	yes	Despite the recent example of a woman prime minister, limited presence in senior posts and in the professions
22	Social and economic equality for women	no	Although changes to the constitution attempt to correct the many inequalities, in practice women remain subservient in most areas. These differ, however, with Hindu, Muslim, and tribal customs – last category including, in Bihar, the killing of "witches"
23	Social and economic equality for ethnic minorities	yes	Affected minorities are usually the victims of traditional social and ethnic discrimination rather than of state policies
24	Independent newspapers	yes	But government exercises some control by administering a monopoly of newsprint sales, a large advertising budget, and the offering of official favors
25	Independent book publishing	yes	Publications which, in the government's opinion, worsen communal tensions may be seized. Some foreign works banned
26	Independent radio and television networks	no	Government monopoly. A new bill will eventually give greater autonomy to radio and television. Networks of illegal stations, cable systems, etc., function with little interference from officials
27	All courts to total independence	yes	Variations from state to state but a high degree of impartial justice. Some corruption at lower levels
28	Independent trade unions	yes	Under the Essential Services Act, strikes may be banned. Majority of workers in agriculture and therefore deprived of union protection. Occasional conflicts between labor and police

	LEGAL RIGHTS:		COMMENTS
29	From deprivation of nationality	**YES**	Rights respected
30	To be considered innocent until proved guilty	no	Abuses by security forces influenced by ethnic, regional, and religious ties. Many instances of wrongful arrests and police dropping investigations for bribes
31	To free legal aid when necessary and counsel of own choice	yes	Free aid under Code of Criminal Procedure but the overburdened system results in limited help
32	From civilian trials in secret	no	Secret trials in Punjab and Assam. Worse in remote areas. Sometimes resulting in convenient "disappearances"

LEGAL RIGHTS: COMMENTS

33 To be brought promptly
    before a judge or court   **no**   Within 24 hours but in turbulent regions or
                                               states the police have arbitrary powers. The
                                               judicial system is also overburdened, with
                                               significant areas of corruption

34 From police searches of
    home without a warrant   **no**   Law frequently set aside if delay affects
                                               police action. Also special powers in
                                               disturbed regions

35 From arbitrary seizure of
    personal property   **yes**   Some ill-disciplined police and military
                                               units use search powers as pretext for
                                               looting. Worse in separatist areas

PERSONAL RIGHTS: COMMENTS

36 To interracial, interreligious,
    or civil marriage   **yes**   Despite the constitution, the persistence of
                                               the caste system, religious and tribal laws,
                                               etc., remain barriers and reflect the
                                               government's limited commitment to its
                                               own laws

37 Equality of sexes during
    marriage and for divorce   **no**   In practice, constitutional equality is
    proceedings   meaningless. Crimes against wives extend
                                               from inheritance disadvantages to child
                                               marriage and the occasional "dowry death"

38 To practice any religion   **yes**   Many religious confrontations, with
                                               intercommunal violence resulting in deaths
                                               and pillaging. Restrictions usually at local
                                               state level, also against missionaries

39 To use contraceptive pills
    and devices   **YES**   Government seeking to reduce
                                               overpopulation

40 To noninterference by state
    in strictly private affairs   **no**   Rapes of women during "house-to-house"
                                               searches. Tensions on families and private
                                               lives where police exceed their lawful
                                               activities

Indonesia

Human rights rating: 34%

YES 3 **yes** 4 **no** 26 NO 7

Population: 184,300,000
Life expectancy: 61.5
Infant mortality (0–5 years)
   per 1,000 births: 100
United Nations covenants ratified:
Convention on Equality for Women

Form of government: Military
presidency
GNP per capita (US$): 440
% of GNP spent by state:
On health: 0.7
On military: 2.5
On education: 2.3

FACTORS AFFECTING HUMAN RIGHTS
The country is governed by a small military group under the leadership for over 20 years of President Soeharto. Under what was claimed to be the "new order government," a national ideology called *Pancasila* was introduced. Among its principles are the need for democracy, social justice, and individual rights. In practice, however, few human rights are honored and separatist movements in three of the major islands face excessive military actions, suffering summary executions and extreme forms of torture. The press, political life, associations, and the legal process are all subject to government pressures or arbitrary controls.

	FREEDOM TO:		COMMENTS
1	Travel in own country	no	After many years of unrest, areas of East Timor, Irian Jaya, Aceh, and elsewhere are still forbidden to most travelers or need permits
2	Travel outside own country	yes	Exit permits may be refused to known political opponents or those who intend to criticize the country while abroad
3	Peacefully associate and assemble	no	No rallies, protest meetings or waving, clandestine flags if purpose is "subversive" – i.e., likely to embarrass the government. Suppression even stricter in the rebellious parts of the country
4	Teach ideas and receive information	no	Universities under general police surveillance. Academics follow understood guidelines. Student protests frequent – and for which they suffer. All seats of learning must follow and teach the national ideology, *Pancasila* (the "five principles")
5	Monitor human rights violations	no	More monitoring is now officially permitted but government regards it as interference in country's internal affairs, particularly when international groups are involved. Harassment, limited access to rebellious areas

FREEDOM TO:		COMMENTS
6	Publish and educate in ethnic language	**NO**	No Chinese schools or newspapers for the minority of 4 million citizens. All public displays of Chinese characters banned. Other minorities affected

FREEDOM FROM:		COMMENTS
7	Serfdom, slavery, forced or child labor	no	Pattern of child labor. Despite efforts by government Manpower Department to control employment of children under 14 years, a figure as high as 2 million has been quoted
8	Extrajudicial killings or "disappearances"	**NO**	Summary killings in the rebellious islands and elsewhere. Dumping of headless bodies as a warning by the military. Also execution of criminal syndicates by state "death squads"
9	Torture or coercion by the state	**NO**	Particularly of secessionist elements during interrogations. Beatings, electric shocks, semi drowning, semi suffocation, slashing by knives, rape, etc. Officials have admitted their concern over extent of torture
10	Compulsory work permits or conscription of labor	no	In the remoter islands well-documented evidence of indigenous peoples forced to work in lumber camps and girls sold to brothels. In some other areas, work permits necessary
11	Capital punishment by the state	**NO**	By firing squad for murder, treason, drug trafficking, etc. But no longer exceeding thousands a year (including executions by the military)
12	Court sentences of corporal punishment	**YES**	Rights respected
13	Indefinite detention without charge	no	Many detentions of political opponents, some over 20 years. Hundreds of "prisoners of conscience" detained in the secessionist areas
14	Compulsory membership of state organizations or parties	no	*Pancasila*, by a 1985 law, must be included as part of the aims of all social groups
15	Compulsory religion or state ideology in schools	**NO**	Both Islam, the state religion, and *Pancasila* are part of school curricula
16	Deliberate state policies to control artistic works	no	New government policy – *keterbukaan* – of greater tolerance toward the arts does not extend to satirical plays and the performing arts. "The art field has become the battlefield of political maneuvers" (film director)

FREEDOM FROM:		COMMENTS
17	Political censorship of press	**no**	Understood guidelines. Limits on criticism of government or *Pancasila*. Occasional arrests and harassment of journalists
18	Censorship of mail or telephone tapping	**no**	Surveillance of political opponents and left-wing sympathizers, said to number 20 million. Even greater surveillance in areas of unrest

FREEDOM FOR OR RIGHTS TO:		COMMENTS
19	Peaceful political opposition	**NO**	The consequences of defying legal guidelines, which include acceptance of *Pancasila* as part of any political program, may be charges of subversion, "communism", etc.
20	Multiparty elections by secret and universal ballot	**no**	Three legal parties but dominance of present ruling party, with the support of the military, is assured and absolute. General election 1992. Presidential election 1993. Political leader states: "democracy would only be misunderstood by the masses"
21	Political and legal equality for women	**no**	Underrepresented in both government and the professions
22	Social and economic equality for women	**no**	Wide pay and employment differences. Much of the agricultural laboring done by women
23	Social and economic equality for ethnic minorities	**no**	As well as disadvantages for the Chinese, dominant positions in government and state services held by Muslims and Javanese
24	Independent newspapers	**no**	Understood guidelines. 1984 decree permits withdrawal of licenses if reporting is not in a "healthy, free and responsible manner." Pressures from military officials to "clean up the press"
25	Independent book publishing	**no**	Strict guidelines. No books on communism, certain interpretations of the corrupt practices of the country's ruling family, and anything that "offends the state", etc.
26	Independent radio and television networks	**no**	Mostly state owned and controlled. Commercial stations concentrate on music and entertainment and must not produce news and current affairs programs
27	All courts to total independence	**no**	Judges are public servants and subject to government pressures and bribery. Corruption accepted as "part of the system"

FREEDOM FOR OR RIGHTS TO:		COMMENTS
28	Independent trade unions	**no**	Despite constitutional rights, unions are frequently suspected of communist leanings. No unions in public services. The union movement is weakened by migration from country areas, providing a constant source of cheap labor

LEGAL RIGHTS:		COMMENTS
29	From deprivation of nationality	**YES**	Rights respected
30	To be considered innocent until proved guilty	**no**	Areas and islands seeking secession are under constant surveillance by military and security forces who arrest arbitrarily, frequently without grounds for suspicion
31	To free legal aid when necessary and counsel of own choice	**yes**	Legal aid for the "most needy." Defense lawyers appointed by court in subversion cases
32	From civilian trials in secret	**no**	Arbitrary. Those taking place, particularly in secessionist areas, would receive no publicity
33	To be promptly before a judge or court	**no**	Long delays in practice. System overburdened and inadequate. Bribery may help pretrial release
34	From police searches of home without a warrant	**no**	The narrow distinction between ordinary crimes and the many categories regarded as subversive permit security forces to ignore the judicial requirement of a warrant. Many abuses
35	From arbitrary seizure of personal property	**yes**	Widespread official corruption. Appropriation of land for resettlement programs, with farmers and villagers complaining of absurd undervaluation of land and homes

PERSONAL RIGHTS:		COMMENTS
36	To interracial, interreligious, or civil marriage	**yes**	Exclusivity usually on ethnic or religious grounds. Males may marry at 19 years, females at 16
37	Equality of sexes during marriage and for divorce proceedings	**no**	In practice affected by regional tradition and by the Islamic faith (that of nearly 90% of the population). Advantages and privileges usually favor husbands
38	To practice any religion	**no**	Smaller religions banned. They include Baha'i, Jehovah's Witnesses, Islamic fundamentalists and "misleading religious cults"

	PERSONAL RIGHTS:		COMMENTS
39	To use contraceptive pills and devices	**YES**	Rights respected
40	To noninterference by state in strictly private affairs	**NO**	Security actions throughout islands and areas demanding secession. Harassment of innocent communities, looting by troops. Private life affected by threats and fear

Iran

Human rights rating: 22%

Population: 54,600,000
Life expectancy: 66.2
Infant mortality (0–5 years)
  per 1,000 births: 64
United Nations covenants ratified:
Civil and Political Rights, Economic,
Social and Cultural Rights

Form of government: Fundamentalist
Islamic
GNP per capita (US$): 1,800
% of GNP spent by state:
On health: 1.4
On military: 20.0
On education: 4.6

FACTORS AFFECTING HUMAN RIGHTS
The Islamic republic was formed after the overthrow of the Shah in 1979. Since the death of the Supreme Leader, Ayatollah Khomeini, in 1989, there has been some moderation in the total obedience to his teachings. Nevertheless, summary executions of those opposing the government and its extreme religious beliefs frequently take place, torture in prisons and during interrogations is widespread, while thousands suffer indefinite detention. Although human rights monitoring is not permitted, a visit by a UN special rapporteur was allowed and, though granted only limited access, his report confirmed many of the human rights crimes.

	FREEDOM TO:		COMMENTS
1	Travel in own country	YES	Rights respected
2	Travel outside own country	no	Exit visas required. Denied for political reasons. Iranian nationals abroad, if condemned by the state, have been assassinated
3	Peacefully associate and assemble	NO	Only rallies and meetings to support state or religion. Others forcibly suppressed. Shops closed by "vice squads" for "neglecting Islamic codes"
4	Teach ideas and receive information	no	A system marginally less rigid than under the late Ayatollah Khomeini. Teaching dominated by religious orthodoxy. Many informers in universities. Official view: "Academic freedom corrupts Islamic youth"
5	Monitor human rights violations	NO	The slight relaxation of permitting a UN special rapporteur to visit and enquire about foreigners in prison was construed as a need for an improved international image
6	Publish and educate in ethnic language	yes	Some limitations on Kurds to educate in own language. Pressures on minorities to conform to majority precepts

FREEDOM FROM:		COMMENTS
7	Serfdom, slavery, forced or child labor	yes	Regional practice of child labor
8	Extrajudicial killings or "disappearances"	NO	Many. Exact numbers unknown but well-documented proof of summary executions
9	Torture or coercion by the state	NO	Widely practiced in Iranian prisons. The visiting UN special rapporteur received many accounts of floggings, foot beatings, long periods of suspension from ceilings, etc.
10	Compulsory work permits or conscription of labor	YES	Rights respected
11	Capital punishment by the state	NO	Many. Until 1989, 1,000 a year. numbers currently reduced except for drug smugglers (in one day in 1990, 46 killed by hanging, shooting, or stoning)
12	Court sentences of corporal punishment	NO	Sentences include floggings, amputation of fingers for theft, etc.
13	Indefinite detention without charge	NO	Thousands. Dissidents, drug offenders, social misfits. Arbitrary action by state and religious authorities
14	Compulsory membership of state organizations or parties	YES	Rights respected
15	Compulsory religion or state ideology in schools	no	Islamic committees established in schools. Religious orthodoxy. Females must wear "Islamic dress"
16	Deliberate state policies to control artistic works	no	Islamic beliefs and values prevail. Ruling: "The cinema must play its own role in propagating Islam, just like a mosque"
17	Political censorship of press	NO	Nothing that conflicts with Islamic orthodoxy. Underground papers (*samizdat*) being circulated
18	Censorship of mail or telephone tapping	NO	Surveillance continues despite slightly less paranoid policies

FREEDOM FOR OR RIGHTS TO:		COMMENTS
19	Peaceful political opposition	NO	Only those acceptable to the Council of Guardians may be nominated for a parliament where religious conformity is compulsory
20	Multiparty elections by secret and universal ballot	NO	Politics still dominated by the late Ayatollah Khomeini's teachings. Parliamentary factions rather than parties compete for power
21	Political and legal equality for women	no	Position strongly influenced by religious attitudes. Inequalities at all levels

FREEDOM FOR OR RIGHTS TO:		COMMENTS
22	Social and economic equality for women	no	Women's conformity with her status under religious orthodoxy an obvious violation of UN human rights. Floggings, fines, or imprisonment for "incorrect" dress, "provocative" wearing of cosmetics, etc.
23	Social and economic equality for ethnic minorities	no	Kurds, the largest minority, suffer degrees of discrimination socially and culturally though position is better than under the monarchy
24	Independent newspapers	NO	Over 1,000 newspapers banned or seized since end of monarchy in 1979
25	Independent book publishing	no	Press and publications office controls all books. A system of "issuing licenses" with fixed prices. Many bannings
26	Independent radio and television networks	NO	State owned and controlled
27	All courts to total independence	NO	Revolutionary courts can overturn decisions of regular civilian courts. In 1990 the visiting UN special rapporteur noted the lack of many elementary procedural guarantees
28	Independent trade unions	NO	Government control through Islamic labor councils

LEGAL RIGHTS:		COMMENTS
29	From deprivation of nationality	no	Complicated situation because some dissidents living abroad are deprived of nationality by official assassination order
30	To be considered innocent until proved guilty	NO	Situation arbitrary. Many official assumptions of guilt without trial
31	To free legal aid when necessary and counsel of own choice	NO	Many cases, particularly those conducted by revolutionary courts, do not permit defending counsel
32	From civilian trials in secret	NO	Arbitrary rulings by both state and religious authorities
33	To be brought promptly before a judge or court	NO	Legal requirements frequently ignored. Special revolutionary courts may try "political" cases to their own timetable
34	From police searches of home without a warrant	NO	Arbitrary situation. Surveillance also for religious nonconformism and banned literature
35	From arbitrary seizure of personal property	NO	Many cases of confiscation of property of various categories of dissidents and "subversives"

PERSONAL RIGHTS:		COMMENTS
36	To interracial, interreligious, or civil marriage	Strict compliance with Islamic law
37	Equality of sexes during marriage and for divorce proceedings	Inequalities include husband's right to divorce at will
38	To practice any religion	Many discriminations, especially against Baha'i religion. No executions of Baha'i, however, since 1989
39	To use contraceptive pills and devices	Limited government approval
40	To noninterference by state in strictly private affairs	Government does not accept the concept of a private life outside its control. Risk of capital punishment for adultery, fornication, and homosexual acts. Example: five reported cases in January 1990

Iraq (As at August 1991)

Human rights rating: 17%

YES 0 yes 11 no 4 NO 25

Population: 18,900,000
Life expectancy: 65.0
Infant mortality (0–5 years)
  per 1,000 births: 89
United Nations covenants ratified:
Civil and Political Rights, Economic,
Social and Cultural Rights, Convention
on Equality for Women

Form of government: Military council
GNP per capita (US$): 3,020
% of GNP spent by state:
On health: 0.8
On military: 32.0
On education: 3.7

FACTORS AFFECTING HUMAN RIGHTS

Iraq's invasion of Kuwait in 1990 was followed at the beginning of 1991 by its military defeat by the USA and its allies. The country was devastated by air attacks, and small areas in the north and south remain outside government control, but President Saddam Hussein and the Ba'ath party continue in power. The human rights position, which had been one of the worst in the world has, if possible, worsened. The international embargo on trade and relations with other countries means considerable suffering for the masses. A meaningful improvement is unlikely – impossible – until the present regime is overthrown.

	FREEDOM TO:		COMMENTS
1	Travel in own country	NO	Countrywide restrictions following 1990–91 gulf war
2	Travel outside own country	NO	Affected by virtual isolation of country following the gulf war and by international embargo
3	Peacefully associate and assemble	NO	Only in support of the regime
4	Teach ideas and receive information	NO	Nothing that conflicts with regime's policies or the president's unpredictable convictions. Intellectuals and academics regarded as antileadership or dissident in their beliefs risk death
5	Monitor human rights violations	NO	Not permitted. The country is effectively cut off from international monitoring
6	Publish and educate in ethnic language	yes	Minority languages generally unrestricted though licenses for ethnic schools may be required

FREEDOM FROM:		COMMENTS
7	Serfdom, slavery, forced or child labor	**NO**	Forced labor as required by the civilian and military authorities. The rebuilding of country and essential services, following aerial bombardment during the gulf war, a major priority
8	Extrajudicial killings or "disappearances"	**NO**	The elimination of opponents of the ruling regime is continuous and merciless. The overall number of deaths may never be known
9	Torture or coercion by the state	**NO**	By beatings, burns, electric shocks, sexual violations, etc. Death frequently follows
10	Compulsory work permits or conscription of labor	**NO**	The need to rebuild the devastated country has meant civilian as well as military conscription
11	Capital punishment by the state	**NO**	By hanging or firing squad
12	Court sentences of corporal punishment	no	The extent of torture with the encouragement of the country's rulers must be regarded as corporal punishment by the state
13	Indefinite detention without charge	**NO**	Many opponents of the regime face indefinite imprisonment. Ethnic groups such as Kurds particularly affected
14	Compulsory membership of state organizations or parties	yes	In a paranoid society affected by the recent military defeat and by the countrywide devastation, membership of the Ba'ath party offers many advantages
15	Compulsory religion or state ideology in schools	yes	Indoctrination of children has continued for many years. First, during the 8-year war against Iran and, more recently, during and following the gulf war
16	Deliberate state policies to control artistic works	yes	Cultural life has been one of the casualties of the recent gulf war. Situation and state policies are currently unverifiable
17	Political censorship of press	**NO**	Total. All dissent and contrary views regarded as a threat to the regime
18	Censorship of mail or telephone tapping	**NO**	Total surveillance. The country, despite the end of the gulf hostilities, remains on a war footing

FREEDOM FOR OR RIGHTS TO:		COMMENTS
19	Peaceful political opposition	**NO**	Opposition faces physical elimination, either by death or "disappearance"
20	Multiparty elections by secret and universal ballot	**NO**	The governing regime has absolute power. Its Ba'ath party gives it "rubber stamp" legitimacy

FREEDOM FOR OR RIGHTS TO: COMMENTS

21 Political and legal equality for women **yes** The decimation of the male population caused by the Iran war has improved the status of women, now an important part of public and military services

22 Social and economic equality for women **yes** The national need for able-bodied citizens, following the Iran and gulf wars, has overcome religious and traditional attitudes toward women

23 Social and economic equality for ethnic minorities **NO** Many discriminations. Particularly against Kurds though many occupy a separate zone in north of country in the transitional phase following the gulf war

24 Independent newspapers **NO** Independence circumscribed. Journalists must belong to the Ba'ath party and closures imposed should the newspaper express disagreement with the regime

25 Independent book publishing **NO** Heavily censored and controlled

26 Independent radio and television networks **NO** State owned and controlled

27 All courts to total independence **no** Many cases regarded as matters of national security tried by revolutionary courts. Little legal protection for the defendant but ordinary civil and criminal cases are fairly conducted

28 Independent trade unions **NO** The chaotic state of society and the economy precludes independence. The 1987 trade union reforms have been overtaken by the recent national emergency

LEGAL RIGHTS: COMMENTS

29 From deprivation of nationality **NO** Many citizens of Iranian descent expelled. Others have been deprived of nationality

30 To be considered innocent until proved guilty **NO** The "state-of-war" attitude of the authorities has meant arbitrary arrests without evidence and on the word of a network of informers

31 To free legal aid when necessary and counsel of own choice **yes** Free aid for the needy but counsel appointed by court

32 From civilian trials in secret **NO** At discretion of the security forces

33 To be brought promptly before a judge or court **NO** Suspects held incommunicado for long periods, particularly in security cases – and sometimes "disappear"

34 From police searches of home without a warrant **NO** In the emergency situation, the military and security forces have arbitrary powers

LEGAL RIGHTS:		COMMENTS
35	From arbitrary seizure of personal property	**no**	The devastation caused by the 1990–91 gulf war has meant seizures and confiscations in the "national interest"

PERSONAL RIGHTS:		COMMENTS
36	To interracial, interreligious, or civil marriage	**yes**	In this area, Iraq is more liberal than neighboring Muslim states. Marriage for both sexes legal at 18 years
37	Equality of sexes during marriage and for divorce proceedings	**yes**	The status of wives is improving; also for divorce. But husband's permission needed for travel abroad
38	To practice any religion	**yes**	Some tolerance of foreign religions but government surveillance of Muslim activists for fundamentalist militancy
39	To use contraceptive pills and devices	**yes**	No government support scheme. Supplies affected by the current embargo
40	To noninterference by state in strictly private affairs	**no**	The continuing emergency of a state devastated by war and under the present absolute ruler means that private affairs have no significance

Irish Republic

Human rights rating: 94%

YES 31 yes 8 no 1 NO 0

Population: 3,700,000
Life expectancy: 74.6
Infant mortality (0–5 years)
   per 1,000 births: 9
United Nations covenants ratified:
Civil and Political Rights, Economic,
Social and Cultural Rights

Form of government: Multiparty
democracy
GNP per capita (US$): 7,750
% of GNP spent by state:
On health: 7.8
On military: 1.9
On education: 6.9

FACTORS AFFECTING HUMAN RIGHTS

The individual respected, the government answerable to the people. In general, democratic practices prevail, with a woman recently elected as the nonexecutive president. The Roman Catholic church, the religion of 94% of the population, exerts a dominant influence in many areas. These include marriage, divorce – which is prohibited – and birth control methods and their availability, which are subject to restrictions. As a member country of the European Community, however, some of these decrees, regarded by some as human rights breaches, may be moderating.

	FREEDOM TO:		COMMENTS
1	Travel in own country	YES	Rights respected
2	Travel outside own country	yes	Anti-abortion law may restrict freedom to travel abroad for women seeking to terminate pregnancy
3	Peacefully associate and assemble	YES	Except for terrorist organizations
4	Teach ideas and receive information	yes	The authority exercised by the powerful Catholic church inhibits inquiry in some areas. Also influential in some university appointments
5	Monitor human rights violations	YES	Rights respected
6	Publish and educate in ethnic language	YES	Rights respected

	FREEDOM FROM:		COMMENTS
7	Serfdom, slavery, forced or child labor	YES	Rights respected
8	Extrajudicial killings or "disappearances"	YES	Rights respected

FREEDOM FROM:		COMMENTS
9	Torture or coercion by the state	**YES**	Rights respected
10	Compulsory work permits or conscription of labor	**YES**	Rights respected
11	Capital punishment by the state	**YES**	Abolished 1990
12	Court sentences of corporal punishment	**YES**	Rights respected
13	Indefinite detention without charge	**YES**	Rights respected
14	Compulsory membership of state organizations or parties	**YES**	Rights respected
15	Compulsory religion or state ideology in schools	yes	Wide degree of state-condoned compulsion in Roman Catholic schools
16	Deliberate state policies to control artistic works	**YES**	But occasional restrictions on grounds of blasphemy
17	Political censorship of press	**YES**	Rights respected except for open support for terrorist groups
18	Censorship of mail or telephone tapping	**YES**	Rights respected

FREEDOM FOR OR RIGHTS TO:		COMMENTS
19	Peaceful political opposition	**YES**	Rights respected
20	Multiparty elections by secret and universal ballot	**YES**	Rights respected. Known terrorist groups excluded from the electoral process
21	Political and legal equality for women	yes	The traditional Catholic view of woman's role a factor in limiting representation at highest level but current president is a woman
22	Social and economic equality for women	no	Difference between women's pay and men's pay is the highest in Western Europe
23	Social and economic equality for ethnic minorities	**YES**	Rights respected
24	Independent newspapers	**YES**	Rights respected
25	Independent book publishing	**YES**	Rights respected
26	Independent radio and television networks	**YES**	The ending of the state monopoly in 1988 has been followed by many independent radio and television stations
27	All courts to total independence	**YES**	Special courts of a small panel of judges may try cases affecting national security

FREEDOM FOR OR RIGHTS TO:		COMMENTS
28	Independent trade unions **YES**	Rights respected

LEGAL RIGHTS:		COMMENTS
29	From deprivation of nationality **YES**	Rights respected
30	To be considered innocent until proved guilty **YES**	Rights respected
31	To free legal aid when necessary and counsel of own choice **yes**	Means test. Choice of counsel from a panel
32	From civilian trials in secret **YES**	Rights respected
33	To be brought promptly before a judge or court **YES**	Within 48 hours
34	From police searches of home without a warrant **YES**	Rights respected
35	From arbitrary seizure of personal property **YES**	Rights respected

PERSONAL RIGHTS:		COMMENTS
36	To interracial, interreligious, or civil marriage **YES**	Rights respected
37	Equality of sexes during marriage and for divorce proceedings **yes**	Strong social and religious influences favor husbands. The question of divorce, still illegal, remains a continuing issue
38	To practice any religion **YES**	Rights respected
39	To use contraceptive pills and devices **yes**	Restrictions on sale of contraceptives. Family planning associations fined for sale of condoms through a nonpharmacy outlet. Suppression of AIDS advice booklet because of church opposition to promotion of use of condoms
40	To noninterference by state in strictly private affairs **yes**	Pressures by influential antiabortion groups, which have led to the court closures of clinics, and government restrictions on the sale and advertising of condoms may be seen as interference with private life

Israel (ex-Occupied Territories)

Human rights rating: 76%

YES 17 **yes** 18 **no** 4 NO 1

Population: 4,600,000
Life expectancy: 75.9
Infant mortality (0–5 years)
  per 1,000 births: 14
United Nations covenants ratified:
None

Form of government: Parliamentary democracy
GNP per capita (US$): 8,650
% of GNP spent by state:
On health: 2.1
On military: 19.2
On education: 7.3

FACTORS AFFECTING HUMAN RIGHTS

From independence in 1948, a persistent formal state of hostilities with most of its Arab neighbors has meant a series of wars and a continuing period of alertness and semimobilization. Certain Arab lands were occupied after the 1967 war and are known as the Occupied Territories. The human rights situation for non-Arabs in Israel (80% of the population) is satisfactory and contrasts, in particular, with the limited rights of Arabs under military rule in the Occupied Territories. The information below applies only to Israel itself. In November 1991, the USA persuaded the parties to the dispute to begin "peace talks." Both the readiness to continue and the outcome of these negotiations are unpredictable.

FREEDOM TO: | COMMENTS

	FREEDOM TO:		COMMENTS
1	Travel in own country	yes	With exception of security areas
2	Travel outside own country	YES	Rights respected (Occ. Terr. not comparable)
3	Peacefully associate and assemble	no	Certain pro-Palestinian rallies are prohibited. Occasional imprisonment of Arabs for displaying Palestinian flag or for anti-Israel graffiti
4	Teach ideas and receive information	yes	Arabs at universities and in education practice circumspection. Hostile views toward Israel may damage career prospects (Occ. Terr. not comparable)
5	Monitor human rights violations	YES	Rights respected (Occ. Terr. not comparable)
6	Publish and educate in ethnic language	YES	Rights respected

FREEDOM FROM: | COMMENTS

	FREEDOM FROM:		COMMENTS
7	Serfdom, slavery, forced or child labor	YES	Rights respected

FREEDOM FROM:		COMMENTS
8	Extrajudicial killings or "disappearances"	yes	Rare excuses by overzealous security forces despite strict government edict (Occ. Terr. not comparable)
9	Torture or coercion by the state	yes	Local abuses by security forces against Arabs but subject to government investigations (Occ. Terr. not comparable)
10	Compulsory work permits or conscription of labor	**YES**	Rights respected
11	Capital punishment by the state	yes	Only for Nazi war crimes. Otherwise abolished from 1954
12	Court sentences of corporal punishment	**YES**	Rights respected
13	Indefinite detention without charge	no	Indefinite detentions of Palestinians from the Occupied Territories in camps actually sited in Israel (evidence of opposition accusations of linkage). Security cases may be prolonged for Israeli citizens (Occ. Terr. not comparable)
14	Compulsory membership of state organizations or parties	**YES**	Rights respected
15	Compulsory religion or state ideology in schools	**YES**	Rights respected
16	Deliberate state policies to control artistic works	yes	Surveillance and occasional banning of plays, songs, etc., on grounds of "security"
17	Political censorship of press	yes	Disputed official interpretation of "security matters." Censorship more severe – and wider – for Arab-language publications
18	Censorship of mail or telephone tapping	no	General surveillance on grounds of security and public order in a country still formally at war with some of its neighbors

FREEDOM FOR OR RIGHTS TO:		COMMENTS
19	Peaceful political opposition	**YES**	Rights respected
20	Multiparty elections by secret and universal ballot	yes	The social and ethnic status of Arabs effectively contributes to their underrepresentation in Parliament
21	Political and legal equality for women	yes	The Equal Opportunities Law is usually enforced but despite a previous prime minister being a woman, most senior posts are held by men
22	Social and economic equality for women	yes	Powerful religious laws frequently to women's disadvantage, particularly among Orthodox Jews and Arabs

FREEDOM FOR OR RIGHTS TO:		COMMENTS
23	Social and economic equality for ethnic minorities	**yes**	Discrimination against Arabs in most areas of society, including government service
24	Independent newspapers	**yes**	Granting of licenses prejudiced if newspaper or journal suspected of supporting Palestinians seeking end to Israeli control of Occupied Territories
25	Independent book publishing	**yes**	But military may press for banning of books considered "subversive"
26	Independent radio and television networks	**yes**	Controlled by a nonpolitical broadcasting authority but this is subject to state of emergency provisions. Occasional restrictions in line with "not serving enemy propaganda" (policy statement)
27	All courts to total independence	**YES**	Rights respected
28	Independent trade unions	**yes**	Discrimination of workers from Occupied Territories working in Israel. Israelis enjoy trade union rights denied to Palestinian workers

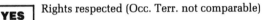

LEGAL RIGHTS:		COMMENTS
29	From deprivation of nationality	**YES**	Rights respected
30	To be considered innocent until proved guilty	**YES**	Rights respected (Occ. Terr. not comparable)
31	To free legal aid when necessary and counsel of own choice	**YES**	Legal aid and choice of counsel, who may refuse the case
32	From civilian trials in secret	**yes**	Security cases under emergency legislation
33	To be brought promptly before a judge or court	**YES**	Rights respected (Occ. Terr. not comparable)
34	From police searches of home without a warrant	**yes**	Police abuses against Arabs who protest at being "second-class citizens" (Occ. Terr. not comparable)
35	From arbitrary seizure of personal property	**YES**	Rights respected (Occ. Terr. not comparable)

PERSONAL RIGHTS:		COMMENTS
36	To interracial, interreligious, or civil marriage	**NO**	No civil marriage. Religious law forbids intermarriage. Nonbelievers forced to marry abroad

PERSONAL RIGHTS: COMMENTS

37 Equality of sexes during
    marriage and for divorce
    proceedings

no

State has followed Ottoman precedent of
allowing religious courts (rabbinical and
Shari'a) to try marriage, divorce, and
alimony cases. Both religions favor men

38 To practice any religion

YES

Rights respected

39 To use contraceptive pills
    and devices

YES

Rights respected

40 To noninterference by state
    in strictly private affairs

yes

Official attitudes and social discrimination
toward Arabs are intimidating factors and
create fear or uncertainty in private life

Italy

Human rights rating: 90%

YES 29 yes 11 no 0 NO 0

Population: 57,100,000
Life expectancy: 76.0
Infant mortality (0–5 years) per 1,000 births: 11
United Nations covenants ratified:
Civil and Political Rights, Economic, Social and Cultural Rights, Convention on Equality for Women

Form of government: Constitutional democracy
GNP per capita (US$): 13,330
% of GNP spent by state:
On health: 4.5
On military: 2.3
On education: 4.0

FACTORS AFFECTING HUMAN RIGHTS
The individual respected. A government answerable to the people. A prosperous economy and one of the more important members of the European Community. The two areas of human rights concern are the strength of organized crime, and its influence on sections of society, and the discrimination against an influx of foreign labor, particularly from North Africa. There is a prosperity gap between the north and the south of the country, with the latter occasionally adhering to feudal customs.

	FREEDOM TO:		COMMENTS
1	Travel in own country	YES	Rights respected
2	Travel outside own country	YES	Rights respected
3	Peacefully associate and assemble	YES	Rights respected
4	Teach ideas and receive information	YES	Rights respected
5	Monitor human rights violations	YES	Rights respected
6	Publish and educate in ethnic language	YES	Rights respected

	FREEDOM FROM:		COMMENTS
7	Serfdom, slavery, forced or child labor	yes	Illegal child labor, especially in less-developed south
8	Extrajudicial killings or "disappearances"	YES	Rights respected
9	Torture or coercion by the state	yes	Illegal local police abuses. Some vicious

FREEDOM FROM:		COMMENTS
10	Compulsory work permits or conscription of labor	**YES**	Rights respected
11	Capital punishment by the state	**YES**	Abolished 1944
12	Court sentences of corporal punishment	**YES**	Rights respected
13	Indefinite detention without charge	yes	Recent 4-year limitation of pretrial detention occasionally disregarded
14	Compulsory membership of state organizations or parties	**YES**	Rights respected
15	Compulsory religion or state ideology in schools	yes	Despite controversy, instances continue of compulsion in many Catholic schools
16	Deliberate state policies to control artistic works	**YES**	Rights respected
17	Political censorship of press	yes	Defamation of state authorities has led to occasional seizures
18	Censorship of mail or telephone tapping	**YES**	Rights respected

FREEDOM FOR OR RIGHTS TO:		COMMENTS
19	Peaceful political opposition	**YES**	But Fascist party banned
20	Multiparty elections by secret and universal ballot	**YES**	Rights respected
21	Political and legal equality for women	yes	Gradual improvement in women's status at all levels though still underrepresented in senior government service
22	Social and economic equality for women	yes	Minor inequalities in pay and employment. More evident in the less-prosperous south
23	Social and economic equality for ethnic minorities	yes	Influx of foreign labor, particularly from North Africa, is encouraging wide but unlawful discrimination
24	Independent newspapers	**YES**	Rights respected
25	Independent book publishing	**YES**	Rights respected
26	Independent radio and television networks	yes	Part state owned with allegations of political bias in favor of government despite guarantees of impartiality
27	All courts to total independence	**YES**	Rights respected
28	Independent trade unions	**YES**	Rights respected. 1990 law limits strikes in essential services

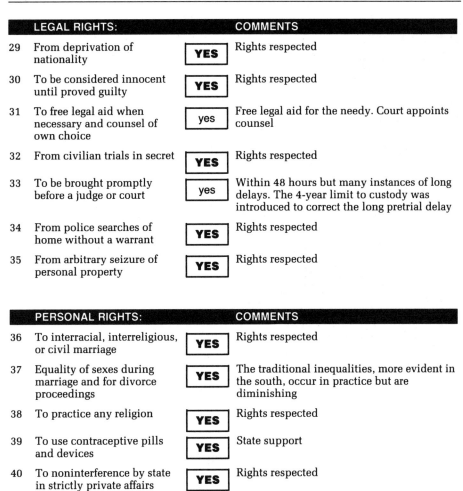

LEGAL RIGHTS:		COMMENTS
29 From deprivation of nationality	**YES**	Rights respected
30 To be considered innocent until proved guilty	**YES**	Rights respected
31 To free legal aid when necessary and counsel of own choice	yes	Free legal aid for the needy. Court appoints counsel
32 From civilian trials in secret	**YES**	Rights respected
33 To be brought promptly before a judge or court	yes	Within 48 hours but many instances of long delays. The 4-year limit to custody was introduced to correct the long pretrial delay
34 From police searches of home without a warrant	**YES**	Rights respected
35 From arbitrary seizure of personal property	**YES**	Rights respected

PERSONAL RIGHTS:		COMMENTS
36 To interracial, interreligious, or civil marriage	**YES**	Rights respected
37 Equality of sexes during marriage and for divorce proceedings	**YES**	The traditional inequalities, more evident in the south, occur in practice but are diminishing
38 To practice any religion	**YES**	Rights respected
39 To use contraceptive pills and devices	**YES**	State support
40 To noninterference by state in strictly private affairs	**YES**	Rights respected

Ivory Coast

Human rights rating: 75%

YES 14 yes 20 no 5 NO 1

Population: 12,000,000
Life expectancy: 53.4
Infant mortality (0–5 years) per 1,000 births: 139
United Nations covenants ratified: None

Form of government: Multiparty system
GNP per capita (US$): 770
% of GNP spent by state:
On health: 1.1
On military: 1.2
On education: 5.0

FACTORS AFFECTING HUMAN RIGHTS

After 30 years of one-party rule, the adoption in 1990 of democratic multiparty elections resulted in the continuation in power of the Democratic party with over 90% of the votes. The opposition protested at various irregularities. Confirmation by the electorate, however, has not persuaded the government to correct a number of human rights violations. The opposition is still under constant surveillance and the media almost entirely controlled by the government. Although the octogenarian president is virtually an absolute ruler, the human rights trend is toward improvement.

	FREEDOM TO:		COMMENTS
1	Travel in own country	yes	System of road checks offers security police an opportunity to augment their low pay
2	Travel outside own country	YES	Rights respected
3	Peacefully associate and assemble	no	The moves toward a multiparty political system have meant a more liberal policy for public meetings but police still break up some pro-democracy rallies
4	Teach ideas and receive information	no	Government pressures on academics to limit criticism. The government's overwhelming victory in the first multiparty election since the country's independence in 1960 may restrict progress toward more academic freedom
5	Monitor human rights violations	YES	Rights respected. No recent evidence of government obstruction
6	Publish and educate in ethnic language	YES	Rights respected. Official language is French

---

	FREEDOM FROM:		COMMENTS
7	Serfdom, slavery, forced or child labor	yes	Regional pattern of child labor

FREEDOM FROM:		COMMENTS
8	Extrajudicial killings or "disappearances"	**YES**	Rights respected
9	Torture or coercion by the state	yes	Occasional minor abuses
10	Compulsory work permits or conscription of labor	**YES**	Rights respected
11	Capital punishment by the state	yes	Sentences regularly commuted to 20 years' imprisonment. No executions of pregnant women until 2 months after giving birth
12	Court sentences of corporal punishment	**YES**	Rights respected
13	Indefinite detention without charge	yes	Pattern of brief detentions for political views represents a considerable improvement over recent years
14	Compulsory membership of state organizations or parties	yes	With the introduction in 1990 of a multiparty system, the law that made all citizens members of the ruling party has been abolished. Advantages, however, in civil service and professions for those continuing their membership
15	Compulsory religion or state ideology in schools	**YES**	Rights respected
16	Deliberate state policies to control artistic works	yes	Provocative creative works risk government seizures. No insulting or satirizing the president
17	Political censorship of press	yes	Understood guidelines, which usually avoid direct intervention by the government
18	Censorship of mail or telephone tapping	yes	Some surveillance of political opponents. The ruling party and aging president have been in power since 1960

FREEDOM FOR OR RIGHTS TO:		COMMENTS
19	Peaceful political opposition	yes	The introduction of multiparty politics, after 30 years of single-party rule, was relatively peaceful
20	Multiparty elections by secret and universal ballot	yes	Many complaints of election irregularities, which favored the ruling party, after introduction of democratic reforms
21	Political and legal equality for women	no	Despite electoral reforms, little participation in politics at senior levels
22	Social and economic equality for women	no	Traditional inequalities. Worse in rural areas. Female circumcision still widely practiced. Also excision
23	Social and economic equality for ethnic minorities	**YES**	Rights respected

FREEDOM FOR OR RIGHTS TO: COMMENTS

24	Independent newspapers	yes	Smaller publications in independent hands but government indirectly controls printing presses as well as the major daily newspapers
25	Independent book publishing	yes	Limited book publication but care advisable in choice of subjects
26	Independent radio and television networks	**NO**	State owned and controlled
27	All courts to total independence	yes	Some interference in cases considered by the government to affect "national security." Occasional instances of corruption
28	Independent trade unions	yes	Union compliance with the governing party which may lessen with the recent democratic reforms

LEGAL RIGHTS: COMMENTS

29	From deprivation of nationality	**YES**	Rights respected
30	To be considered innocent until proved guilty	yes	Random arrests by local police for the purpose of extracting bribes in exchange for release
31	To free legal aid when necessary and counsel of own choice	yes	Court appoints counsel for the needy
32	From civilian trials in secret	**YES**	Rights respected
33	To be brought promptly before a judge or court	yes	Within 48 hours. May be followed by 4 months' detention. A few violations and instances of bribery to gain release
34	From police searches of home without a warrant	yes	Law usually respected. Searches forbidden between 9 pm and 4 am
35	From arbitrary seizure of personal property	yes	A few violations of the law by underpaid police. Seizures may be on pretext of recovering "stolen" property

PERSONAL RIGHTS: COMMENTS

36	To interracial, interreligious, or civil marriage	**YES**	Rights respected
37	Equality of sexes during marriage and for divorce proceedings	**no**	Customary tribal laws govern many marriages. Such laws usually favor husbands
38	To practice any religion	**YES**	Rights respected. Animists 45% of population
39	To use contraceptive pills and devices	**YES**	Rights respected

40 To noninterference by state
in strictly private affairs

YES

Rights respected

Jamaica

Human rights rating: 72%

YES 23 yes 12 no 4 NO 1

Population: 2,500,000
Life expectancy: 73.1
Infant mortality (0–5 years) per 1,000 births: 21
United Nations covenants ratified:
Civil and Political Rights, Economic, Social and Cultural Rights, Convention on Equality for Women

Form of government: Parliamentary democracy
GNP per capita (US$): 1,070
% of GNP spent by state:
On health: 2.8
On military: 1.5
On education: 5.6

FACTORS AFFECTING HUMAN RIGHTS

The major human rights violations concern the social realities of a relatively small island with a high population density and an economy relying on limited resources. This has meant a sustained degree of violence, usually between criminal gangs and the police. In 1990 the police were responsible for over 100 deaths, many explained as being "in the course of duty" or "killed while trying to escape." Prison conditions remain bad with overcrowding and maltreatment of the inmates. Complaints to the courts about police conduct have resulted in many awards of damages.

FREEDOM TO:		COMMENTS
1 Travel in own country	YES	Rights respected
2 Travel outside own country	YES	Rights respected
3 Peacefully associate and assemble	YES	Rights respected
4 Teach ideas and receive information	YES	Rights respected
5 Monitor human rights violations	YES	Rights respected
6 Publish and educate in ethnic language	YES	Rights respected

FREEDOM FROM:		COMMENTS
7 Serfdom, slavery, forced or child labor	yes	Regional pattern of child labor
8 Extrajudicial killings or "disappearances"	no	Over 100 killings in 1990 in police operations against violent gangs and sundry lawbreakers. Charges of precipitate shootings by police

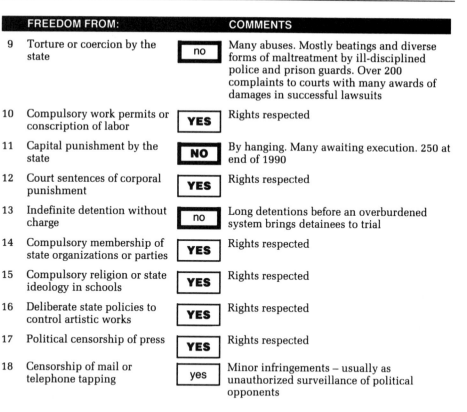

	FREEDOM FROM:		COMMENTS
9	Torture or coercion by the state	no	Many abuses. Mostly beatings and diverse forms of maltreatment by ill-disciplined police and prison guards. Over 200 complaints to courts with many awards of damages in successful lawsuits
10	Compulsory work permits or conscription of labor	YES	Rights respected
11	Capital punishment by the state	NO	By hanging. Many awaiting execution. 250 at end of 1990
12	Court sentences of corporal punishment	YES	Rights respected
13	Indefinite detention without charge	no	Long detentions before an overburdened system brings detainees to trial
14	Compulsory membership of state organizations or parties	YES	Rights respected
15	Compulsory religion or state ideology in schools	YES	Rights respected
16	Deliberate state policies to control artistic works	YES	Rights respected
17	Political censorship of press	YES	Rights respected
18	Censorship of mail or telephone tapping	yes	Minor infringements – usually as unauthorized surveillance of political opponents

	FREEDOM FOR OR RIGHTS TO:		COMMENTS
19	Peaceful political opposition	YES	Rights respected
20	Multiparty elections by secret and universal ballot	yes	Elections are marked by degrees of fraud and lawlessness. Intimidation of voters, thefts of ballot boxes, etc.
21	Political and legal equality for women	yes	Minor inequalities in practice. The domination by men of positions of authority, in a society featuring excesses of male aggression, limits women's involvement in politics
22	Social and economic equality for women	yes	Pay and employment inequality, worsened by current economic recession
23	Social and economic equality for ethnic minorities	YES	90% of population is of African origin
24	Independent newspapers	YES	Rights respected
25	Independent book publishing	YES	Rights respected

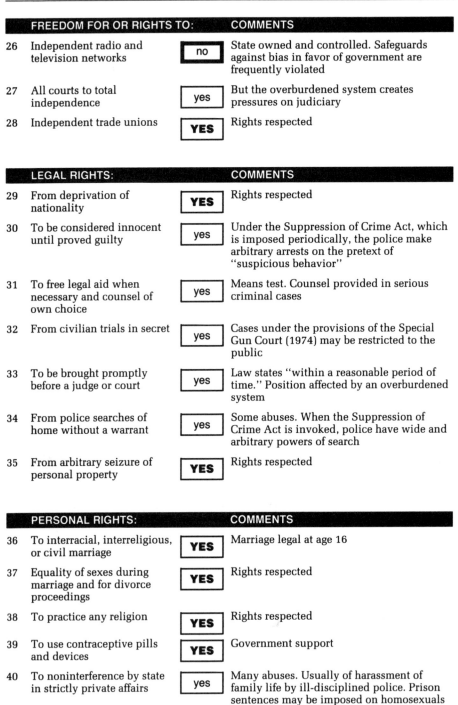

FREEDOM FOR OR RIGHTS TO:		COMMENTS
26	Independent radio and television networks	**no**	State owned and controlled. Safeguards against bias in favor of government are frequently violated
27	All courts to total independence	yes	But the overburdened system creates pressures on judiciary
28	Independent trade unions	**YES**	Rights respected

LEGAL RIGHTS:		COMMENTS
29	From deprivation of nationality	**YES**	Rights respected
30	To be considered innocent until proved guilty	yes	Under the Suppression of Crime Act, which is imposed periodically, the police make arbitrary arrests on the pretext of "suspicious behavior"
31	To free legal aid when necessary and counsel of own choice	yes	Means test. Counsel provided in serious criminal cases
32	From civilian trials in secret	yes	Cases under the provisions of the Special Gun Court (1974) may be restricted to the public
33	To be brought promptly before a judge or court	yes	Law states "within a reasonable period of time." Position affected by an overburdened system
34	From police searches of home without a warrant	yes	Some abuses. When the Suppression of Crime Act is invoked, police have wide and arbitrary powers of search
35	From arbitrary seizure of personal property	**YES**	Rights respected

PERSONAL RIGHTS:		COMMENTS
36	To interracial, interreligious, or civil marriage	**YES**	Marriage legal at age 16
37	Equality of sexes during marriage and for divorce proceedings	**YES**	Rights respected
38	To practice any religion	**YES**	Rights respected
39	To use contraceptive pills and devices	**YES**	Government support
40	To noninterference by state in strictly private affairs	yes	Many abuses. Usually of harassment of family life by ill-disciplined police. Prison sentences may be imposed on homosexuals

Japan

Human rights rating: 82%

YES 25 yes 12 no 2 NO 1

Population: 123,500,000
Life expectancy: 78.6
Infant mortality (0–5 years) per 1,000 births: 8
United Nations covenants ratified:
Civil and Political Rights, Economic,
Social and Cultural Rights, Convention
on Equality for Women

Form of government: Parliamentary
monarchy
GNP per capita (US$): 21,020
% of GNP spent by state:
On health: 4.9
On military: 1.0
On education: 5.0

FACTORS AFFECTING HUMAN RIGHTS
A prosperous economy, democratic institutions, a free press, and a cohesive society.
The new constitution of 1947, which replaced the Meiji constitution of 1889, affirmed
Japan's postwar commitment to human rights violations. Nevertheless, some serious
violations continue. Behind the facade of a formalized society with a high regard for
conservative values and behavior, the status of women is still affected by a traditional
subservience, the *Burakumin* ("outcasts") and Korean minorities suffer major
disadvantages, the police are guilty of abuses, particularly during lengthy interrogation
of suspects, and individual judges have been known to complain about the
consequences to career prospects of not upholding conformist values. From time to time
senior government ministers and the financial establishment are involved in major
corruption scandals, though penalties for the misdemeanors of the most powerful
individuals appear to be minimal.

	FREEDOM TO:		COMMENTS
1	Travel in own country	YES	Rights respected
2	Travel outside own country	YES	Rights respected
3	Peacefully associate and assemble	YES	Rights respected
4	Teach ideas and receive information	YES	Rights respected but government and some historians encourage indirect censorship of Japan's role in Second World War. Protests of "whitewashing"
5	Monitor human rights violations	yes	Restrictions on monitoring treatment in mental hospitals and of delinquent children in institutional care. No official monitoring of discrimination against Buraku people ("outcasts")
6	Publish and educate in ethnic language	YES	Rights respected

FREEDOM FROM:		COMMENTS
7	Serfdom, slavery, forced or child labor	**YES**	Rights respected
8	Extrajudicial killings or "disappearances"	**YES**	Rights respected
9	Torture or coercion by the state	yes	Police abuses. Beatings and maltreatment, particularly during harsh and lengthy interrogation while in police custody (*daiyo kangoku*)
10	Compulsory work permits or conscription of labor	**YES**	Rights respected
11	Capital punishment by the state	**NO**	By hanging. For 18 different offenses. No executions since 1990 but approximately 50 cases await death
12	Court sentences of corporal punishment	**YES**	Rights respected
13	Indefinite detention without charge	yes	Interrogation of detainees in police cells can be extended to a maximum of 300 days without charge. Lawyers frequently obstructed
14	Compulsory membership of state organizations or parties	**YES**	Rights respected
15	Compulsory religion or state ideology in schools	**YES**	Rights respected
16	Deliberate state policies to control artistic works	**YES**	Rights respected
17	Political censorship of press	yes	Degree of self-censorship in a society highly sensitive to criticism of emperor, treatment of *Burakumin* minority, and business corruption. Threats against journalists by extremist groups
18	Censorship of mail or telephone tapping	yes	Illegal wiretapping, usually of Communists. Resignations of implicated senior police officers

FREEDOM FOR OR RIGHTS TO:		COMMENTS
19	Peaceful political opposition	**YES**	Rights respected
20	Multiparty elections by secret and universal ballot	**YES**	Rights respected
21	Political and legal equality for women	yes	Despite a number of government measures to improve status of women, traditional subservience continues. House of Representatives has 12 women out of 504

FREEDOM FOR OR RIGHTS TO:		COMMENTS
22 Social and economic equality for women	**no**	Major differences in pay and employment opportunities. Traditional attitudes persist. A more stressful society has generated a category of women known as "kitchen drinkers"
23 Social and economic equality for ethnic minorities	yes	Korean and *Burakumin*, 2 million in number, suffer discrimination in a country concerned with "Japaneseness." Pay and employment disadvantages. Proof of assimiliation of Japanese culture required for naturalization
24 Independent newspapers	**YES**	Rights respected
25 Independent book publishing	**YES**	But certain taboo subjects. These include the need to respect the emperor, prudence toward violent crime syndicates, and exposure of corruption "in high places"
26 Independent radio and television networks	**YES**	Rights respected. Of 180 television and radio companies, only one is government controlled
27 All courts to total independence	yes	Administrative pressures on individual judges to conform to prevailing conservative values. Complaints by Japanese Federation of Bar Associates submitted to United Nations
28 Independent trade unions	**YES**	Rights respected

LEGAL RIGHTS:		COMMENTS
29 From deprivation of nationality	**YES**	Rights respected
30 To be considered innocent until proved guilty	yes	Many forced confessions. Japanese conviction rate three times that of USA, twice that of France. Police prejudice cases by making disclosures to the media
31 To free legal aid when necessary and counsel of own choice	**no**	Denial of counsel during lengthy interrogation while in police custody, in practice violating law (*Kenpo*, Article 37). Limited aid after indictment
32 From civilian trials in secret	**YES**	Rights respected
33 To be brought promptly before a judge or court	yes	72 hours may be followed by a preindictment 20-day period of interrogation (sometimes longer) in police isolation cells (*daiyo kangoku*), in practice a successful way of extracting "confessions"
34 From police searches of home without a warrant	**YES**	Rights respected

LEGAL RIGHTS:		COMMENTS
35	From arbitrary seizure of personal property	**YES**	Rights respected

PERSONAL RIGHTS:		COMMENTS
36	To interracial, interreligious, or civil marriage	**YES**	Rights respected
37	Equality of sexes during marriage and for divorce proceedings	yes	Despite constitutional equality, traditional subservience of wives remains the social reality. This influences official attitudes toward marriage and divorce – and toward a significant increase of violence against wives
38	To practice any religion	**YES**	Rights respected
39	To use contraceptive pills and devices	yes	Moves to liberalize prescribing birth control pills. Abortion legal and sometimes favored by doctors who may therefore prosper financially from prohibition of "the pill"
40	To noninterference by state in strictly private affairs	**YES**	Rights respected

Jordan

Human rights rating: 65%

Population: 4,000,000
Life expectancy: 66.9
Infant mortality (0–5 years)
  per 1,000 births: 55
United Nations covenants ratified:
Civil and Political Rights, Economic,
Social and Cultural Rights

Form of government: Nominated by
monarch
GNP per capita (US$): 1,500
% of GNP spent by state:
On health: 2.7
On military: 13.8
On education: 6.5

FACTORS AFFECTING HUMAN RIGHTS
The king is an absolute ruler and political life is severely circumscribed for the
population. Tentative plans for a national charter which would grant his subjects
greater participation in political and public life are being considered. From time to time
the king's authority is reinforced by the declaration of martial law. The country's
economy has been affected by its association with Iraq during the recent gulf war.

	FREEDOM TO:		COMMENTS
1	Travel in own country	**YES**	Rights respected except in specific military zones
2	Travel outside own country	yes	Passports for wives need husband's consent in writing
3	Peacefully associate and assemble	**no**	Permission to hold rallies depends on following understood guidelines. Prohibited issues may be political, religious, or criticism of the king
4	Teach ideas and receive information	yes	Understood guidelines on politics, religions, and the king
5	Monitor human rights violations	yes	In practice, no significant monitoring by local groups. Some official reluctance to cooperate with visiting international investigators
6	Publish and educate in ethnic language	**YES**	Rights respected

	FREEDOM FROM:		COMMENTS
7	Serfdom, slavery, forced or child labor	yes	Regional pattern of child labor
8	Extrajudicial killings or "disappearances"	**YES**	Rights respected

FREEDOM FROM:		**COMMENTS**
9	Torture or coercion by the state	yes	Increased supervision by the authorities has meant fewer abuses at local level
10	Compulsory work permits or conscription of labor	YES	Rights respected
11	Capital punishment by the state	NO	By hanging or firing squad. Pregnant women may not be executed before 3 months following birth of child
12	Court sentences of corporal punishment	YES	Rights respected
13	Indefinite detention without charge	no	State of emergency, when in force, permits indefinite detention. Many held incommunicado
14	Compulsory membership of state organizations or parties	YES	Rights respected
15	Compulsory religion or state ideology in schools	no	In most Muslim schools, the religion of over 90% of population
16	Deliberate state policies to control artistic works	yes	Subject to the beliefs and restraints of the area and the national religion
17	Political censorship of press	yes	Mostly self-censorship. Government "advises" rather than censors directly
18	Censorship of mail or telephone tapping	yes	Limited surveillance. Increased during the recent gulf war

FREEDOM FOR OR RIGHTS TO:		**COMMENTS**
19	Peaceful political opposition	no	Political opposition and Parliament may influence ministerial policies but ultimate executive power rests with the king
20	Multiparty elections by secret and universal ballot	no	For a parliament with very limited powers; and for local elections. A national charter to increase participation of the population in political and public life is being considered by a royal commission
21	Political and legal equality for women	no	Recent progress in the professions and limited involvement in politics, though not at senior government level. Women may now serve in the armed forces
22	Social and economic equality for women	no	Wide inequalities perpetuated by regional, ethnic, and religious factors
23	Social and economic equality for ethnic minorities	yes	The end of the gulf war added to the social difficulties of absorbing Palestinians from Iraq. The country's economic plight has meant an increase in discrimination
24	Independent newspapers	no	Government control of leading newspapers through shareholdings of over 50%

FREEDOM FOR OR RIGHTS TO:		COMMENTS
25	Independent book publishing	yes	No publishing of works likely to offend the king, Islamic sensibilities, or the military
26	Independent radio and television networks	**NO**	State owned and controlled
27	All courts to total independence	yes	Rare instance of courts being influenced by government pressures. Shari'a courts for family and personal cases
28	Independent trade unions	yes	Small proportion of workers unionized, probably 10%. Some restrictions on strike action

LEGAL RIGHTS:		COMMENTS
29	From deprivation of nationality	**YES**	Rights respected
30	To be considered innocent until proved guilty	**YES**	Rights respected
31	To free legal aid when necessary and counsel of own choice	yes	But only for the most serious cases – i.e., murder charges, etc.
32	From civilian trials in secret	yes	The king has executive powers to order secret trials
33	To be brought promptly before a judge or court	yes	Within 48 hours. The elimination of most martial law provisions will reduce delays and arbitrary detentions
34	From police searches of home without a warrant	yes	Local abuses usually relate to "hot pursuit"
35	From arbitrary seizure of personal property	**YES**	Rights respected

PERSONAL RIGHTS:		COMMENTS
36	To interracial, interreligious, or civil marriage	no	Inter-marriage rare because of the constraints of Islamic religion, that of over 90% of the population
37	Equality of sexes during marriage and for divorce proceedings	no	Koranic law favors husbands in almost all matters, including divorce, inheritance, and the right to "discipline" wives
38	To practice any religion	**YES**	Rights respected
39	To use contraceptive pills and devices	**YES**	Rights respected
40	To noninterference by state in strictly private affairs	yes	The state may support intrusion into private affairs by coercive religious zealots

Kenya

Human rights rating: 46%

Population: 24,000,000
Life expectancy: 59.7
Infant mortality (0–5 years)
per 1,000 births: 111
United Nations covenants ratified:
Civil and Political Rights, Economic,
Social and Cultural Rights, Convention
on Equality for Women

Form of government: One-party state
GNP per capita (US$): 370
% of GNP spent by state:
On health: 1.7
On military: 1.2
On education: 6.1

FACTORS AFFECTING HUMAN RIGHTS

A one-party state with a powerful president. Authority is ensured by close surveillance and occasional arrests of opponents, press restrictions, and virtual control of broadcasting. In July 1990, demonstrations and rioting calling for change were forcibly suppressed with over 20 deaths and 1,000 arrests. Western countries which have given generous aid to a regime they regarded as one of the most deserving in Africa have tried to persuade the president to improve the human rights situation and to introduce a multiparty system. In December 1991, the president and his Kenya African National Union appeared to yield to this and promised amendments to the constitution to permit democratic elections.

	FREEDOM TO:		COMMENTS
1	Travel in own country	**YES**	Rights respected
2	Travel outside own country	yes	Passports of political opposition occasionally withdrawn or issue refused
3	Peacefully associate and assemble	**NO**	Demonstrations against government in favor of multiparty democracy forcibly suppressed. Many dead, including children. Pattern continues
4	Teach ideas and receive information	no	Universities frequently closed. Restrictions on academic subjects seen to create opposition to government. Surveillance by system of informers
5	Monitor human rights violations	no	President insists that no monitoring is necessary. Occasional visits by international groups face obstacles rather than ban on activities
6	Publish and educate in ethnic language	**YES**	Rights respected

FREEDOM FROM:		COMMENTS
7 Serfdom, slavery, forced or child labor	yes	Child labor in rural areas
8 Extrajudicial killings or "disappearances"	no	Usually excessive force by security services to put down public protests of one-party rule
9 Torture or coercion by the state	no	Well-documented evidence of torture. Frequently to extract confessions. Intellectuals held on bogus charges of treason and tortured
10 Compulsory work permits or conscription of labor	YES	Rights respected
11 Capital punishment by the state	NO	By hanging. For treason, murder, robbery with violence, etc.
12 Court sentences of corporal punishment	no	Floggings by rod or switch
13 Indefinite detention without charge	NO	Public Security Act permits indefinite detention
14 Compulsory membership of state organizations or parties	yes	But civil servants expected to belong to the ruling Kenya African National Union
15 Compulsory religion or state ideology in schools	YES	Rights respected
16 Deliberate state policies to control artistic works	no	Arrests of satirists and songwriters producing lyrics critical of government. "Folk songs can be seditious"
17 Political censorship of press	yes	Self-imposed guidelines, to ensure freedom from arrest
18 Censorship of mail or telephone tapping	no	Constant surveillance of security suspects, liberal academics, and those making illegal currency transfers overseas

FREEDOM FOR OR RIGHTS TO:		COMMENTS
19 Peaceful political opposition	NO	Severe punishments for some political opponents. President is virtually an absolute ruler. KANU the single authorized party. But see Factors
20 Multiparty elections by secret and universal ballot	NO	Government justifies opposition to multiparty elections on grounds that it would encourage "tribal conflicts". But see Factors
21 Political and legal equality for women	no	Modest progress from customary subservient status but discrimination remains widespread
22 Social and economic equality for women	no	High proportion of women work as farm laborers and as cheap labor. Unlawful female circumcision still practiced by some tribes

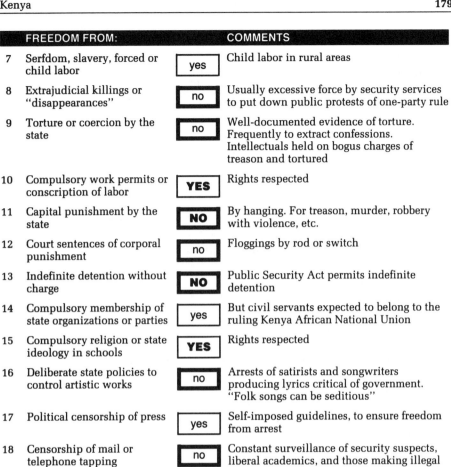

FREEDOM FOR OR RIGHTS TO:		COMMENTS
23	Social and economic equality for ethnic minorities	yes	The ethnic Somalis of the north suffer official discrimination over identity papers and registration requirements
24	Independent newspapers	no	Newspapers follow prudent guidelines. Occasional seizures. Government not obliged to explain actions
25	Independent book publishing	no	Publishers have suffered closures and arrests when charged with sedition or seditious intent
26	Independent radio and television networks	**NO**	Owned or controlled by government or the ruling party. The single American cable network is sometimes censored
27	All courts to total independence	no	All senior court appointments by the president
28	Independent trade unions	yes	Overall control of unions rests with the Central Organization of Trade Unions, which in turn is influenced by the government

LEGAL RIGHTS:		COMMENTS
29	From deprivation of nationality	**YES**	Rights respected
30	To be considered innocent until proved guilty	no	Many suspected political dissidents held in police custody on flimsy or nonexistent evidence
31	To free legal aid when necessary and counsel of own choice	no	Only for capital offenses. Counsel appointed by court
32	From civilian trials in secret	**NO**	Political detainees may be tried *in camera*
33	To be brought promptly before a judge or court	no	Within 14 days but long delays permitted under the Public Security Act
34	From police searches of home without a warrant	no	In practice arbitrary powers, particularly when police claim to be in "hot pursuit"
35	From arbitrary seizure of personal property	**YES**	Rights respected

PERSONAL RIGHTS:		COMMENTS
36	To interracial, interreligious, or civil marriage	**YES**	Legal age 18 years for both sexes
37	Equality of sexes during marriage and for divorce proceedings	yes	Many marriages under customary (tribal) law with advantages to husband. Some polygamy still practiced

PERSONAL RIGHTS:		COMMENTS
38	To practice any religion	**YES**	Churches risk "deregistration" if involved in politics but little interference in practice of verbal statements by church leaders
39	To use contraceptive pills and devices	**YES**	State support
40	To noninterference by state in strictly private affairs	no	"Umbrella" searches without warrants. Psychological pressures on government critics affect personal life. Police informers active

Korea, North

Human rights rating: 20%

YES 6 yes 2 no 6 NO 26

Population: 21,800,000
Life expectancy: 70.4
Infant mortality (0–5 years) per 1,000 births: 36
United Nations covenants ratified: Civil and Political Rights, Economic, Social and Cultural Rights

Form of government: One-party Communist state
GNP per capita (US$): 1,240
% of GNP spent by state:
On health: 1.0
On military: 10.0
On education: n/a

FACTORS AFFECTING HUMAN RIGHTS
Individual rights are subordinated to the aims and the ideology of the Korean Workers party. This Marxist-Leninist party has been dominated for over 40 years by the president, whose other titles are General Secretary of the Party, Chairman of the Central People's Committee and Supreme Commander of the Armed Forces. All aspects of life, political, social, and economic, are controlled by the regime, and opposition is prohibited. The honoring of human rights as contained in the instruments of the United Nations is disregarded. In September 1991, the country became a member of that organization.

FREEDOM TO:		COMMENTS
1 Travel in own country	NO	Not beyond one's town or village without a permit
2 Travel outside own country	NO	Permission for a few students and relatives to visit family abroad represents a remarkable liberalization
3 Peacefully associate and assemble	NO	Only in support of, or with the approval of, the state, the president, or the Korean Worker's party
4 Teach ideas and receive information	NO	All teaching within the framework of support for the leader and party. Listening to foreign radio may be a crime
5 Monitor human rights violations	NO	Nothing. No concessions to international monitors
6 Publish and educate in ethnic language	YES	No significant minorities

FREEDOM FROM:		COMMENTS
7 Serfdom, slavery, forced or child labor	no	Long periods in reeducation camps. Forced labor for an estimated 150,000

FREEDOM FROM:		COMMENTS
8	Extrajudicial killings or "disappearances"	**NO**	Killings of purported "enemies" have extended to enemies beyond the national frontier
9	Torture or coercion by the state	**NO**	Internal security forces have arbitrary powers of interrogation. Torture routine. Beatings, starvation, sleep deprivation
10	Compulsory work permits or conscription of labor	**NO**	Directed labor throughout the economy
11	Capital punishment by the state	**NO**	By shooting. For political opposition, "economic sabotage", etc.
12	Court sentences of corporal punishment	no	Courts are an extension of the executive government which means that torture and state-approved violence equate with corporal punishment
13	Indefinite detention without charge	**NO**	Exact numbers in detention unknown. Defectors escaping country give estimates of around 100,000
14	Compulsory membership of state organizations or parties	yes	Social and career prospects helped by membership in the party
15	Compulsory religion or state ideology in schools	**NO**	Ideological instruction centered on the cult figure of President Kim Il Sung in all schools
16	Deliberate state policies to control artistic works	**NO**	Only works of cultural and political orthodoxy. Satire banned, as is most Western music
17	Political censorship of press	**NO**	Total party control. Journalists are all party members
18	Censorship of mail or telephone tapping	**NO**	Surveillance of all communications, particularly infiltration of anything from South Korea

FREEDOM FOR OR RIGHTS TO:		COMMENTS
19	Peaceful political opposition	**NO**	Consequences may be the death penalty
20	Multiparty elections by secret and universal ballot	**NO**	All candidates to Supreme People's Assembly belong to ruling party. Government claims 100% poll attendance
21	Political and legal equality for women	no	Despite constitutional guarantees, little representation at senior levels
22	Social and economic equality for women	yes	But discrimination in employment, pay, and opportunities
23	Social and economic equality for ethnic minorities	**YES**	No significant minorities
24	Independent newspapers	**NO**	State owned and controlled

FREEDOM FOR OR RIGHTS TO: COMMENTS

25 Independent book publishing — **NO** — State owned and controlled. A major purpose of publications is to extol the president

26 Independent radio and television networks — **NO** — State owned and controlled

27 All courts to total independence — **NO** — Courts' responsibility "are to the people, not to the defendant." Sentences may be decided before trial takes place

28 Independent trade unions — **NO** — Party controlled. No rights to strike

LEGAL RIGHTS: COMMENTS

29 From deprivation of nationality — **YES** — Kinship with fatherland an important concept

30 To be considered innocent until proved guilty — **NO** — Arbitrary police powers. A decision on innocence may precede court trial

31 To free legal aid when necessary and counsel of own choice — **no** — Free legal aid but state appoints one of its own staff as defense counsel

32 From civilian trials in secret — **NO** — All trials may be held in secret

33 To be brought promptly before a judge or court — **NO** — Authorities decide timetable arbitrarily. The only publicity may be an announcement of the sentence passed – as a warning to others

34 From police searches of home without a warrant — **NO** — Police have arbitrary powers

35 From arbitrary seizure of personal property — **no** — A wide definition of "economic corruption" includes possessing "unauthorized" wealth, which may be seized

PERSONAL RIGHTS: COMMENTS

36 To interracial, interreligious, or civil marriage — **YES** — Rights respected, but late marriages (28–30) part of government policy

37 Equality of sexes during marriage and for divorce proceedings — **YES** — Rights respected

38 To practice any religion — **no** — No proselytizing by the small groups of practicing Christians. Harassment rather than suppression

39 To use contraceptive pills and devices — **YES** — Rights respected

40 To noninterference by state in strictly private affairs — **NO** — Intrusion. System of neighborhood informers. A badge of leader Kim Il Sung must be worn by everyone and his picture displayed in home

Korea, South

Human rights rating: 59%

YES 14 yes 9 no 14 NO 3

Population: 42,800,000
Life expectancy: 70.1
Infant mortality (0–5 years)
per 1,000 births: 31
United Nations covenants ratified:
Civil and Political Rights, Economic,
Social and Cultural Rights, Convention
on Equality for Women

Form of government: Executive
presidency
GNP per capita (US$): 3,600
% of GNP spent by state:
On health: 0.4
On military: 5.2
On education: 3.0

FACTORS AFFECTING HUMAN RIGHTS

The continuing division of the country since 1948, with the North governed by a Communist regime, has made security a major priority of South Korea. Although there are multiparty elections, the governing party enjoys certain electoral advantages. The human rights situation over the last few years has alternated between surveillance and arrests of perceived opponents of the government and periods of greater tolerance. The economy is one of the most prosperous in Asia, a large US military force has been present since the 1950–52 war, but contacts with North Korea are prohibited and calls for unification of the peninsula are discouraged by the government. Recent information places the number of political prisoners at over 1,000.

FREEDOM TO:		COMMENTS
1	Travel in own country	YES	Rights respected
2	Travel outside own country	yes	But prison for visiting North Korea without special permission
3	Peacefully associate and assemble	no	Many violent police actions against demonstrations threatening "law and order," particularly those of students, trade unions, and dissidents. Some protesters driven to self-immolation in public
4	Teach ideas and receive information	no	Teachers may not form unions. Careers jeopardized for those supporting unification with North Korea. Network of informers in universities
5	Monitor human rights violations	no	Surveillance by military intelligence. Sometimes resulting in imprisonment for human rights activists
6	Publish and educate in ethnic language	YES	Rights respected

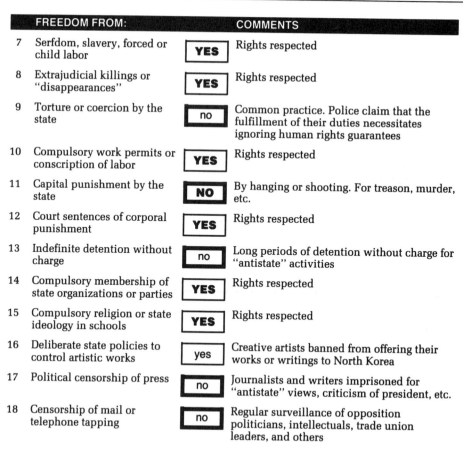

	FREEDOM FROM:		**COMMENTS**
7	Serfdom, slavery, forced or child labor	YES	Rights respected
8	Extrajudicial killings or "disappearances"	YES	Rights respected
9	Torture or coercion by the state	no	Common practice. Police claim that the fulfillment of their duties necessitates ignoring human rights guarantees
10	Compulsory work permits or conscription of labor	YES	Rights respected
11	Capital punishment by the state	NO	By hanging or shooting. For treason, murder, etc.
12	Court sentences of corporal punishment	YES	Rights respected
13	Indefinite detention without charge	no	Long periods of detention without charge for "antistate" activities
14	Compulsory membership of state organizations or parties	YES	Rights respected
15	Compulsory religion or state ideology in schools	YES	Rights respected
16	Deliberate state policies to control artistic works	yes	Creative artists banned from offering their works or writings to North Korea
17	Political censorship of press	no	Journalists and writers imprisoned for "antistate" views, criticism of president, etc.
18	Censorship of mail or telephone tapping	no	Regular surveillance of opposition politicians, intellectuals, trade union leaders, and others

	FREEDOM FOR OR RIGHTS TO:		**COMMENTS**
19	Peaceful political opposition	no	Many who sympathize with North Korea or come within a wide definition of "subversives" have been arrested
20	Multiparty elections by secret and universal ballot	yes	Voting age at 20. The powerful ruling party maintains certain electoral advantages
21	Political and legal equality for women	yes	Despite constitutional equality, women have only 2% of parliamentary seats and minimal representation in senior ministries
22	Social and economic equality for women	no	Professions and commerce male dominated. Traditional Confucian ideas of women's status still widely held
23	Social and economic equality for ethnic minorities	YES	Rights respected

FREEDOM FOR OR RIGHTS TO:		COMMENTS
24	Independent newspapers	**no**	If understood guidelines for publications are not followed, owners may be arrested and papers closed. No "aiding the cause of the enemy"
25	Independent book publishing	**no**	As question 24
26	Independent radio and television networks	**no**	Networks indirectly owned by the state. Close supervision for subversive broadcasts and communist influences
27	All courts to total independence	**yes**	But sensitive political cases may provoke pressures from the government
28	Independent trade unions	**no**	Though freely formed, unions are under constant surveillance by police and a network of informers. Some demonstrations and strike actions violently suppressed

LEGAL RIGHTS:		COMMENTS
29	From deprivation of nationality	**YES**	Rights respected
30	To be considered innocent until proved guilty	**yes**	Assumption of guilt more likely when links with North Korea suspected
31	To free legal aid when necessary and counsel of own choice	**yes**	Means test but lawyer may be appointed by court
32	From civilian trials in secret	**NO**	The May 1991 amendments to the National Security Law do not abolish secret trials for wide interpretation of antistate activities
33	To be brought promptly before a judge or court	**NO**	The National Security Law may hold back case for lengthy interrogation
34	From police searches of home without a warrant	**no**	Raids on homes of leading opposition figures justified under special "stability" laws
35	From arbitrary seizure of personal property	**YES**	Rights respected

PERSONAL RIGHTS:		COMMENTS
36	To interracial, interreligious, or civil marriage	**YES**	Rights respected
37	Equality of sexes during marriage and for divorce proceedings	**yes**	An amended family law has not totally corrected inequalities of property claims for divorced wives
38	To practice any religion	**YES**	Rights respected
39	To use contraceptive pills and devices	**YES**	State support

PERSONAL RIGHTS:		COMMENTS
40	To noninterference by state in strictly private affairs	**yes**	Wide surveillance of political suspects exerts psychological and social pressures on family life

Kuwait

Human rights rating: 33%

YES 5 yes 10 no 13 NO 12

Population: 2,000,000
Life expectancy: 73.4
Infant mortality (0–5 years)
per 1,000 births: 20
United Nations covenants ratified:
None

Form of government: Absolute rule by emir
GNP per capita (US$): 13,400
% of GNP spent by state:
On health: 2.7
On military: 5.8
On education: 5.1

FACTORS AFFECTING HUMAN RIGHTS

The country is being rebuilt after the devastation of the 1990–91 gulf war. The absolute rule of the al-Sabah family has been restored though there have been promises of permitting greater participation of the people in government and political life. Before the war, over half the population was of non-Kuwaiti origin, mostly foreign workers, but many of these have been expelled or have returned to their own country. The end of the gulf war was followed by reprisals against suspected Iraqi sympathizers, with many summary killings and reports of torture. This phase appears to have exhausted itself.

	FREEDOM TO:		COMMENTS
1	Travel in own country	yes	Within the limits of a country recovering from the Iraqi occupation, security zones, the clearing of mined areas, and restoration of the destroyed oil fields
2	Travel outside own country	yes	Except for those suspected of cooperating with the Iraqi invaders
3	Peacefully associate and assemble	no	No rallies or expressions of hostility toward the ruling emir or against the wave of nationalism that has followed the gulf war
4	Teach ideas and receive information	no	The country's academic life is being rebuilt on traditional lines of conservatism and loyalty to the emir. Censorship of books that refer to the existence of Israel
5	Monitor human rights violations	yes	No direct bannings. Obstruction by government of investigations into post-gulf war excesses against suspected collaborators, traitors, etc.
6	Publish and educate in ethnic language	YES	Rights respected

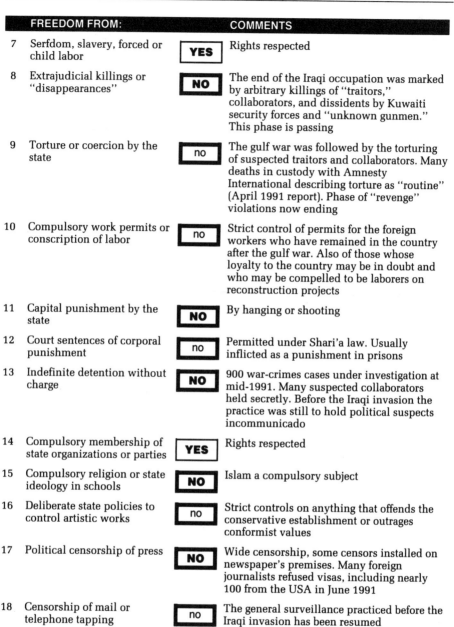

FREEDOM FROM:		COMMENTS
7	Serfdom, slavery, forced or child labor	**YES**	Rights respected
8	Extrajudicial killings or "disappearances"	**NO**	The end of the Iraqi occupation was marked by arbitrary killings of "traitors," collaborators, and dissidents by Kuwaiti security forces and "unknown gunmen." This phase is passing
9	Torture or coercion by the state	no	The gulf war was followed by the torturing of suspected traitors and collaborators. Many deaths in custody with Amnesty International describing torture as "routine" (April 1991 report). Phase of "revenge" violations now ending
10	Compulsory work permits or conscription of labor	no	Strict control of permits for the foreign workers who have remained in the country after the gulf war. Also of those whose loyalty to the country may be in doubt and who may be compelled to be laborers on reconstruction projects
11	Capital punishment by the state	**NO**	By hanging or shooting
12	Court sentences of corporal punishment	no	Permitted under Shari'a law. Usually inflicted as a punishment in prisons
13	Indefinite detention without charge	**NO**	900 war-crimes cases under investigation at mid-1991. Many suspected collaborators held secretly. Before the Iraqi invasion the practice was still to hold political suspects incommunicado
14	Compulsory membership of state organizations or parties	**YES**	Rights respected
15	Compulsory religion or state ideology in schools	**NO**	Islam a compulsory subject
16	Deliberate state policies to control artistic works	no	Strict controls on anything that offends the conservative establishment or outrages conformist values
17	Political censorship of press	**NO**	Wide censorship, some censors installed on newspaper's premises. Many foreign journalists refused visas, including nearly 100 from the USA in June 1991
18	Censorship of mail or telephone tapping	no	The general surveillance practiced before the Iraqi invasion has been resumed

FREEDOM FOR OR RIGHTS TO:		COMMENTS
19	Peaceful political opposition	**NO**	No opposition to the emir. The opposition chooses to operate from abroad

FREEDOM FOR OR RIGHTS TO:		COMMENTS
20	Multiparty elections by secret and universal ballot	**NO**	Introduction of limited parliamentary democracy planned for 1992. Meanwhile, the country's small, all-male National Assembly survives, obedient to the absolute ruler
21	Political and legal equality for women	**NO**	Women have limited involvement in public life. The Western influences associated with the gulf war may be establishing a trend to greater equality
22	Social and economic equality for women	no	The departure or expulsion of much of the essential force of foreign labor is offering Kuwaiti women the opportunity to take previously nonavailable jobs
23	Social and economic equality for ethnic minorities	**NO**	Before the Iraqi war, over half the population consisted of immigrant labor. These ethnic groups, some long established in the country, have become targets for grievances or have been expelled from Kuwait
24	Independent newspapers	no	Mostly privately owned. Licenses may be cancelled, owners imprisoned if emir criticized. Similar fate for open criticism of the royal family or neighboring Arab leaders. Islamic beliefs must be respected
25	Independent book publishing	no	Vigilant censorship of books. Subject matter must conform to question 24
26	Independent radio and television networks	**NO**	Government controlled
27	All courts to total independence	yes	All senior appointments by the emir. Nationalistic pressures, following the country's liberation, may affect independence of the courts
28	Independent trade unions	yes	Unions being reconstituted after gulf war. Limits on strike action. Most union members in government service and follow understood guidelines

LEGAL RIGHTS:		COMMENTS
29	From deprivation of nationality	yes	Restrictions on foreigners remaining after gulf war stricter than ever. Non-Muslims barred from becoming citizens. Political opponents remain abroad
30	To be considered innocent until proved guilty	yes	The anti-Iraqi fury which accompanied the liberation of Kuwait, with arbitrary arrests and trials of "traitors," has subsided
31	To free legal aid when necessary and counsel of own choice	no	Court appoints counsel and question of aid is at the discretion of the judge

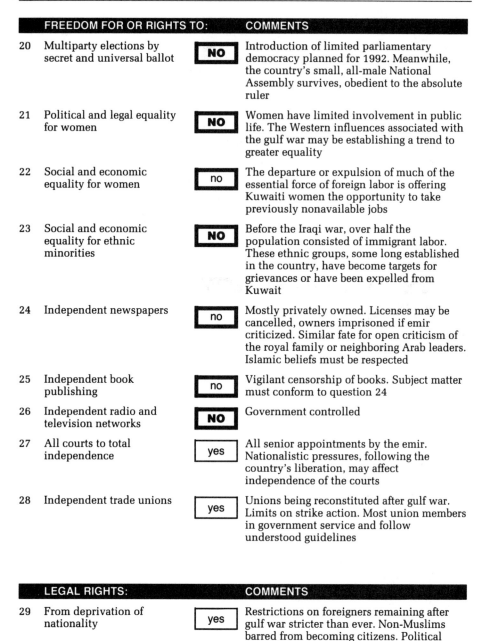

LEGAL RIGHTS: COMMENTS

32	From civilian trials in secret	yes	The ending of the brief period of martial law, which lasted until June 1991, does not prevent the occasional trial by a military court
33	To be brought promptly before a judge or court	yes	Pre-gulf war procedures not yet fully restored. These require special authorization for detentions longer than 24 hours
34	From police searches of home without a warrant	no	The end of the gulf war brought widespread searches for spies, traitors, and collaborators. This phase has passed but police still have increased authority to search
35	From arbitrary seizure of personal property	no	Seizures of property of foreign workers and noncitizens suspected of pro-Iraqi sympathies as they are expelled from the country

PERSONAL RIGHTS: COMMENTS

36	To interracial, interreligious, or civil marriage	NO	No civil law. Marriage only under Shari'a law
37	Equality of sexes during marriage and for divorce proceedings	NO	Shari'a marriage law may be based on a contract which is invariably in husband's favor. He may also practice polygamy and freely divorce though wife may litigate for custody of children
38	To practice any religion	yes	No proselytizing of Muslims. Small communities of expatriates choose to worship in their own homes
39	To use contraceptive pills and devices	YES	Rights respected
40	To noninterference by state in strictly private affairs	YES	Rights respected

Libya

Human rights rating: 24%

YES 2 **yes** 5 **no** 11 NO 22

Population: 4,500,000
Life expectancy: 61.8
Infant mortality (0–5 years)
  per 1,000 births: 116
United Nations covenants ratified:
Civil and Political Rights, Economic,
Social and Cultural Rights

Form of government: Islamic socialist
regime
GNP per capita (US$): 5,420
% of GNP spent by state:
On health: 3.0
On military: 12.0
On education: 10.1

FACTORS AFFECTING HUMAN RIGHTS
The country has been governed since 1969 by Colonel Qadhafi and his revolutionary
committees. A national policy of a "popular democracy" is based on the Islamic
religion, socialism, and the leader's philosophy as set out in his "Green Book." Human
rights are severely circumscribed, opposition to the government and system is
forbidden, and the country's security apparatus extends throughout society and
includes people's committees which act as a network of informers. The leader's
enemies abroad are occasionally assassinated.

	FREEDOM TO:		COMMENTS
1	Travel in own country	**YES**	Rights respected
2	Travel outside own country	yes	Exit permits compulsory. Some easing of recent restrictions though more difficult to obtain foreign currency, resulting in significant corruption
3	Peacefully associate and assemble	**NO**	Meetings only in support of the government or in association with "people's committees"
4	Teach ideas and receive information	**NO**	In conformity with the guidelines set out by the leader and the countrywide network of the revolutionary committees
5	Monitor human rights violations	**NO**	International groups refused entry. A local human rights group is understood to be a political gesture by the government and is not independent
6	Publish and educate in ethnic language	yes	In many instances previous use of foreign languages is being replaced by compulsory Arabic

FREEDOM FROM:		COMMENTS
7 Serfdom, slavery, forced or child labor	**YES**	Rights respected
8 Extrajudicial killings or "disappearances"	**NO**	Until 1990, when the situation began to improve, many enemies of the revolution were "liquidated" by one means or another, including some living overseas
9 Torture or coercion by the state	**NO**	A great number of methods including beatings, electric shocks, breaking bones, foot flogging, and semisuffocation
10 Compulsory work permits or conscription of labor	yes	Labor mobilized for major projects. Considered to be a well-paid form of conscription
11 Capital punishment by the state	**NO**	Methods have included hanging in public squares. For political opposition, murder, economic crimes, etc.
12 Court sentences of corporal punishment	no	The extent of torture with the approval of the government can be equated with corporal punishment by the state
13 Indefinite detention without charge	**NO**	Both political prisoners and criminals. Long incommunicado periods. Not only in prisons but in detention centers
14 Compulsory membership of state organizations or parties	no	All the population must be involved in the system of "People's committees"
15 Compulsory religion or state ideology in schools	**NO**	Teachings of the "Green Book" (by the country's leader) and from the Koran
16 Deliberate state policies to control artistic works	no	Rigid ideas of propriety and Islamic principles inhibit artistic creation in all fields
17 Political censorship of press	**NO**	Total censorship. One daily newspaper in capital. Circulation 40,000
18 Censorship of mail or telephone tapping	**NO**	Widespread censorship and surveillance

FREEDOM FOR OR RIGHTS TO:		COMMENTS
19 Peaceful political opposition	**NO**	No opposition permitted. The country's leader appoints all senior ministers and officials
20 Multiparty elections by secret and universal ballot	**NO**	The country is a "republic of the masses" (*Jamahirija*) and political parties are forbidden
21 Political and legal equality for women	no	Traditional role of women slowly improving, encouraged by the leadership
22 Social and economic equality for women	no	Women imprisoned for adultery. Husband's consent needed to travel abroad. Excision of clitoris still practiced in remote areas of the south

FREEDOM FOR OR RIGHTS TO:		COMMENTS
23	Social and economic equality for ethnic minorities	yes	Dependent on their support for the system and the leadership
24	Independent newspapers	NO	Owned and controlled by the state
25	Independent book publishing	NO	Nothing contrary to official policy and the leader's teachings
26	Independent radio and television networks	NO	Owned and controlled by the state
27	All courts to total independence	NO	Security forces and officials have authority to transfer cases to other courts. These may be "People's Courts," revolutionary courts, or military courts
28	Independent trade unions	NO	Despite membership in the International Labour Organization, unions are controlled by the regime. Strikes unlawful

LEGAL RIGHTS:		COMMENTS
29	From deprivation of nationality	NO	Leadership may make arbitrary ruling
30	To be considered innocent until proved guilty	NO	Security forces have arbitrary powers, particularly with "traitors of the people"
31	To free legal aid when necessary and counsel of own choice	no	"People's courts" would not usually incur costs. Threats against lawyers if they fail to cooperate with the authorities
32	From civilian trials in secret	NO	When considered necessary, in secret
33	To be brought promptly before a judge or court	NO	The irregular "People's Courts" may involve long delays
34	From police searches of home without a warrant	no	Law frequently ignored. Area of corruption, bribery, harassment
35	From arbitrary seizure of personal property	no	Seizures "when in the interests of the state." Also abuses in the interests of corrupt officials

PERSONAL RIGHTS:		COMMENTS
36	To interracial, interreligious, or civil marriage	NO	No civil or intermarriage for Muslims (99% of population). Leadership attempting to assimilate Muslim Berbers through intermarriage
37	Equality of sexes during marriage and for divorce proceedings	no	Traditional role of women, despite declarations to the contrary by the leadership, continues. Islamic law ensures advantages for husbands

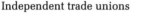

PERSONAL RIGHTS:		COMMENTS
38	To practice any religion	no	Intolerance is usually directed at schismatic Muslims and fundamentalists
39	To use contraceptive pills and devices	yes	Available but not encouraged
40	To noninterference by state in strictly private affairs	no	The extent of surveillance by both the authorities and a network of "people's informers" discourages nonconformism, including strictly private behavior

Malawi

Human rights rating: 33%

Population: 8,800,000
Life expectancy: 48.1
Infant mortality (0–5 years)
per 1,000 births: 258
United Nations covenants ratified:
Convention on Equality for Women

Form of government: One-party state
GNP per capita (US$): 170
% of GNP spent by state:
On health: 1.9
On military: 2.3
On education: 3.2

FACTORS AFFECTING HUMAN RIGHTS

The country has been ruled for over 30 years by the president-for-life. He and his government have absolute power and there is little regard for human rights, which are considered a "forbidden subject." The media are controlled by the government, foreign correspondents in the country are discouraged or expelled, and intellectuals and academics frequently choose to live abroad and are then considered to be exiles. The security services monitor the population's activities at all levels and there is an extensive network of informers.

	FREEDOM TO:		COMMENTS
1	Travel in own country	yes	Limitations on movements by known dissidents
2	Travel outside own country	no	Migrant workers to neighboring countries, including South Africa, receive passports but dissidents and those likely to be critics abroad refused
3	Peacefully associate and assemble	NO	Nothing that may be considered hostile to the president or the ruling Malawi Congress party. Groups of three may be charged with unlawful assembly
4	Teach ideas and receive information	NO	Outspoken opposition to the ruling authorities may lead to arrest, dismissal, forfeiture of privileges, etc.
5	Monitor human rights violations	NO	No monitoring. Human rights a seditious concept – despite UN membership
6	Publish and educate in ethnic language	yes	But the means and the opportunites limited by poverty

	FREEDOM FROM:		COMMENTS
7	Serfdom, slavery, forced or child labor	yes	Regional pattern of child labor

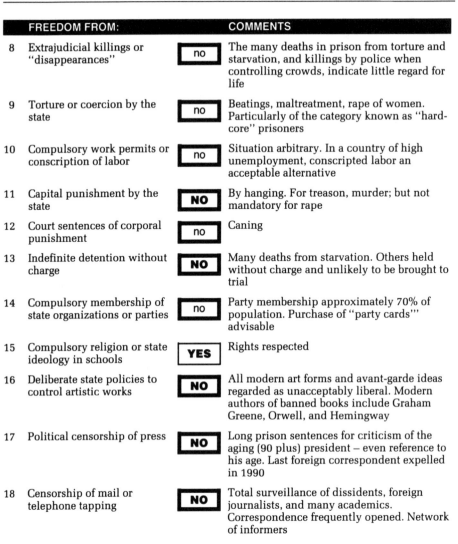

	FREEDOM FROM:		**COMMENTS**
8	Extrajudicial killings or "disappearances"	no	The many deaths in prison from torture and starvation, and killings by police when controlling crowds, indicate little regard for life
9	Torture or coercion by the state	no	Beatings, maltreatment, rape of women. Particularly of the category known as "hard-core" prisoners
10	Compulsory work permits or conscription of labor	no	Situation arbitrary. In a country of high unemployment, conscripted labor an acceptable alternative
11	Capital punishment by the state	NO	By hanging. For treason, murder; but not mandatory for rape
12	Court sentences of corporal punishment	no	Caning
13	Indefinite detention without charge	NO	Many deaths from starvation. Others held without charge and unlikely to be brought to trial
14	Compulsory membership of state organizations or parties	no	Party membership approximately 70% of population. Purchase of "party cards'" advisable
15	Compulsory religion or state ideology in schools	YES	Rights respected
16	Deliberate state policies to control artistic works	NO	All modern art forms and avant-garde ideas regarded as unacceptably liberal. Modern authors of banned books include Graham Greene, Orwell, and Hemingway
17	Political censorship of press	NO	Long prison sentences for criticism of the aging (90 plus) president – even reference to his age. Last foreign correspondent expelled in 1990
18	Censorship of mail or telephone tapping	NO	Total surveillance of dissidents, foreign journalists, and many academics. Correspondence frequently opened. Network of informers

	FREEDOM FOR OR RIGHTS TO:		**COMMENTS**
19	Peaceful political opposition	NO	Forbidden. Severe penalties. The opposition resides abroad
20	Multiparty elections by secret and universal ballot	NO	As question 19
21	Political and legal equality for women	yes	Government seeking to improve women's status in a male-dominated society
22	Social and economic equality for women	no	Traditional discrimination. Most agricultural laboring done by women. Some circumcision of females

FREEDOM FOR OR RIGHTS TO:		COMMENTS
23	Social and economic equality for ethnic minorities	yes	The Asian community, largely in business, suffers limited restrictions. Also from government attempts to give greater share to Malawians
24	Independent newspapers	**NO**	Mostly owned and controlled by the state. Attempts to be independent described as "hazardous"
25	Independent book publishing	no	Little independent book publishing. Prudence required and government approval advisable
26	Independent radio and television networks	**NO**	State owned and controlled
27	All courts to total independence	**NO**	"Traditional courts" have priority in trying major cases. No defending lawyers. President may intervene arbitrarily
28	Independent trade unions	yes	Restrictions and harassment by authorities. No unions for state employees. Overall membership about 20% of work force

LEGAL RIGHTS:		COMMENTS
29	From deprivation of nationality	yes	Political opponents living abroad suffer de facto deprivation
30	To be considered innocent until proved guilty	yes	Arbitrary. Innocence less likely to be recognized if political dissidence suspected
31	To free legal aid when necessary and counsel of own choice	**NO**	No meaningful system. Cases before "traditional courts" may be denied defense lawyer
32	From civilian trials in secret	no	President has absolute powers
33	To be brought promptly before a judge or court	no	Police officers free to hold suspect. Up to 28 days without further authority. Longer under the Public Security Act
34	From police searches of home without a warrant	no	Police may indulge in arbitrary searches which, even without the required authority, will not be subject to discipline
35	From arbitrary seizure of personal property	**NO**	Under the Forfeiture Act, all property and possessions may be seized. Understood that this law is to deter offenses by the more prosperous Asian community

PERSONAL RIGHTS:		COMMENTS
36	To interracial, interreligious, or civil marriage	**YES**	No minimum age for marriage

PERSONAL RIGHTS:		COMMENTS

37 Equality of sexes during marriage and for divorce proceedings — **yes** — Husbands enjoy advantages in tribal areas under customary law

38 To practise any religion — **yes** — But Jehovah's Witnesses refused official registration and some followers arrested

39 To use contraceptive pills and devices — **YES** — Rights respected

40 To noninterference by state in strictly private affairs — **no** — The countrywide network of informers and the degree of fear of the authorities threaten private life. A compulsion to conform

Malaysia

Human rights rating: 61%

YES 10 yes 21 no 7 NO 2

Population: 17,900,000
Life expectancy: 70.1
Infant mortality (0–5 years) per 1,000 births: 19
United Nations covenants ratified: None

Form of government: Parliamentary democracy
GNP per capita (US$): 1,940
% of GNP spent by state:
On health: 1.8
On military: 6.1
On education: 7.9

FACTORS AFFECTING HUMAN RIGHTS

The National Front coalition has been in power since 1957. Elections in 1990 extended this for a further period though there were allegations of irregularities, including government dominance of the media. The maintenance of the 1960 Internal Security Act and others enable the government to arrest and detain without charge and to restrict judicial reviews. A number of former Communist party members remain in detention and are understood to be undergoing "rehabilitation." The population is almost 55% Malays, 35% Chinese and nearly 10% Indian. Although a member of the United Nations since 1957, the prime minister sees its human rights instruments as an aspect of Western domination.

	FREEDOM TO:		COMMENTS
1	Travel in own country	**YES**	Rights respected
2	Travel outside own country	yes	Following lifting of restrictions in 1989, only a few applications for passports refused. No permits for South Africa or Israel
3	Peacefully associate and assemble	yes	Rallies regarded as socially divisive banned. Students need approval to form associations. Peaceful protests may suffer police violence and be charged with illegal assembly
4	Teach ideas and receive information	no	Limitations on political activities in universities. Academics and students may have privileges withdrawn. No Marxist ideologies
5	Monitor human rights violations	yes	Monitoring by foreigners regarded as intrusion. Despite Malaysia's voluntary membership of United Nations since 1957, that organization's human rights treaties still seen as Western imposition. Local monitors tolerated but not supported
6	Publish and educate in ethnic language	**YES**	Rights respected

	FREEDOM FROM:		COMMENTS
7	Serfdom, slavery, forced or child labor	yes	Degree of traditional child labor
8	Extrajudicial killings or "disappearances"	yes	But a few suspicious deaths of prisoners held in police custody
9	Torture or coercion by the state	yes	Unlawful abuses, usually by prison wardens
10	Compulsory work permits or conscription of labor	YES	Rights respected
11	Capital punishment by the state	NO	By hanging. For drug trafficking, murder, possession of guns, etc.
12	Court sentences of corporal punishment	no	Flogging. By a half-inch thick cane. Sometimes added to a prison sentence
13	Indefinite detention without charge	NO	Over 100 currently detained under the Internal Security Act. Also possible under two other laws
14	Compulsory membership of state organizations or parties	YES	Rights respected
15	Compulsory religion or state ideology in schools	yes	Orthodox Islamic schools adhere to strict religious instruction. *Dakwah* (Muslim proselytizing movement) remains active
16	Deliberate state policies to control artistic works	yes	Severe interpretation of "obscenity" subjects. Bannings of artworks, concerts, etc., emanating from "Zionist sources"
17	Political censorship of press	yes	Understood guidelines. Heavy fines for the offense of "false news." No comment under Sedition Act of favored position in society of Malays
18	Censorship of mail or telephone tapping	yes	With the official ending of action against the Communist party, surveillance lessened but government is still alert to what it regards as security threats

	FREEDOM FOR OR RIGHTS TO:		COMMENTS
19	Peaceful political opposition	YES	Rights respected
20	Multiparty elections by secret and universal ballot	yes	Preponderance of media controlled or owned by parties supporting the government. Others inhibited by the need for self-censorship. Charges of irregularities at counting of votes
21	Political and legal equality for women	yes	Women's position improving though still limited representation at highest levels
22	Social and economic equality for women	no	Disadvantages in pay and employment. The majority of women, being Muslim, accept a role in accordance with their beliefs

FREEDOM FOR OR RIGHTS TO:		COMMENTS
23	Social and economic equality for ethnic minorities	**no**	Government policy of advancing the interests of the previously handicapped Malay majority, particularly economically, may mean disadvantages for the minority ethnic groups
24	Independent newspapers	**yes**	Permits renewed annually. Seizures under a wide ruling of "prejudice to public order and national security." The national news agency, Bernama, controls distribution of foreign news, photos, etc.
25	Independent book publishing	**yes**	Bannings and fines when books are seen as a threat to public order and national security. This ruling includes many scholarly works on Islam
26	Independent radio and television networks	**yes**	State controlled and private. Licenses of latter revoked if considered to offend or violate what are termed "Malaysian values"
27	All courts to total independence	**no**	Amendments to the Internal Security Act restrict judicial reviews of those held in preventive detention. Protests by Bar Council at attempts to limit powers of court
28	Independent trade unions	**yes**	Considerable government powers to control strikes and to suspend union's registration

LEGAL RIGHTS:		COMMENTS
29	From deprivation of nationality	**YES**	Rights respected
30	To be considered innocent until proved guilty	**YES**	Rights respected
31	To free legal aid when necessary and counsel of own choice	**yes**	Means test. Court appoints counsel
32	From civilian trials in secret	**yes**	Attorney general has discretion under the 1975 Essential (Security Cases) Regulations
33	To be brought promptly before a judge or court	**yes**	Within 24 hours for ordinary crimes but early appearance in court may be delayed by arbitrary powers under the Internal Security Act
34	From police searches of home without a warrant	**no**	Considerable search powers under the Internal Security Act. Seizures of books and documents of those suspected under a wide definition of "subversion"
35	From arbitrary seizure of personal property	**yes**	Usually books and papers of suspected subversives (see question 34)

PERSONAL RIGHTS:		COMMENTS
36	To interracial, interreligious, or civil marriage	**YES**	In practice there is little intermarriage between the different ethnic and religious groups – Malay, Chinese, and Indians
37	Equality of sexes during marriage and for divorce proceedings	no	The revised 1989 law improved rights of Muslim wives but disadvantages remain. Muslim males may still practice polygamy
38	To practice any religion	**YES**	Rights respected
39	To use contraceptive pills and devices	**YES**	But official hopes of trebling population by year 2010
40	To noninterference by state in strictly private affairs	yes	Homosexuals conduct themselves with prudence. Maximum sentence 20 years in prison

Mexico

Human rights rating: 64%

YES 14 **yes** 19 **no** 4 NO 3

Population: 88,600,000
Life expectancy: 69.7
Infant mortality (0–5 years)
  per 1,000 births: 51
United Nations covenants ratified:
Civil and Political Rights, Economic,
Social and Cultural Rights, Convention
on Equality for Women

Form of government: Multiparty
system (nominally)
GNP per capita (US$): 1,760
% of GNP spent by state:
On health: 2.2
On military: 0.6
On education: 2.8

FACTORS AFFECTING HUMAN RIGHTS
The governing party has been in power since 1929 under what is claimed to be a
multiparty democratic system. There is considerable evidence of continuing electoral
irregularities, fraud, corruption, patronage, and, at the local level, physical threats.
Behind the superficial claim to be a democracy, there are widespread human rights
crimes by the police and security forces. Extrajudicial killings, disappearances, torture
and arbitrary detentions take place over most of the country. Among the victims are
human rights activists, peasants in dispute with powerful landowners, and political
militants. From time to time the government, without much success, has tried to correct
the excesses and to introduce reforms.

	FREEDOM TO:		COMMENTS
1	Travel in own country	**YES**	Rights respected
2	Travel outside own country	**YES**	Rights respected
3	Peacefully associate and assemble	yes	Occasional police restrictions, especially in areas of peasant agitation pressing for land reform
4	Teach ideas and receive information	yes	Academics and progressives investigating extrajudicial killings and "disappearances" receive death threats. Source of these are frequently anonymous security agents
5	Monitor human rights violations	yes	Following the formation of the National Human Rights Commission in 1990, the authorities are permitting freer inquiries. Monitors, however, still receive anonymous death threats
6	Publish and educate in ethnic language	**YES**	Rights respected

FREEDOM FROM:		COMMENTS
7	Serfdom, slavery, forced or child labor	yes	Child labor in rural areas. Guatemalan refugees employed for subsistence wages
8	Extrajudicial killings or "disappearances"	NO	Including human rights activists. Many extrajudicial killings in rural areas where security forces openly support large landowners against peaceful peasant protesters
9	Torture or coercion by the state	NO	Despite government claim to be eliminating culprits, torture widely practiced. Also against a few US citizens. Beatings, electric shocks, etc.
10	Compulsory work permits or conscription of labor	YES	Rights respected
11	Capital punishment by the state	yes	Recently abolished but see question 8. Degree of government culpability because of apathy toward extrajudicial killings
12	Court sentences of corporal punishment	YES	Rights respected
13	Indefinite detention without charge	NO	Long incommunicado detentions. Of criminals, drug traffickers, and militant peasant leaders
14	Compulsory membership of state organizations or parties	YES	Rights respected
15	Compulsory religion or state ideology in schools	YES	Rights respected
16	Deliberate state policies to control artistic works	YES	Rights respected
17	Political censorship of press	yes	Despite constitutional freedom, intimidation of journalists rather than direct censorship is frequently practiced. In 1990, some journalists murdered
18	Censorship of mail or telephone tapping	yes	Irregular unofficial surveillance. Motives include bribery, political corruption, and investigating financial crimes

FREEDOM FOR OR RIGHTS TO:		COMMENTS
19	Peaceful political opposition	yes	New reforms introduced in 1990 but political assassinations during elections still occur
20	Multiparty elections by secret and universal ballot	yes	The Institutional Revolutionary party has been in power since 1929. Regular allegations of electoral fraud, including those during the 1991 election
21	Political and legal equality for women	yes	Progress toward equality but traditional attitudes impede change

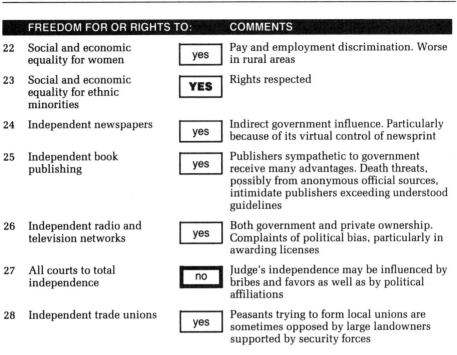

FREEDOM FOR OR RIGHTS TO:		COMMENTS
22	Social and economic equality for women	yes	Pay and employment discrimination. Worse in rural areas
23	Social and economic equality for ethnic minorities	YES	Rights respected
24	Independent newspapers	yes	Indirect government influence. Particularly because of its virtual control of newsprint
25	Independent book publishing	yes	Publishers sympathetic to government receive many advantages. Death threats, possibly from anonymous official sources, intimidate publishers exceeding understood guidelines
26	Independent radio and television networks	yes	Both government and private ownership. Complaints of political bias, particularly in awarding licenses
27	All courts to total independence	no	Judge's independence may be influenced by bribes and favors as well as by political affiliations
28	Independent trade unions	yes	Peasants trying to form local unions are sometimes opposed by large landowners supported by security forces

LEGAL RIGHTS:		COMMENTS
29	From deprivation of nationality	YES	Rights respected
30	To be considered innocent until proved guilty	no	Many arbitrary arrests. Bribes a persuasive factor in proving innocence. Reforms proposed to stop practice of "confessions" (sometimes forced) being accepted by courts as proof
31	To free legal aid when necessary and counsel of own choice	yes	Means test. Court appoints "public defender"
32	From civilian trials in secret	yes	Security cases only. Decision for higher authority
33	To be brought promptly before a judge or court	yes	48 hours but this period frequently violated
34	From police searches of home without a warrant	no	Arbitrary harassment, particularly in remote rural areas or in land tenure disputes
35	From arbitrary seizure of personal property	YES	Rights respected

PERSONAL RIGHTS:		COMMENTS
36	To interracial, interreligious, or civil marriage	YES	From age 18, both sexes

PERSONAL RIGHTS: COMMENTS

37 Equality of sexes during
    marriage and for divorce | yes | Traditional and religious influences in a
    proceedings   strongly Roman Catholic country inhibit
                                            divorce and occasionally place wives at a
                                            disadvantage

38 To practice any religion | **YES** | Rights respected

39 To use contraceptive pills | **YES** | State support
    and devices

40 To noninterference by state | no | Police actions in support of oppressive
    in strictly private affairs   landowners include harassment of peasants
                                            and intrusion into homes

Morocco

Human rights rating: 56%

Population: 25,100,000
Life expectancy: 62.0
Infant mortality (0–5 years)
per 1,000 births: 116
United Nations covenants ratified:
Civil and Political Rights, Economic,
Social and Cultural Rights

Form of government: Nominated by
monarch
GNP per capita (US$): 830
% of GNP spent by state:
On health: 0.9
On military: 5.1
On education: 5.0

FACTORS AFFECTING HUMAN RIGHTS
The king is virtually an absolute ruler and although there is an electoral multiparty
system, the Parliament has little authority and the prime minister is appointed by the
"head of state." Promised reforms have not materialized. In response to international
complaints about such human rights abuses as torture and indefinite detention, a
commission has been set up to report to the king and to recommend changes. The
proposals so far put forward do not include freedom to criticize the monarchy, Islam,
the political system, and the dispute over the western Sahara, which remain prohibited
subjects.

FREEDOM TO: / COMMENTS

1	Travel in own country	YES	Only frontier area subject to restrictions
2	Travel outside own country	no	Passports withheld from suspected political subversives and Baha'i believers
3	Peacefully associate and assemble	no	Permits provided for meetings and activities within accepted political and religious guidelines. But not otherwise
4	Teach ideas and receive information	no	Surveillance of universities. No criticism of the king and certain sensitive issues. Careers depend on prudence
5	Monitor human rights violations	no	Foreign monitors occasionally expelled. Local human rights groups function but with limited effectiveness
6	Publish and educate in ethnic language	YES	Rights respected

FREEDOM FROM: / COMMENTS

7	Serfdom, slavery, forced or child labor	yes	Child labor widespread, particularly in country areas, and many under legal age of 12 years

FREEDOM FROM:		COMMENTS
8	Extrajudicial killings or "disappearances"	no	Usually while in police custody or of rioters seized by security forces. But few cases in recent years
9	Torture or coercion by the state	no	Use of torture is usually to encourage confessions. Sleep and food deprivation favored methods
10	Compulsory work permits or conscription of labor	YES	Rights respected
11	Capital punishment by the state	no	No officially admitted executions since 1973 but the death sentence still legal
12	Court sentences of corporal punishment	YES	Rights respected
13	Indefinite detention without charge	no	Many political detainees, some from 1971–72 assassination attempts on king
14	Compulsory membership of state organizations or parties	YES	Rights respected
15	Compulsory religion or state ideology in schools	no	State-supported compulsion in most orthodox Muslim schools
16	Deliberate state policies to control artistic works	yes	Self-censorship the preferred alternative to state interference
17	Political censorship of press	NO	The monarchy, Islam, territorial claims on neighbors are among subjects regarded as "forbidden"
18	Censorship of mail or telephone tapping	no	Extensive security network

FREEDOM FOR OR RIGHTS TO:		COMMENTS
19	Peaceful political opposition	no	Severe limitations on opposition to king, Islam, and the present system
20	Multiparty elections by secret and universal ballot	yes	Despite claims that party elections are free, ministers appointed by king. Any constitutional changes need royal approval
21	Political and legal equality for women	no	Little evidence of women in senior positions. Islamic orthodoxy about woman's place in politics supported by king
22	Social and economic equality for women	no	Permission of father or husband needed for passport. Women suffer inequalities in most areas of society
23	Social and economic equality for ethnic minorities	yes	But Berber minority complain of many disadvantages
24	Independent newspapers	yes	Prudent self-censorship required to ensure limited independence. Official press agency owned by government

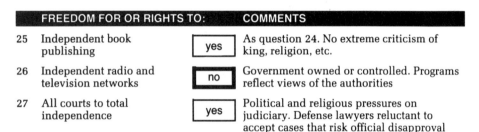

FREEDOM FOR OR RIGHTS TO:		COMMENTS
25	Independent book publishing	yes	As question 24. No extreme criticism of king, religion, etc.
26	Independent radio and television networks	no	Government owned or controlled. Programs reflect views of the authorities
27	All courts to total independence	yes	Political and religious pressures on judiciary. Defense lawyers reluctant to accept cases that risk official disapproval
28	Independent trade unions	yes	Strong union movement. Occasional attempts to influence policies

LEGAL RIGHTS:		COMMENTS
29	From deprivation of nationality	**YES**	Rights respected
30	To be considered innocent until proved guilty	no	Arbitrary arrests. Onus of proof may be on those suspected of opposition to the monarchy and Islamic establishment
31	To free legal aid when necessary and counsel of own choice	yes	Means test in civil courts with counsel, when considered necessary, appointed by judge
32	From civilian trials in secret	**YES**	Rights respected
33	To be brought promptly before a judge or court	no	Within 4 to 8 days. Frequently violated. Prolonged interrogation by police without access to counsel
34	From police searches of home without a warrant	no	Constitutional right ignored by security forces. Political opposition, writers, and students under surveillance
35	From arbitrary seizure of personal property	**YES**	Rights respected

PERSONAL RIGHTS:		COMMENTS
36	To interracial, interreligious, or civil marriage	no	Shari'a law ensures Islamic marriage for 99% of population
37	Equality of sexes during marriage and for divorce proceedings	**NO**	Husband can repudiate marriage but wife cannot. If wife committing adultery is murdered by husband, no charge. If wife kills an adulterous husband, she faces trial
38	To practice any religion	yes	But Baha'i severely restricted. Islamic fundamentalists also face government suppression
39	To use contraceptive pills and devices	**YES**	Rights respected

PERSONAL RIGHTS:		COMMENTS

40 To noninterference by state in strictly private affairs   **yes**   State supports religious and social pressures to ensure conformity in private life. Adultery and homosexuality between consenting adults illegal

Mozambique

Human rights rating: 53%

YES 10 yes 14 no 10 NO 6

Population: 15,700,000
Life expectancy: 47.5
**Infant mortality (0–5 years)
  per 1,000 births**: 297
United Nations covenants ratified:
None

Form of government: One-party system
GNP per capita (US$): 100
% of GNP spent by state:
On health: 1.8
On military: 7.0
On education: n/a

FACTORS AFFECTING HUMAN RIGHTS

For most of the period since independence in 1975, Mozambique has been a one-party state governed by FRELIMO. It has also been involved in a civil war with RENAMO but the ending of aid and military supplies to that group by the South African government, and intervention by neighboring states supporting FRELIMO, have resulted in an agreed cease-fire. These developments have encouraged the government to prepare for the introduction of a multiparty system, to release from detention many political opponents, to adopt a new constitution, and to revive relations with neighboring South Africa. Current improvements in the human rights situation depend on the honoring of the cease-fire by all parties.

	FREEDOM TO:		COMMENTS
1	Travel in own country	no	The system of permits is related to military action against the insurgent movement, RENAMO, and the need for surveillance within security zones
2	Travel outside own country	yes	Cross-frontier movements of over a million refugees without passports or permits, seeking safety, food, or work, remain the reality. Travel abroad depends on purpose of journey and the availability of foreign currency
3	Peacefully associate and assemble	yes	The plans for a new multiparty political system have meant public rallies and assemblies which are usually permitted
4	Teach ideas and receive information	yes	Situation improving since the adoption of the new constitution at the end of 1990 but academics exercise considerable restraint
5	Monitor human rights violations	YES	Rights respected but continuing military actions against RENAMO rebels limit on-the-spot monitoring
6	Publish and educate in ethnic language	YES	Rights respected

FREEDOM FROM:		COMMENTS
7	Serfdom, slavery, forced or child labor	**no**	Traditional pattern of child labor. Many instances of forced labor in insurgency zones – by all factions
8	Extrajudicial killings or "disappearances"	**NO**	Military operations against the major rebel movement, RENAMO, have meant indiscriminate killings on a large scale. Many "disappearances." Worse excesses, including massacres, by rebels. With the current cease-fire, these may be ending
9	Torture or coercion by the state	**NO**	By both the government and insurgents. Also outside the immediate war areas. Victims include journalists
10	Compulsory work permits or conscription of labor	**no**	Military and civilian conscription on a reduced scale. Situation changeable and subject to local factors and the civil war cease-fire
11	Capital punishment by the state	**YES**	Abolished 1990
12	Court sentences of corporal punishment	**YES**	Rights respected
13	Indefinite detention without charge	**no**	Many releases since the adoption of the new constitution but hundreds still detained without trial, not all RENAMO rebels
14	Compulsory membership of state organizations or parties	**YES**	Rights respected
15	Compulsory religion or state ideology in schools	**YES**	Rights respected
16	Deliberate state policies to control artistic works	**YES**	Rights respected
17	Political censorship of press	**yes**	Greater freedom encouraged by government as preparation for a multiparty political system. But prudence advisable on security-sensitive subjects
18	Censorship of mail or telephone tapping	**NO**	Wide surveillance, a legacy from the previous Marxist regime. The current cease-fire with the RENAMO rebels has not yet lessened the monitoring of suspects

FREEDOM FOR OR RIGHTS TO:		COMMENTS
19	Peaceful political opposition	**no**	Although a new constitution proposes major changes, the realities of a continuing military operation against RENAMO insurgents and emergency security measures limit political opposition
20	Multiparty elections by secret and universal ballot	**no**	Proposed changes from a single-party state to multiparty government, with a strong president, are still being formulated

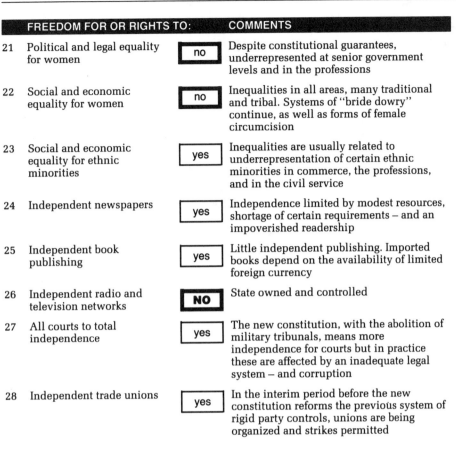

FREEDOM FOR OR RIGHTS TO:		COMMENTS
21	Political and legal equality for women	no	Despite constitutional guarantees, underrepresented at senior government levels and in the professions
22	Social and economic equality for women	no	Inequalities in all areas, many traditional and tribal. Systems of "bride dowry" continue, as well as forms of female circumcision
23	Social and economic equality for ethnic minorities	yes	Inequalities are usually related to underrepresentation of certain ethnic minorities in commerce, the professions, and in the civil service
24	Independent newspapers	yes	Independence limited by modest resources, shortage of certain requirements – and an impoverished readership
25	Independent book publishing	yes	Little independent publishing. Imported books depend on the availability of limited foreign currency
26	Independent radio and television networks	NO	State owned and controlled
27	All courts to total independence	yes	The new constitution, with the abolition of military tribunals, means more independence for courts but in practice these are affected by an inadequate legal system – and corruption
28	Independent trade unions	yes	In the interim period before the new constitution reforms the previous system of rigid party controls, unions are being organized and strikes permitted

LEGAL RIGHTS:		COMMENTS
29	From deprivation of nationality	YES	Rights respected
30	To be considered innocent until proved guilty	yes	The improvement under the new constitution is still limited. Much depends on political and social stability
31	To free legal aid when necessary and counsel of own choice	yes	Means test for accused but free choice of counsel from an approved panel
32	From civilian trials in secret	yes	Revolutionary military tribunals abolished in 1988 but considerations of "national security" still permit secret trials
33	To be brought promptly before a judge or court	NO	Up to 84 days before formal charges but this may be extended. Many violations. Large-scale corruption

LEGAL RIGHTS:		COMMENTS
34	From police searches of home without a warrant	**NO**	The civil war and general instability are responsible for many breaches of the law. Arbitrary searches, sometimes a pretext for looting, seizures, etc.
35	From arbitrary seizure of personal property	no	Counterinsurgency operations have meant the forced displacement of whole communities. Situation confused, with some abuses by government

PERSONAL RIGHTS:		COMMENTS
36	To interracial, interreligious, or civil marriage	**YES**	Rights respected. Tribal marriages from age 10
37	Equality of sexes during marriage and for divorce proceedings	no	Traditional practices still prevail in most rural areas – usually to husband's advantage
38	To practice any religion	yes	Many previous restrictions being lifted with the adoption of the new constitution. 60% follow animist beliefs
39	To use contraceptive pills and devices	**YES**	State aided since 1978
40	To noninterference by state in strictly private affairs	yes	Military operations in large areas of the country have meant many abuses, threats, etc., to private life. Interference is constant but indirect

Nepal

Human rights rating: 69%

YES 13 **yes** 20 **no** 6 NO 1

Population: 19,100,000
Life expectancy: 52.2
Infant mortality (0–5 years) per 1,000 births: 193
United Nations covenants ratified: None

Form of government: Multiparty monarchy
GNP per capita (US$): 180
% of GNP spent by state:
On health: 0.9
On military: 1.5
On education: 2.1

FACTORS AFFECTING HUMAN RIGHTS

Significant changes favoring respect for human rights took place in 1990. The absolute ruler, King Birendra, following widespread demonstrations which culminated in the security forces shooting into a crowd with between 50 and 100 deaths, abolished the national *Panchayat* (legislature) and finally agreed to multiparty elections. A new constitution was adopted in 1990 and the transition to a working democracy has been relatively peaceful. Human rights reservations concern the continuing powers of the military forces and the police practice of resorting to torture, usually beatings, when interrogating suspected criminals.

	FREEDOM TO:		COMMENTS
1	Travel in own country	yes	Occasional restrictions of movements at times of social unrest
2	Travel outside own country	yes	Passports of ministers of previous government impounded in 1991. Otherwise general freedom to travel
3	Peacefully associate and assemble	yes	Previous *Panchayet* government restrictions lifted in 1990 but rallies and protests may be banned if "law and order" are threatened
4	Teach ideas and receive information	yes	Academics are adapting to a freer intellectual environment after restraints previously regarded as prudent
5	Monitor human rights violations	yes	The introduction of democratic rule has been accompanied by relative freedom for both national and international monitors
6	Publish and educate in ethnic language	**YES**	Rights respected

FREEDOM FROM: COMMENTS

7 Serfdom, slavery, forced or child labor — **no** — Bonded and child labor traditional. Also trafficking in young women to become prostitutes in India, with little government interference

8 Extrajudicial killings or "disappearances" — **yes** — With the change to democratic government in May 1991, the police and security forces have ceased shooting into mass demonstrations with many deaths. Too early to assume a stable society

9 Torture or coercion by the state — **no** — Torture has been reduced since the change to democracy but still continues, mostly as beatings during interrogation of prisoners

10 Compulsory work permits or conscription of labor — **no** — Bonded labor still practiced

11 Capital punishment by the state — **YES** — Abolished 1990 for "ordinary offenses"

12 Court sentences of corporal punishment — **YES** — Rights respected. The practice of arbitrary beatings by the police has ceased

13 Indefinite detention without charge — **yes** — Laws permitting detention revoked in 1990 or not used. Most detainees now released

14 Compulsory membership of state organizations or parties — **YES** — Rights respected

15 Compulsory religion or state ideology in schools — **yes** — Compulsion in orthodox religious schools

16 Deliberate state policies to control artistic works — **yes** — Some "subject" censorship of Western films and art forms on grounds of "Hindu susceptibilities"

17 Political censorship of press — **yes** — Despite the more open press following the end of the *Panchayat* system, comments on the royal family must be restrained and respectful

18 Censorship of mail or telephone tapping — **YES** — Rights respected

FREEDOM FOR OR RIGHTS TO: COMMENTS

19 Peaceful political opposition — **YES** — Following the 1990 constitutional changes, the country is adapting to a democratic system – so far peacefully

20 Multiparty elections by secret and universal ballot — **YES** — The first multiparty elections for 30 years, in May 1991, gives government a narrow parliamentary majority

21 Political and legal equality for women — **no** — Traditional role for women. Little representation in politics and professions

22 Social and economic equality for women — **no** — Inequalities worst in rural areas where women provide much of the agricultural labor

	FREEDOM FOR OR RIGHTS TO:		**COMMENTS**
23	Social and economic equality for ethnic minorities	yes	Caste distinctions still evident, particularly affecting "untouchables," which is against government policy
24	Independent newspapers	yes	Independence being reasserted in the freer society though major newspapers still state owned
25	Independent book publishing	yes	A degree of self-censorship still exercised in view of the recentness of democratic change
26	Independent radio and television networks	**NO**	State owned and controlled
27	All courts to total independence	yes	The king has retained limited powers under the new constitution, which may allow him to overturn sentences
28	Independent trade unions	yes	New trade union laws, under consideration, will legalize the greater freedom that has followed the end of the *Panchayat* system

	LEGAL RIGHTS:		**COMMENTS**
29	From deprivation of nationality	**YES**	Rights respected
30	To be considered innocent until proved guilty	**YES**	Rights respected
31	To free legal aid when necessary and counsel of own choice	yes	Limited by inadequate funds and system. Reforms under the new constitution
32	From civilian trials in secret	**YES**	Rights respected
33	To be brought promptly before a judge or court	yes	Within 24 hours by law. Some delays but the worst abuses of the previous regime have ceased
34	From police searches of home without a warrant	yes	Occasional violations though fewer with the establishment of democratic rights
35	From arbitrary seizure of personal property	**YES**	Rights respected

	PERSONAL RIGHTS:		**COMMENTS**
36	To interracial, interreligious, or civil marriage	yes	Strong Hindu caste system. Marriage, in practice, follows tradition and religion – though no legal restrictions
37	Equality of sexes during marriage and for divorce proceedings	no	Marriage and divorce practices favor husbands. Husband may commit adultery but not the wife; favored in child custody laws
38	To practice any religion	yes	But no proselytizing. No conversions. Prison for offenders

PERSONAL RIGHTS:		COMMENTS
39	To use contraceptive pills and devices	**YES**	Rights respected
40	To noninterference by state in strictly private affairs	**YES**	Rights respected

Netherlands

Human rights rating: 98%

Population: 15,000,000
Life expectancy: 77.2
Infant mortality (0–5 years)
  per 1,000 births: 8
United Nations covenants ratified:
Civil and Political Rights, Economic,
Social and Cultural Rights, Convention
on Equality for Women

Form of government: Democratic
monarchy
GNP per capita (US$): 14,520
% of GNP spent by state:
On health: 7.5
On military: 3.1
On education: 6.6

FACTORS AFFECTING HUMAN RIGHTS
The individual respected. Government answerable to the people. A prosperous
economy, a free press, and a cohesive society. The government and its citizens are
among the most committed to human rights causes.

	FREEDOM TO:		COMMENTS
1	Travel in own country	YES	Rights respected
2	Travel outside own country	YES	Rights respected
3	Peacefully associate and assemble	YES	Rights respected
4	Teach ideas and receive information	YES	Rights respected
5	Monitor human rights violations	YES	An international leader in encouraging monitoring
6	Publish and educate in ethnic language	YES	Rights respected

	FREEDOM FROM:		COMMENTS
7	Serfdom, slavery, forced or child labor	YES	Rights respected
8	Extrajudicial killings or "disappearances"	YES	Rights respected
9	Torture or coercion by the state	YES	Rights respected
10	Compulsory work permits or conscription of labor	YES	Rights respected

FREEDOM FROM: COMMENTS

11	Capital punishment by the state	YES	Rights respected
12	Court sentences of corporal punishment	YES	Rights respected
13	Indefinite detention without charge	YES	Rights respected
14	Compulsory membership of state organizations or parties	YES	Rights respected
15	Compulsory religion or state ideology in schools	YES	Rights respected
16	Deliberate state policies to control artistic works	YES	Rights respected
17	Political censorship of press	YES	Rights respected
18	Censorship of mail or telephone tapping	YES	Rights respected

FREEDOM FOR OR RIGHTS TO: COMMENTS

19	Peaceful political opposition	YES	Rights respected
20	Multiparty elections by secret and universal ballot	YES	National elections every 4 years. By proportional representation. Approximately nine parties
21	Political and legal equality for women	yes	Underrepresented in senior government posts but 20% of Second Chamber of Parliament are women
22	Social and economic equality for women	yes	Minor inequalities in pay and employment
23	Social and economic equality for ethnic minorities	yes	Despite determined government efforts, some non-European immigrants suffer discrimination. And used as cheap labor
24	Independent newspapers	YES	Rights respected. Extremely liberal in practice
25	Independent book publishing	YES	Rights respected
26	Independent radio and television networks	YES	Rights respected
27	All courts to total independence	YES	Rights respected
28	Independent trade unions	YES	Rights respected

LEGAL RIGHTS:		COMMENTS
29	From deprivation of nationality	**YES**	Rights respected
30	To be considered innocent until proved guilty	**YES**	Rights respected
31	To free legal aid when necessary and counsel of own choice	**YES**	Rights respected
32	From civilian trials in secret	**YES**	Rights respected
33	To be brought promptly before a judge or court	**YES**	Usually held for limit of 6 hours. Serious crimes up to 48 hours
34	From police searches of home without a warrant	**YES**	Rights respected
35	From arbitrary seizure of personal property	**YES**	Rights respected

PERSONAL RIGHTS:		COMMENTS
36	To interracial, interreligious, or civil marriage	**YES**	Rights respected
37	Equality of sexes during marriage and for divorce proceedings	**YES**	Significant government help and protection for victimized and "battered" wives
38	To practice any religion	**YES**	Rights respected
39	To use contraceptive pills and devices	**YES**	Rights respected
40	To noninterference by state in strictly private affairs	**YES**	Rights respected

New Zealand

Human rights rating: 98%

YES 37 yes 3 no 0 NO 0

Population: 3,400,000
Life expectancy: 75.2
Infant mortality (0–5 years)
  per 1,000 births: 12
United Nations covenants ratified:
Civil and Political Rights, Economic,
Social and Cultural Rights, Convention
on Equality for Women

Form of government: Parliamentary
democracy
GNP per capita (US$): 10,000
% of GNP spent by state:
On health: 5.6
On military: 2.2
On education: 4.8

FACTORS AFFECTING HUMAN RIGHTS
The individual respected. Government answerable to the people. A high standard of
living, democratic institutions, and a free press ensure respect for human rights and the
country's involvement in international efforts to improve them elsewhere.

	FREEDOM TO:		COMMENTS
1	Travel in own country	YES	Rights respected
2	Travel outside own country	YES	Rights respected
3	Peacefully associate and assemble	YES	Rights respected
4	Teach ideas and receive information	YES	Rights respected
5	Monitor human rights violations	YES	Leading supporter of human rights monitoring. A national Human Rights Commission investigates violations
6	Publish and educate in ethnic language	YES	Rights respected

	FREEDOM FROM:		COMMENTS
7	Serfdom, slavery, forced or child labor	YES	Rights respected
8	Extrajudicial killings or "disappearances"	YES	Rights respected
9	Torture or coercion by the state	YES	Rights respected
10	Compulsory work permits or conscription of labor	YES	Rights respected

FREEDOM FROM:		COMMENTS
11	Capital punishment by the state	**YES**	Rights respected
12	Court sentences of corporal punishment	**YES**	Rights respected
13	Indefinite detention without charge	**YES**	Rights respected
14	Compulsory membership of state organizations or parties	**YES**	Rights respected
15	Compulsory religion or state ideology in schools	**YES**	Rights respected
16	Deliberate state policies to control artistic works	**YES**	Rights respected
17	Political censorship of press	**YES**	Rights respected
18	Censorship of mail or telephone tapping	**YES**	Rights respected

FREEDOM FOR OR RIGHTS TO:		COMMENTS
19	Peaceful political opposition	**YES**	Rights respected
20	Multiparty elections by secret and universal ballot	**YES**	Four members of Maori minority in House of Representatives of 92 members
21	Political and legal equality for women	yes	Underrepresented in Parliament and at senior government levels. Vote granted in 1893
22	Social and economic equality for women	yes	Minor inequalities in pay and employment
23	Social and economic equality for ethnic minorities	yes	Maoris approximately 10% of population suffer discrimination in pay, employment, and socially, which government seeks to rectify
24	Independent newspapers	**YES**	Rights respected
25	Independent book publishing	**YES**	Rights respected
26	Independent radio and television networks	**YES**	Vigilant safeguards ensure independence of state owned and -operated networks. No bias
27	All courts to total independence	**YES**	Rights respected
28	Independent trade unions	**YES**	Rights respected

LEGAL RIGHTS: COMMENTS

29	From deprivation of nationality	**YES**	Rights respected
30	To be considered innocent until proved guilty	**YES**	Rights respected
31	To free legal aid when necessary and counsel of own choice	**YES**	Rights respected
32	From civilian trials in secret	**YES**	Rights respected
33	To be brought promptly before a judge or court	**YES**	Rights respected
34	From police searches of home without a warrant	**YES**	Rights respected
35	From arbitrary seizure of personal property	**YES**	Rights respected

PERSONAL RIGHTS: COMMENTS

36	To interracial, interreligious, or civil marriage	**YES**	Rights respected
37	Equality of sexes during marriage and for divorce proceedings	**YES**	Determined government policies to offer active protection to victimized and "battered" wives
38	To practice any religion	**YES**	Rights respected
39	To use contraceptive pills and devices	**YES**	Government supported
40	To noninterference by state in strictly private affairs	**YES**	Rights respected

Nicaragua

Human rights rating: 75%

YES 20 **yes** 14 **no** 5 NO 1

Population: 3,900,000
Life expectancy: 64.8
Infant mortality (0–5 years)
  per 1,000 births: 92
United Nations covenants ratified:
Civil and Political Rights, Economic,
Social and Cultural Rights, Convention
on Equality for Women

Form of government: Multiparty
system
GNP per capita (US$): 830
% of GNP spent by state:
On health: 6.6
On military: 16.0
On education: 6.1

FACTORS AFFECTING HUMAN RIGHTS

The present, more democratic, society has evolved through a number of conflicting
phases. The oppressive Somoza dynasty ruled the country from 1933 to 1979, when it
was overthrown by an armed revolution. This was followed by 10 years of government
by the left-wing Sandinista National Liberation Front. Its rule was soon contested by a
counterrevolutionary guerrilla movement (the *contras*) which, supported and supplied
by the USA, continued a civil war until the Sandinistas accepted democratic elections
in 1990. These elections were marked by violence and irregularities but resulted in a
14-party coalition taking power. As part of the political settlement, the Sandinistas
were granted key appointments in the army and security forces. Human rights
violations, despite efforts to consolidate democracy, are most evident in the scattered
killings by well-armed supporters of both sides.

	FREEDOM TO:		COMMENTS
1	Travel in own country	yes	Gangs termed *recontras* threaten travelers and set up roadblocks because of local breakdowns of law and order
2	Travel outside own country	yes	Formality of paying a fee for exit visa
3	Peacefully associate and assemble	YES	Sporadic violence at public meetings between rival political factions, some armed
4	Teach ideas and receive information	YES	Rights respected
5	Monitor human rights violations	YES	Rights respected
6	Publish and educate in ethnic language	YES	Rights respected

	FREEDOM FROM:		COMMENTS
7	Serfdom, slavery, forced or child labor	yes	Regional pattern of child labor

FREEDOM FROM:		COMMENTS
8 Extrajudicial killings or "disappearances"	**NO**	The conflict between the former ruling Sandinistas and the newly elected government, despite the national accord, persists as scattered killings by well-armed supporters of both sides. Arbitrary shootings by police not always investigated by higher authority
9 Torture or coercion by the state	no	Charges and countercharges against police and military units enjoying own internal independence. Beatings during interrogation of detainees of opposing factions
10 Compulsory work permits or conscription of labor	**YES**	Rights respected
11 Capital punishment by the state	**YES**	Abolished 1979
12 Court sentences of corporal punishment	**YES**	Rights respected
13 Indefinite detention without charge	yes	But the Police Functions Law occasionally abused. Detentions beyond legal limits. Sandinistas still have a dominant role in security forces
14 Compulsory membership of state organizations or parties	**YES**	Rights respected
15 Compulsory religion or state ideology in schools	**YES**	Rights respected
16 Deliberate state policies to control artistic works	**YES**	Rights respected
17 Political censorship of press	**YES**	Rights respected, new publications appearing but old rivalries between democratic and Sandinista factions, with the possibility of violence, encourages some self-censorship
18 Censorship of mail or telephone tapping	yes	Security services still controlled by senior officers of the previous Sandinista government. Unofficial surveillance

FREEDOM FOR OR RIGHTS TO:		COMMENTS
19 Peaceful political opposition	yes	Since the change to democratic rule after the election of 1990, opposition permitted and practiced. But a legacy of sporadic violence after 10 years of Sandinista rule perpetuates instability
20 Multiparty elections by secret and universal ballot	yes	The government policy of reconciliation that has followed the 1990 elections has not prevented many areas of dispute. Democratic system under stress

	FREEDOM FOR OR RIGHTS TO:		COMMENTS
21	Political and legal equality for women	yes	Position improving, despite traditional prejudice. Increasing representation in government and professions
22	Social and economic equality for women	no	Society and economy male dominated. Major pay and employment inequalities for women
23	Social and economic equality for ethnic minorities	YES	Rights respected
24	Independent newspapers	YES	Situation transformed since the democratic election of 1990. Independence relates to commercial considerations
25	Independent book publishing	YES	As question 24
26	Independent radio and television networks	no	Transition from total state ownership to part private ownership. The changes are affected by violence, strikes, and threats, as rival factions seek to influence transfers or new licences
27	All courts to total independence	yes	Independence being established though previous political loyalties, with a shortage of qualified justices, may still influence courts
28	Independent trade unions	YES	Rights respected as the union movement adapts to a more democratic society. Decree of previous government seeking to control union agreements now abolished

	LEGAL RIGHTS:		COMMENTS
29	From deprivation of nationality	YES	The amnesty law, as part of the national reconciliation, removes threat of deprivation
30	To be considered innocent until proved guilty	yes	Some abuses by police still influenced by previous political loyalties
31	To free legal aid when necessary and counsel of own choice	yes	Means test. Judge appoints "public defender"
32	From civilian trials in secret	yes	Complaints that crimes against the military by civilians should not be tried in military secret courts
33	To be brought promptly before a judge or court	yes	Within 9 days. Occasional violations at local level
34	From police searches of home without a warrant	no	Many politically motivated forced entries by police, frequently supporting Sandinistas, to harass opponents. On many false pretexts

LEGAL RIGHTS:		COMMENTS
35	From arbitrary seizure of personal property	**no**	Police and military, usually loyal to previous Sandinista government, forcibly enter homes of political opponents. Looting, intimidation, etc.

PERSONAL RIGHTS:		COMMENTS
36	To interracial, interreligious, or civil marriage	**YES**	Rights respected
37	Equality of sexes during marriage and for divorce proceedings	yes	Traditional male domination persists despite constitutional equality
38	To practice any religion	**YES**	Rights respected
39	To use contraceptive pills and devices	**YES**	Rights respected
40	To noninterference by state in strictly private affairs	**YES**	Rights respected

Nigeria

Human rights rating: 49%

YES 9 yes 14 no 12 NO 5

Population: 108,500,000
Life expectancy: 51.5
Infant mortality (0–5 years)
  per 1,000 births: 170
United Nations covenants ratified:
Convention on Equality for Women

Form of government: Military council
GNP per capita (US$): 290
% of GNP spent by state:
On health: 0.2
On military: 1.0
On education: 1.4

FACTORS AFFECTING HUMAN RIGHTS

Since gaining independence from British colonial in 1960, the government has alternated between civilian and military administrations. The present military government took power in 1985 and is composed almost entirely of members of the armed forces. The heads of the 21 states are also from the military. A series of decrees ensures absolute rule by the council. A coup attempt in 1990 by military dissidents was quickly defeated and around 70 of the plotters were executed with many others placed under detention. The regime has recently promised a new constitution in 1992 and agreed to consider multiparty elections. The population has three major ethnic groups and traditional rivalries continue. Corruption is endemic in most areas of social, political, legal, and financial spheres.

	FREEDOM TO:		COMMENTS
1	Travel in own country	**YES**	Rights respected
2	Travel outside own country	yes	Occasional withdrawal of passports of senior political opponents. Airport officials notorious for intimidating passengers, frequently for gain
3	Peacefully associate and assemble	yes	Rallies need permits but many ignore this requirement. Associations and clubs need not register
4	Teach ideas and receive information	yes	Academics exercise prudence. Some enforced "retirements," students union banned, brief closures of turbulent universities
5	Monitor human rights violations	yes	Some government harassment of local group but international monitors allowed visits
6	Publish and educate in ethnic language	**YES**	Rights respected

FREEDOM FROM:		COMMENTS
7	Serfdom, slavery, forced or child labor	no	Traditional pattern of child labor and serfdom. Young girls occasionally sold in remote tribal areas
8	Extrajudicial killings or "disappearances"	NO	Summary executions of approximately 70 after failed military coup in April 1990. Many deaths of criminals while in custody
9	Torture or coercion by the state	no	Many beatings. Frequent reports on torture by human rights groups. Practiced by some police to extract money from victims
10	Compulsory work permits or conscription of labor	YES	Rights respected
11	Capital punishment by the state	NO	By shooting. Sometimes in public squares
12	Court sentences of corporal punishment	no	Caning: males and females up to age 45. Maximum of 12 strokes
13	Indefinite detention without charge	NO	Many held after 1990 coup attempt. Arbitrary police powers under Decree Two. Bribes may secure release
14	Compulsory membership of state organizations or parties	YES	Rights respected
15	Compulsory religion or state ideology in schools	YES	Rights respected
16	Deliberate state policies to control artistic works	YES	Rights respected
17	Political censorship of press	no	Subjects banned under Military Council's Decree Two. Self-censorship a necessary precaution. Occasional arrests of journalists
18	Censorship of mail or telephone tapping	yes	Limited surveillance of suspected subversives

FREEDOM FOR OR RIGHTS TO:		COMMENTS
19	Peaceful political opposition	no	Intention to consider multiparty politics has begun with two parties permitted by president. Future unpredictable but the military will dominate any change
20	Multiparty elections by secret and universal ballot	NO	Promise of a new constitution in 1992 may be affected by 1990 coup attempt. "Democracy will be encouraged by a multiparty system" (Military Council)
21	Political and legal equality for women	yes	Position improving in the professions but little representation in senior ministry posts
22	Social and economic equality for women	no	Widespread discrimination in pay and employment, and degrees of ill-treatment suffered from traditional customs. Female circumcision in some tribes

FREEDOM FOR OR RIGHTS TO:		COMMENTS
23	Social and economic equality for ethnic minorities	yes	But the numerous ethnic groups complain of different forms and degrees of discrimination
24	Independent newspapers	yes	Many national dailies. Understood guidelines. Provoking the authorities may lead to arrests, interrogation, or closure
25	Independent book publishing	yes	Circumspection required to avoid seizures or harassment
26	Independent radio and television networks	NO	State owned and controlled. "Voice of the government is the voice of the people"
27	All courts to total independence	no	Military Council may have civil cases transferred. Shari'a religious courts, when choice is offered, frequently chosen by Muslims of the northern states
28	Independent trade unions	yes	But government may ban strikes and impose mediation

LEGAL RIGHTS:		COMMENTS
29	From deprivation of nationality	YES	Rights respected
30	To be considered innocent until proved guilty	no	The catchall Decree Two permits individual's rights to be suspended, leading to detention
31	To free legal aid when necessary and counsel of own choice	yes	Means test. Legal Aid Society offers services. May be set aside in security cases
32	From civilian trials in secret	no	Cases may be arbitrarily transferred from civil to military courts, and in secret, if important political opponent involved
33	To be brought promptly before a judge or court	no	Consitutional safeguards ignored with complicity of senior government ministers. Corruption widespread
34	From police searches of home without a warrant	no	Frequent abuses by police officers with wide powers of arrest and detention. And to extract bribes
35	From arbitrary seizure of personal property	YES	Rights respected

PERSONAL RIGHTS:		COMMENTS
36	To interracial, interreligious, or civil marriage	yes	From age 16. 45% Muslims, usually in northern states, conform to Shari'a law
37	Equality of sexes during marriage and for divorce proceedings	no	Husband is "head of the family." Wives may not own property in some states or inherit on death of husband. Wives need spouse's permission to obtain passport

PERSONAL RIGHTS:		COMMENTS
38	To practice any religion	**yes**	But complaints by Christian denominations at interference by the Islamic authorities of the Muslim north
39	To use contraceptive pills and devices	**YES**	State support
40	To noninterference by state in strictly private affairs	**yes**	Except for occasional arbitrary intrusions by ill-disciplined police, perhaps hoping to receive bribes to ignore infringements

Norway

Human rights rating: 97%

YES 36 yes 4 no 0 NO 0

Population: 4,200,000
Life expectancy: 77.1
Infant mortality (0–5 years)
  per 1,000 births: 10
United Nations covenants ratified:
Civil and Political Rights, Economic,
Social and Cultural Rights, Convention
on Equality for Women

Form of government: Parliamentary
monarchy
GNP per capita (US$): 19,990
% of GNP spent by state:
On health: 5.5
On military: 3.2
On education: 6.8

FACTORS AFFECTING HUMAN RIGHTS
The individual respected. Government answerable to the people. A prosperous
economy, a free press, and a cohesive society. A country committed to human rights
causes.

	FREEDOM TO:		COMMENTS
1	Travel in own country	YES	Rights respected
2	Travel outside own country	YES	Rights respected
3	Peacefully associate and assemble	YES	Police permits invariably granted
4	Teach ideas and receive information	YES	Rights respected
5	Monitor human rights violations	YES	Strong supporter of human rights monitoring. Alleged breaches investigated by independent ombudsman
6	Publish and educate in ethnic language	YES	Government support for Lapp minority culture and survival of ethnic language

	FREEDOM FROM:		COMMENTS
7	Serfdom, slavery, forced or child labor	YES	Rights respected
8	Extrajudicial killings or "disappearances"	YES	Rights respected
9	Torture or coercion by the state	YES	Rights respected
10	Compulsory work permits or conscription of labor	YES	Rights respected

FREEDOM FROM:		COMMENTS
11	Capital punishment by the state	**YES**	Rights respected
12	Court sentences of corporal punishment	**YES**	Rights respected
13	Indefinite detention without charge	**YES**	Rights respected
14	Compulsory membership of state organizations or parties	**YES**	Rights respected
15	Compulsory religion or state ideology in schools	yes	The state religion, Evangelical Lutheran, is nominally that of over 90% of population and is given dominance in schools
16	Deliberate state policies to control artistic works	**YES**	Rights respected
17	Political censorship of press	**YES**	Rights respected
18	Censorship of mail or telephone tapping	**YES**	Court consent required in rare instances of high-risk security offences

FREEDOM FOR OR RIGHTS TO:		COMMENTS
19	Peaceful political opposition	**YES**	Rights respected
20	Multiparty elections by secret and universal ballot	**YES**	By proportional representation
21	Political and legal equality for women	yes	Limited parliamentary representation but in 1986 a woman became prime minister
22	Social and economic equality for women	yes	Ombudsman under the Equal Rights Law investigates occasional discrimination
23	Social and economic equality for ethnic minorities	yes	Increased non-Nordic refugee groups are forming minorities and complain of inequalities. Also a source of "cheap labor"
24	Independent newspapers	**YES**	Rights respected
25	Independent book publishing	**YES**	Brief prison sentences for rare publications that incite racism
26	Independent radio and television networks	**YES**	State owned but independence guaranteed by strict controls against government interference
27	All courts to total independence	**YES**	Rights respected
28	Independent trade unions	**YES**	Rights respected

LEGAL RIGHTS:		COMMENTS
29	From deprivation of nationality	**YES**	Rights respected
30	To be considered innocent until proved guilty	**YES**	Rights respected
31	To free legal aid when necessary and counsel of own choice	**YES**	No means test. All eligible for free counsel in criminal cases
32	From civilian trials in secret	**YES**	Rights respected
33	To be brought promptly before a judge or court	**YES**	To be charged within 24 hours. No prison sentence longer than 21 years. Judge may extend period of pretrial detention
34	From police searches of home without a warrant	**YES**	Only when in "hot pursuit" or when state security genuinely endangered
35	From arbitrary seizure of personal property	**YES**	Rights respected

PERSONAL RIGHTS:		COMMENTS
36	To interracial, interreligious, or civil marriage	**YES**	Rights respected
37	Equality of sexes during marriage and for divorce proceedings	**YES**	Special care, concern, and facilities for "battered" wives
38	To practice any religion	**YES**	Rights respected
39	To use contraceptive pills and devices	**YES**	Rights respected
40	To noninterference by state in strictly private affairs	**YES**	Rights respected

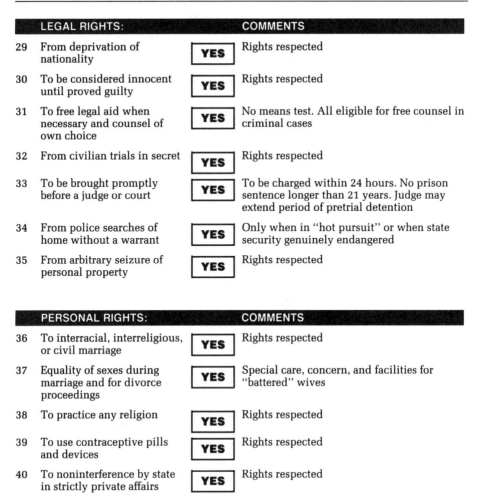

Oman

Human rights rating: 49%

YES 10 yes 7 no 11 NO 12

Population: 1,500,000
Life expectancy: 65.9
**Infant mortality (0–5 years)
  per 1,000 births**: 53
United Nations covenants ratified:
None

Form of government: Absolute monarchy
GNP per capita (US$): 5,000
% of GNP spent by state:
On health: 2.3
On military: 27.6
On education: 5.3

FACTORS AFFECTING HUMAN RIGHTS
The sultan is an absolute ruler. Criticism of the monarch and matters of national consequence are forbidden. There is no codified constitution and the State Consultative Council is appointed by the ruler. The media are either severely censored or state owned and controlled. Political parties or a more democratic system are not under consideration. Oman is a member of the United Nations. It has not ratified any of the instruments.

	FREEDOM TO:		COMMENTS
1	Travel in own country	**YES**	Rights respected
2	Travel outside own country	yes	Wives and daughters need permission from husband or guardian to travel abroad
3	Peacefully associate and assemble	**NO**	No meetings contrary to the interests and the policies of the sultan and government
4	Teach ideas and receive information	no	Strict supervision of curricula. Nothing that contradicts or challenges the rule or the policies of the sultan or government. Single university has banned the subject of politics
5	Monitor human rights violations	**NO**	Visits by international investigators not permitted. Local monitoring would be clandestine
6	Publish and educate in ethnic language	yes	Government pressures favor Arabic in all areas

	FREEDOM FROM:		COMMENTS
7	Serfdom, slavery, forced or child labor	no	Regional patterns of child labor. A few survivors of earlier slavery
8	Extrajudicial killings or "disappearances"	**YES**	Rights respected

	FREEDOM FROM:		COMMENTS
9	Torture or coercion by the state	yes	Occasional abuses at local level
10	Compulsory work permits or conscription of labor	**YES**	Rights respected
11	Capital punishment by the state	**NO**	No information on recent executions but the death penalty is retained and requires The sultan's ratification
12	Court sentences of corporal punishment	no	Canings within Shari'a law jurisdiction
13	Indefinite detention without charge	yes	A few cases of brief detentions on security suspicions. Usually for interrogation
14	Compulsory membership of state organizations or parties	**YES**	Rights respected
15	Compulsory religion or state ideology in schools	no	Islamic instruction compulsory in all Muslim schools
16	Deliberate state policies to control artistic works	no	Close surveillance of all creative art forms for alien or nonconformist ideas, Western vulgarity, etc.
17	Political censorship of press	**NO**	No criticisms of the sultan and matters of national consequence. Foreign publications censored before distribution
18	Censorship of mail or telephone tapping	no	Surveillance of security or dissident threats. Also in criminal investigations

	FREEDOM FOR OR RIGHTS TO:		COMMENTS
19	Peaceful political opposition	**NO**	Not permitted
20	Multiparty elections by secret and universal ballot	**NO**	Not permitted. The sultan is an absolute ruler
21	Political and legal equality for women	no	Limited involvement in politics and the professions
22	Social and economic equality for women	no	Outside government employment, pay and opportunity discrimination in accordance with traditional ideas of a woman's role. Some female circumcision
23	Social and economic equality for ethnic minorities	**YES**	Rights respected
24	Independent newspapers	no	Most are government owned or controlled
25	Independent book publishing	no	Publications must remain within strict guidelines. No nonconformist works, books critical of Islam, or publications that spread or profit from Western vulgarity
26	Independent radio and television networks	**NO**	State owned or controlled

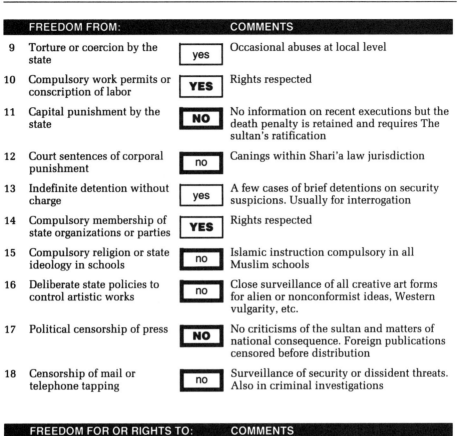

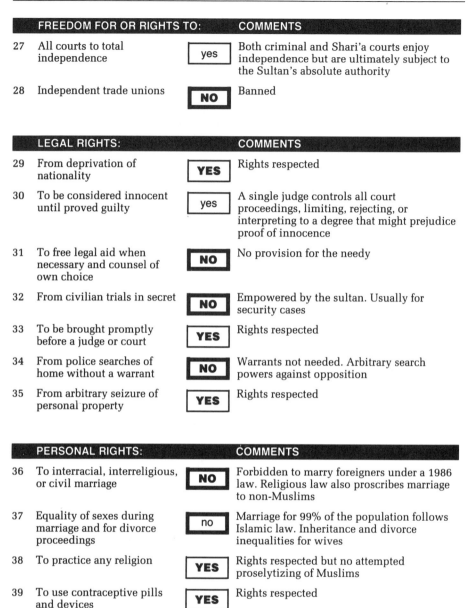

FREEDOM FOR OR RIGHTS TO:		COMMENTS
27	All courts to total independence	yes	Both criminal and Shari'a courts enjoy independence but are ultimately subject to the Sultan's absolute authority
28	Independent trade unions	NO	Banned

LEGAL RIGHTS:		COMMENTS
29	From deprivation of nationality	YES	Rights respected
30	To be considered innocent until proved guilty	yes	A single judge controls all court proceedings, limiting, rejecting, or interpreting to a degree that might prejudice proof of innocence
31	To free legal aid when necessary and counsel of own choice	NO	No provision for the needy
32	From civilian trials in secret	NO	Empowered by the sultan. Usually for security cases
33	To be brought promptly before a judge or court	YES	Rights respected
34	From police searches of home without a warrant	NO	Warrants not needed. Arbitrary search powers against opposition
35	From arbitrary seizure of personal property	YES	Rights respected

PERSONAL RIGHTS:		COMMENTS
36	To interracial, interreligious, or civil marriage	NO	Forbidden to marry foreigners under a 1986 law. Religious law also proscribes marriage to non-Muslims
37	Equality of sexes during marriage and for divorce proceedings	no	Marriage for 99% of the population follows Islamic law. Inheritance and divorce inequalities for wives
38	To practice any religion	YES	Rights respected but no attempted proselytizing of Muslims
39	To use contraceptive pills and devices	YES	Rights respected
40	To noninterference by state in strictly private affairs	yes	The absolute powers of the sultan and the religious authority of Islam inhibits nonconformism in private life

Pakistan

Human rights rating: 42%

YES 7 **yes** 13 **no** 11 NO 9

Population: 122,600,000
Life expectancy: 57.7
Infant mortality (0–5 years)
per 1,000 births: 162
United Nations covenants ratified:
None

Form of government: Multiparty system
GNP per capita (US$): 350
% of GNP spent by state:
On health: 0.2
On military: 6.7
On education: 2.2

FACTORS AFFECTING HUMAN RIGHTS

The death in 1988 of a military president with virtually absolute powers was followed by a multiparty election. This resulted in the country's first woman prime minister. In the period of her government, which lasted nearly 2 years, the judiciary became more independent, academic life freer, and the press allowed to report on national affairs without being inhibited by "understood guidelines." In 1990 the president used his constitutional powers and dismissed the elected prime minister, and her government has been followed by one adhering to more traditional policies. In response to violence and terrorism in some of the provinces, there have been many killings by the security forces, some arbitrary, and large-scale torture is practiced. After the more liberal interregnum, orthodox Islamic influences are increasing and severe punishments under Shari'a law may be enforced.

	FREEDOM TO:		COMMENTS
1	Travel in own country	**YES**	Rights respected but free travel may be affected by regional law-and-order emergencies
2	Travel outside own country	yes	A form of travel clearance may be used by the government. Obligatory for state employees but otherwise limited use
3	Peacefully associate and assemble	no	The degree of violence has meant close scrutiny of permits, with many bannings. Curfews periodically in force
4	Teach ideas and receive information	yes	But self-imposed restraints on academics whose ideas may conflict with the changing policies of political and religious authorities. Also intimidation by opposing factions of militant students
5	Monitor human rights violations	**YES**	Rights respected
6	Publish and educate in ethnic language	**YES**	Rights respected

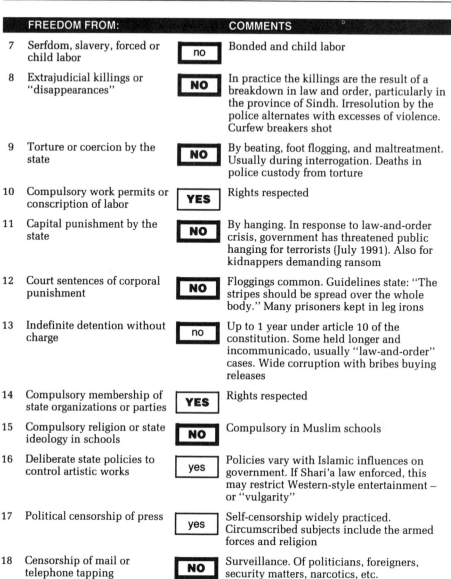

	FREEDOM FROM:		COMMENTS
7	Serfdom, slavery, forced or child labor	**no**	Bonded and child labor
8	Extrajudicial killings or "disappearances"	**NO**	In practice the killings are the result of a breakdown in law and order, particularly in the province of Sindh. Irresolution by the police alternates with excesses of violence. Curfew breakers shot
9	Torture or coercion by the state	**NO**	By beating, foot flogging, and maltreatment. Usually during interrogation. Deaths in police custody from torture
10	Compulsory work permits or conscription of labor	**YES**	Rights respected
11	Capital punishment by the state	**NO**	By hanging. In response to law-and-order crisis, government has threatened public hanging for terrorists (July 1991). Also for kidnappers demanding ransom
12	Court sentences of corporal punishment	**NO**	Floggings common. Guidelines state: "The stripes should be spread over the whole body." Many prisoners kept in leg irons
13	Indefinite detention without charge	**no**	Up to 1 year under article 10 of the constitution. Some held longer and incommunicado, usually "law-and-order" cases. Wide corruption with bribes buying releases
14	Compulsory membership of state organizations or parties	**YES**	Rights respected
15	Compulsory religion or state ideology in schools	**NO**	Compulsory in Muslim schools
16	Deliberate state policies to control artistic works	**yes**	Policies vary with Islamic influences on government. If Shari'a law enforced, this may restrict Western-style entertainment – or "vulgarity"
17	Political censorship of press	**yes**	Self-censorship widely practiced. Circumscribed subjects include the armed forces and religion
18	Censorship of mail or telephone tapping	**NO**	Surveillance. Of politicians, foreigners, security matters, narcotics, etc.

	FREEDOM FOR OR RIGHTS TO:		COMMENTS
19	Peaceful political opposition	**yes**	Subject to intimidation and violence at local or regional level
20	Multiparty elections by secret and universal ballot	**yes**	Charges of vote rigging, unlawful detention of opposition candidates, and intimidation

FREEDOM FOR OR RIGHTS TO:		COMMENTS
21	Political and legal equality for women	**NO**	Tradition and religion prevent equality in politics and professions. The appointment of a recent woman prime minister should be contrasted with the many obstacles preventing women from voting at all
22	Social and economic equality for women	**NO**	Constant discrimination. Worse in rural areas. Little apparent progress
23	Social and economic equality for ethnic minorities	no	Discrimination against Baluchis, Pathans, etc. Also against Ahmedi religious sect and Christians
24	Independent newspapers	yes	Profitable survival may depend on the goodwill of the government, a major source of advertising revenue as well as requiring registration of newspapers
25	Independent book publishing	yes	Bannings vary from province to province. Publishers usually follow a conservative course on political themes
26	Independent radio and television networks	no	In 1990 the first licenses for independent stations were issued but major stations remain state owned and controlled
27	All courts to total independence	no	The civil judicial system may be increasingly affected by the Shari'a bill provisions being introduced in 1991. The Federal Shari'a Court has ordered the government to make the death penalty obligatory for insulting the Prophet Mohammad
28	Independent trade unions	no	Limited rights. Fewer than 10% of workers in unions. Strike restrictions

LEGAL RIGHTS:		COMMENTS
29	From deprivation of nationality	**YES**	Rights respected
30	To be considered innocent until proved guilty	yes	The wide powers of arrest enable police to take arbitrary action. Random arrests during public protests. Paramilitaries accused of deaths of innocent civilians
31	To free legal aid when necessary and counsel of own choice	yes	Courts may grant aid, and in Shari'a court defendant may choose own lawyer
32	From civilian trials in secret	yes	Permitted under special powers, when necessary, to restore law and order
33	To be brought promptly before a judge or court	no	An overburdened and understaffed judicial system causes long delays. Some arbitrary actions by police, such as "on-the-spot" floggings, avoid time-consuming court appearances

LEGAL RIGHTS:		COMMENTS
34	From police searches of home without a warrant	yes	But under special powers many violations, sometimes with corrupt motives
35	From arbitrary seizure of personal property	yes	Confiscation can be ordered by court. Instances of unlawful seizures – looting, etc. – by security and military forces during law-and-order operations

PERSONAL RIGHTS:		COMMENTS
36	To interracial, interreligious, or civil marriage	no	In a country 99% Muslim, marriage to a non-Muslim involves a written application to renounce religion. Shari'a court may reject such applications
37	Equality of sexes during marriage and for divorce proceedings	NO	The 1991 bill to extend Shari'a law perpetuates the subordinate status of wives. Worse in rural areas
38	To practice any religion	no	Ahmadi sect suffers severe harassment, occasional imprisonment. Forced closures of Ahmadi mosques. Occasional violence against Hindus with some forced conversions of women
39	To use contraceptive pills and devices	YES	Direct government support
40	To noninterference by state in strictly private affairs	no	The requirements of Shari'a law demand conformism in private affairs. Muslims disregarding them face severe punishments. Such as death for adultery

Panama

Human rights rating: 81%

YES 22 yes 17 no 1 NO 0

Population: 3,900,000
Life expectancy: 72.4
Infant mortality (0–5 years)
per 1,000 births: 33
United Nations covenants ratified:
Civil and Political Rights, Economic,
Social and Cultural Rights, Convention
on Equality for Women

Form of government: Multiparty
system
GNP per capita (US$): 2,120
% of GNP spent by state:
On health: 5.7
On military: 1.4
On education: 5.4

FACTORS AFFECTING HUMAN RIGHTS
At the end of 1989 the country was invaded by the USA and a military dictator
suspected of being deeply involved in drug trafficking was overthrown. He was
replaced by a civilian government which claimed legitimacy as a result of winning the
1989 election. While the constitution recognizes most human rights, most violations
relate to arbitrary behavior by the police and security forces, corruption at many levels
of society and of the administration, and complaints that detentions and trials of some
of those suspected of supporting the deposed dictator rest on inadequate evidence and
political bias.

	FREEDOM TO:		COMMENTS
1	Travel in own country	YES	Rights respected
2	Travel outside own country	YES	Rights respected
3	Peacefully associate and assemble	YES	Rights respected
4	Teach ideas and receive information	YES	Rights respected
5	Monitor human rights violations	YES	Rights respected
6	Publish and educate in ethnic language	YES	Rights respected

	FREEDOM FROM:		COMMENTS
7	Serfdom, slavery, forced or child labor	yes	Regional pattern of casual child labor
8	Extrajudicial killings or "disappearances"	yes	A US military presence helps to restrain excesses by a police and security force that served under the previous dictator

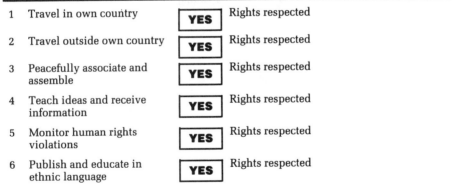

FREEDOM FROM:		COMMENTS
9	Torture or coercion by the state	yes	Unlawful abuses by an ill-disciplined security service, continuing some of the practices of the previous regime
10	Compulsory work permits or conscription of labor	**YES**	Limited use of convicts for manual work
11	Capital punishment by the state	**YES**	Abolished. Last execution claimed to be in 1906
12	Court sentences of corporal punishment	**YES**	Rights respected
13	Indefinite detention without charge	yes	But thousands of previous regime's military supporters are awaiting trial. An inadequate judicial system is therefore further overburdened
14	Compulsory membership of state organizations or parties	**YES**	Rights respected
15	Compulsory religion or state ideology in schools	**YES**	Rights respected
16	Deliberate state policies to control artistic works	**YES**	Rights respected
17	Political censorship of press	yes	New press laws, drafted since the end of the previous regime, set guidelines for journalists, who must now have a university degree
18	Censorship of mail or telephone tapping	yes	Surveillance of supporters of the previous dictator and other violent factions

FREEDOM FOR OR RIGHTS TO:		COMMENTS
19	Peaceful political opposition	yes	The return to democracy is helped by the influence of a considerable US presence and must therefore be seen as unpredictable
20	Multiparty elections by secret and universal ballot	**YES**	Rights respected following the ousting of the previous military dictator
21	Political and legal equality for women	yes	In practice women do not enjoy the equality granted in the amended constitution of 1983
22	Social and economic equality for women	no	Regional attitudes toward women. Worse in rural areas
23	Social and economic equality for ethnic minorities	yes	Limited discrimination against blacks and Amerindians. Lighter skins enjoy many advantages
24	Independent newspapers	**YES**	Rights respected
25	Independent book publishing	**YES**	Rights respected

FREEDOM FOR OR RIGHTS TO:		COMMENTS
26	Independent radio and television networks	**YES**	Mostly private stations. Occasional pressures from officials during periods of authoritarian governments
27	All courts to total independence	yes	Independence is affected by an underfunded system with underqualified judicial officials. Position worsened by a degree of corruption
28	Independent trade unions	yes	But in the period following the previous oppressive regime, many areas of dispute are emerging with harassment of unions

LEGAL RIGHTS:		COMMENTS
29	From deprivation of nationality	**YES**	Rights respected
30	To be considered innocent until proved guilty	yes	Bribes to court officials may be the quickest way to establish innocence
31	To free legal aid when necessary and counsel of own choice	yes	Means test but court may impose own counsel
32	From civilian trials in secret	**YES**	Rights respected
33	To be brought promptly before a judge or court	yes	Officially 24 hours. But may be followed by 90-day review period. Many additional delays
34	From police searches of home without a warrant	yes	Many police sweeps searching for weapons, drugs, and traffickers, which offer opportunities of gain for more corrupt members of police and security forces
35	From arbitrary seizure of personal property	yes	Criminal elements among the police collaborate with lawless gangs in burglaries, seizures, etc.

PERSONAL RIGHTS:		COMMENTS
36	To interracial, interreligious, or civil marriage	**YES**	Males may marry at 14 years, females at 12
37	Equality of sexes during marriage and for divorce proceedings	yes	Traditional regional attitude toward marriage and divorce, usually to husband's advantage
38	To practice any religion	**YES**	Country devoutly Roman Catholic (95% of population)
39	To use contraceptive pills and devices	**YES**	Government support
40	To noninterference by state in strictly private affairs	**YES**	Rights respected

Papua New Guinea

Human rights rating: 70%

YES 21 **yes** 12 **no** 5 NO 2

Population: 3,900,000
Life expectancy: 54.9
Infant mortality (0–5 years)
per 1,000 births: 83
United Nations covenants ratified:
None

Form of government: Multiparty democracy
GNP per capita (US$): 810
% of GNP spent by state:
On health: 3.0
On military: 1.4
On education: 5.0

FACTORS AFFECTING HUMAN RIGHTS

The country gained its independence in 1975. The multiparty democracy, however, suffered its first major crisis when rebels seized control of Bougainville Island, part of the PNG archipelago. The response of ill-trained and ill-disciplined government forces has not only been to oppose the rebels but to commit atrocities against innocent inhabitants of the region. At mid-1991, having failed to suppress the rebellion, the government had resorted to a blockade of the area. Repercussions of the situation, however, have included limited censorship of the press, government sensitivity to criticism, and an increase in abuses by the police.

	FREEDOM TO:		COMMENTS
1	Travel in own country	yes	Bougainville Island seized by rebels. PNG government using blockade as strategy to restore its authority
2	Travel outside own country	YES	Rights respected
3	Peacefully associate and assemble	YES	Permits required only when violence threatened
4	Teach ideas and receive information	yes	Some minor restrictions on the freedom of expatriate university staff to criticize PNG government
5	Monitor human rights violations	yes	Government reluctant to cooperate with human rights groups investigating atrocities by security forces in Bougainville. But no bannings of monitoring
6	Publish and educate in ethnic language	YES	Rights respected

	FREEDOM FROM:		COMMENTS
7	Serfdom, slavery, forced or child labor	yes	Traditional child labor. In some remote mountainous regions, status of women that of near-serfdom

FREEDOM FROM:		COMMENTS
8	Extrajudicial killings or "disappearances"	**no**	In confronting the insurrectionist Bougainville Revolutionary Army, undisciplined government forces used excessive violence on innocent villagers of the region. Many killed
9	Torture or coercion by the state	**NO**	Circle of torture and brutality extended from the initial clash with the insurrectionists to atrocities against whole villages
10	Compulsory work permits or conscription of labor	**YES**	Rights respected
11	Capital punishment by the state	**NO**	Death penalty restored in August 1991 for violent crimes, including rape
12	Court sentences of corporal punishment	**YES**	Rights respected but special measures to control violent crime include proposal to tattoo the foreheads of ex-convicts
13	Indefinite detention without charge	yes	Lengthy detentions usually caused by slow and inadequate systems
14	Compulsory membership of state organizations or parties	**YES**	Rights respected
15	Compulsory religion or state ideology in schools	**YES**	Rights respected
16	Deliberate state policies to control artistic works	yes	Censorship Act of 1990 gives government powers over artistic works as well as over the media
17	Political censorship of press	yes	The Bougainville insurrection encouraged limited censorship. Position anomalous but deteriorating
18	Censorship of mail or telephone tapping	**YES**	Rights respected

FREEDOM FOR OR RIGHTS TO:		COMMENTS
19	Peaceful political opposition	**YES**	Rights respected
20	Multiparty elections by secret and universal ballot	**YES**	Rights respected
21	Political and legal equality for women	no	Outside the few urban areas, customary law prevails. Woman's role continues to be subservient. Little involvement in government
22	Social and economic equality for women	no	The tribal pattern of the population perpetuates woman's lowly status. During tribal wars women are exposed to mass rape by the victors
23	Social and economic equality for ethnic minorities	**YES**	The population consists of nearly 1,000 tribes which prevents a recognizable pattern of discrimination

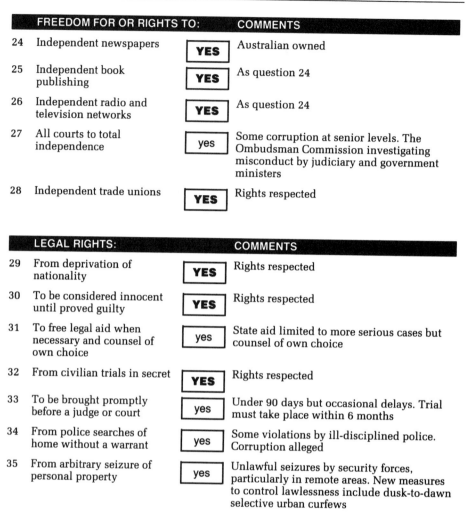

FREEDOM FOR OR RIGHTS TO:		COMMENTS
24	Independent newspapers	**YES**	Australian owned
25	Independent book publishing	**YES**	As question 24
26	Independent radio and television networks	**YES**	As question 24
27	All courts to total independence	yes	Some corruption at senior levels. The Ombudsman Commission investigating misconduct by judiciary and government ministers
28	Independent trade unions	**YES**	Rights respected

LEGAL RIGHTS:		COMMENTS
29	From deprivation of nationality	**YES**	Rights respected
30	To be considered innocent until proved guilty	**YES**	Rights respected
31	To free legal aid when necessary and counsel of own choice	yes	State aid limited to more serious cases but counsel of own choice
32	From civilian trials in secret	**YES**	Rights respected
33	To be brought promptly before a judge or court	yes	Under 90 days but occasional delays. Trial must take place within 6 months
34	From police searches of home without a warrant	yes	Some violations by ill-disciplined police. Corruption alleged
35	From arbitrary seizure of personal property	yes	Unlawful seizures by security forces, particularly in remote areas. New measures to control lawlessness include dusk-to-dawn selective urban curfews

PERSONAL RIGHTS:		COMMENTS
36	To interracial, interreligious, or civil marriage	**YES**	Rights respected
37	Equality of sexes during marriage and for divorce proceedings	no	The constitution and the exercise of law are remote from isolated tribes. Women are subservient in and out of marriage
38	To practice any religion	**YES**	Rights respected
39	To use contraceptive pills and devices	**YES**	Rights respected
40	To noninterference by state in strictly private affairs	no	Ill-disciplined police forces in remoter areas and in tribal lands plunder villages and destroy homes

Paraguay

Human rights rating: 70%

Population: 4,300,000
Life expectancy: 67.1
Infant mortality (0–5 years)
  per 1,000 births: 61
United Nations covenants ratified:
Convention on Equality for Women

Form of government: Party system
being introduced
GNP per capita (US$): 1,180
% of GNP spent by state:
On health: 0.2
On military: 1.0
On education: 1.0

FACTORS AFFECTING HUMAN RIGHTS

A military president who had ruled the country since 1954, and his puppet Colorado party, were overthrown in 1989. The transition to democracy will be completed in 1992, when presidential elections will be held. Many reforms are being introduced, as is a new constitution. In the interim period, however, the armed forces retain considerable political power, as do many appointees of the previous dictator and large landowners. Peasants in dispute with these landowners are frequently assaulted by the police.

FREEDOM TO:　　COMMENTS

1	Travel in own country	**YES**	Rights respected
2	Travel outside own country	**YES**	Rights respected. Previous restrictions on travel to communist countries no longer enforced
3	Peacefully associate and assemble	yes	Prohibitions include demonstrations outside military or police buildings and the president's residence
4	Teach ideas and receive information	**YES**	Rights respected
5	Monitor human rights violations	**YES**	Rights respected
6	Publish and educate in ethnic language	**YES**	Rights respected. Both Spanish and Guarani Indian are official languages

FREEDOM FROM:　　COMMENTS

7	Serfdom, slavery, forced or child labor	yes	Regional pattern of child labor
8	Extrajudicial killings or "disappearances"	yes	Mercenaries hired by large landowners to evict peasant squatters commit killings which are regularly disregarded by the security police

FREEDOM FROM: COMMENTS

9	Torture or coercion by the state	no	Mostly beatings and maltreatment though government is now investigating worst abuses
10	Compulsory work permits or conscription of labor	YES	Rights respected
11	Capital punishment by the state	YES	De facto abolition 1928
12	Court sentences of corporal punishment	YES	Rights respected
13	Indefinite detention without charge	yes	Cases of incommunicado detention continue to be reported but on a diminishing scale
14	Compulsory membership of state organizations or parties	YES	Rights respected
15	Compulsory religion or state ideology in schools	YES	Rights respected
16	Deliberate state policies to control artistic works	YES	Rights respected
17	Political censorship of press	yes	Understood guidelines observed. The transformation to a free press, after the Stroessner dictatorship, still calls for a degree of restraint
18	Censorship of mail or telephone tapping	YES	Rights respected

FREEDOM FOR OR RIGHTS TO: COMMENTS

19	Peaceful political opposition	yes	The change from a dictatorship to democracy is making significant progress
20	Multiparty elections by secret and universal ballot	yes	Total reform of the constitution to be completed by 1992. Parties represented in Congress by proportional representation
21	Political and legal equality for women	yes	After the male-oriented Stroessner regime, some progress toward promotion to senior levels
22	Social and economic equality for women	yes	Unequal pay and employment opportunities. Prospects improving under the new system
23	Social and economic equality for ethnic minorities	yes	Many discriminations against Chinese and Korean immigrant minorities
24	Independent newspapers	yes	The relatively sudden change from dictatorial rule means that a degree of self-censorship is still practiced
25	Independent book publishing	yes	Prudence required when books directly defame or appear to insult the president, who was an associate of the previous leader

FREEDOM FOR OR RIGHTS TO: COMMENTS

26	Independent radio and television networks	yes	State-owned and private stations. Gradual progress toward greater freedom and independence
27	All courts to total independence	yes	Many appointees of the Stroessner regime remain in office and their previous affiliations may influence arbitrary judgments
28	Independent trade unions	yes	The democratic changes of the last 2 years are transforming union practices and independence

LEGAL RIGHTS: COMMENTS

29	From deprivation of nationality	YES	Rights respected
30	To be considered innocent until proved guilty	yes	Some abuses of assumption of innocence. Many high officials and police from the previous dictatorship continue old practices
31	To free legal aid when necessary and counsel of own choice	YES	Means test. Free choice of counsel
32	From civilian trials in secret	no	Parts of court proceedings held in secret when case regarded as politically provocative
33	To be brought promptly before a judge or court	yes	Within 48 hours but many violations – usually because of inadequate judicial system
34	From police searches of home without a warrant	yes	Occasional violations by security forces still influenced by arbitrary practices of previous regime
35	From arbitrary seizure of personal property	YES	Rights respected

PERSONAL RIGHTS: COMMENTS

36	To interracial, interreligious, or civil marriage	YES	Rights respected. Legal age for males 14 years, for females 12 years
37	Equality of sexes during marriage and for divorce proceedings	no	Old forms of discrimination still persist, including wife's need for husband's permission to work, travel abroad, etc.
38	To practice any religion	YES	Rights respected
39	To use contraceptive pills and devices	YES	Limited state support

PERSONAL RIGHTS:		COMMENTS

40   To noninterference by state in strictly private affairs

no

Bitter confrontations between peasants and landowners have meant forced evictions and burning of dwellings by private guards assisted by police. Considerable corruption of officials

Peru	Human rights rating:	54%

YES 14 **yes** 10 **no** 9 NO 7

Population: 21,600,000
Life expectancy: 63.0
Infant mortality (0–5 years)
per 1,000 births: 85
United Nations covenants ratified:
Civil and Political Rights, Economic,
Social and Cultural Rights, Convention
on Equality for Women

Form of government: Multiparty
democracy
GNP per capita (US$): 1,300
% of GNP spent by state:
On health: 0.8
On military: 1.6
On education: 2.2

FACTORS AFFECTING HUMAN RIGHTS
An orderly election in 1985 resulted in a democratic change of government. But the
country has since faced the continuing and disruptive challenge of the violent terrorist
campaign of the *Sendero Luminoso* (Shining Path), a self-styled Marxist organization.
The human rights situation is dominated by a bitter and bloody struggle which, if
anything, has worsened. An annual figure of those killed is estimated at around 3,000,
nearly half being civilians. Peasants in remote areas, and sometimes whole villages, are
frequently the targets of both government and terrorist forces. A new government,
installed during 1990, with policies for improvement, has failed to implement its
program.

	FREEDOM TO:		COMMENTS
1	Travel in own country	yes	But not in emergency zones under military rule or where security operations against the *Sendero Luminoso* are taking place
2	Travel outside own country	YES	Rights respected
3	Peacefully associate and assemble	no	Many instances of excessive violence by police against demonstrators. By water cannon, tear gas, and buckshot
4	Teach ideas and receive information	yes	Surveillance by authorities for supporters of terrorists or left-wing militants. Students arrested for demonstrations against academic restraints
5	Monitor human rights violations	no	Government allows monitoring groups greater freedom to investigate crimes by terrorists than to probe "disappearances," detentions, and torture charges against security forces
6	Publish and educate in ethnic language	YES	Quecha, spoken by 27% of population, made the second official language in 1975

FREEDOM FROM:		COMMENTS
7	Serfdom, slavery, forced or child labor	yes	Child labor a regional practice
8	Extrajudicial killings or "disappearances"	**NO**	An overall annual total of approximately 3,000 killings. *Sendero Luminoso* terrorists and pro-government forces equally guilty. Worst in provinces under emergency laws. Bloodbaths in remote areas
9	Torture or coercion by the state	**NO**	By all factions. Extreme brutalities by both the terrorists and police. The new president, installed in 1990, has begun to reduce torture practices by summarily dismissing hundreds of police officers
10	Compulsory work permits or conscription of labor	**YES**	Rights respected
11	Capital punishment by the state	**NO**	Restored in 1982 for murders by terrorists but executions greatly outnumbered by summary killings by security forces
12	Court sentences of corporal punishment	**YES**	Rights respected
13	Indefinite detention without charge	**NO**	State emergency permits indefinite detention under Decree 046 (1981). Worst excesses in provinces. Many detainees have "disappeared"
14	Compulsory membership of state organizations or parties	**YES**	Rights respected
15	Compulsory religion or state ideology in schools	**YES**	Rights respected
16	Deliberate state policies to control artistic works	**YES**	Rights respected
17	Political censorship of press	yes	Only for supporting or giving media space to *Sendero Luminoso* views or propaganda. Numerous murders of journalists, with some evidence indicating that police as well as terrorists were responsible
18	Censorship of mail or telephone tapping	no	Surveillance most evident in state of emergency provinces, covering 25% of population

FREEDOM FOR OR RIGHTS TO:		COMMENTS
19	Peaceful political opposition	no	Essentially fair elections are affected by efforts of guerrilla groups to disrupt them. Many killings of candidates and voters
20	Multiparty elections by secret and universal ballot	**YES**	Presidential elections every 5 years. Democracy restored in 1980 after 12 years of military rule

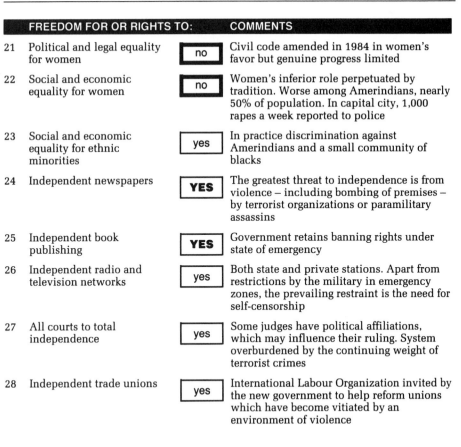

FREEDOM FOR OR RIGHTS TO:		COMMENTS
21	Political and legal equality for women	no	Civil code amended in 1984 in women's favor but genuine progress limited
22	Social and economic equality for women	no	Women's inferior role perpetuated by tradition. Worse among Amerindians, nearly 50% of population. In capital city, 1,000 rapes a week reported to police
23	Social and economic equality for ethnic minorities	yes	In practice discrimination against Amerindians and a small community of blacks
24	Independent newspapers	YES	The greatest threat to independence is from violence – including bombing of premises – by terrorist organizations or paramilitary assassins
25	Independent book publishing	YES	Government retains banning rights under state of emergency
26	Independent radio and television networks	yes	Both state and private stations. Apart from restrictions by the military in emergency zones, the prevailing restraint is the need for self-censorship
27	All courts to total independence	yes	Some judges have political affiliations, which may influence their ruling. System overburdened by the continuing weight of terrorist crimes
28	Independent trade unions	yes	International Labour Organization invited by the new government to help reform unions which have become vitiated by an environment of violence

LEGAL RIGHTS:		COMMENTS
29	From deprivation of nationality	YES	Rights respected
30	To be considered innocent until proved guilty	NO	Many assumptions of guilt in emergency areas. Perhaps followed by equally arbitrary executions by ill-disciplined security forces. Particularly in remote communities of *campesinos*
31	To free legal aid when necessary and counsel of own choice	yes	System of "defender of the poor" for the most needy. Some freedom to choose counsel
32	From civilian trials in secret	no	The military authorities may "persuade" the Supreme Court to conduct "security" trials in secret
33	To be brought promptly before a judge or court	no	Within 24 hours, extended to 15 days for terrorists, drug, and similar cases. Corruption at all levels. Indefinite delays in emergency zones

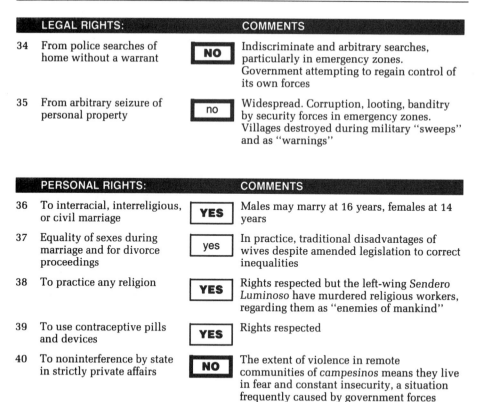

LEGAL RIGHTS:		COMMENTS
34	From police searches of home without a warrant	**NO**	Indiscriminate and arbitrary searches, particularly in emergency zones. Government attempting to regain control of its own forces
35	From arbitrary seizure of personal property	no	Widespread. Corruption, looting, banditry by security forces in emergency zones. Villages destroyed during military "sweeps" and as "warnings"

PERSONAL RIGHTS:		COMMENTS
36	To interracial, interreligious, or civil marriage	**YES**	Males may marry at 16 years, females at 14 years
37	Equality of sexes during marriage and for divorce proceedings	yes	In practice, traditional disadvantages of wives despite amended legislation to correct inequalities
38	To practice any religion	**YES**	Rights respected but the left-wing Sendero Luminoso have murdered religious workers, regarding them as "enemies of mankind"
39	To use contraceptive pills and devices	**YES**	Rights respected
40	To noninterference by state in strictly private affairs	**NO**	The extent of violence in remote communities of campesinos means they live in fear and constant insecurity, a situation frequently caused by government forces

Philippines

Human rights rating: 72%

YES 19 **yes** 17 **no** 3 NO 1

Population: 62,400,000
Life expectancy: 64.2
Infant mortality (0–5 years)
per 1,000 births: 72
United Nations covenants ratified:
Civil and Political Rights, Economic,
Social and Cultural Rights, Convention
on Equality for Women

Form of government: Democratic
presidential system
GNP per capita (US$): 630
% of GNP spent by state:
On health: 0.7
On military: 1.7
On education: 2.4

FACTORS AFFECTING HUMAN RIGHTS

A peaceful revolution described as a demonstration of "people power" ended the 14-year rule of a president who had become a virtual dictator. The new president, a woman, has been democratically elected but faces major difficulties which affect the stability of the country. An active Communist insurgency movement is established in many of the provinces, a large army and security forces have been mobilized to contain the threat, which is a burden on the limited resources of a poor country, and armed Muslim separatists continue to operate in the extreme south of the country. From time to time there are coup attempts which call for a state of emergency. Despite the many extrajudicial killings and "disappearances," principally by the armed forces and the Communists, and the government's failure to control its own military, human rights away from the areas of fighting are usually respected. A destabilizing factor, however, is the widespread corruption, particularly among police and officials.

	FREEDOM TO:		COMMENTS
1	Travel in own country	**YES**	Rights respected but areas may be occupied by rebel forces or be the scene of military operations
2	Travel outside own country	**YES**	Rights respected
3	Peacefully associate and assemble	**YES**	Provided rallies not advocating violent overthrow of the government or threatening public order
4	Teach ideas and receive information	**YES**	Rights respected
5	Monitor human rights violations	yes	International and national monitors relatively free but sometimes obstructed by military. Many death threats from right-wing pressure groups
6	Publish and educate in ethnic language	**YES**	Rights respected

	FREEDOM FROM:		COMMENTS
7	Serfdom, slavery, forced or child labor	yes	Child labor in rural areas. Widespread child prostitution despite efforts to prohibit it
8	Extrajudicial killings or "disappearances"	**NO**	Many summary executions by government forces and paramilitary "vigilantes" confronting Communist rebels. Also many "disappearances"
9	Torture or coercion by the state	no	Torture during interrogation of insurgents and suspects. Extent of violations by security forces beyond government control
10	Compulsory work permits or conscription of labor	**YES**	Rights respected
11	Capital punishment by the state	yes	Abolished 1987 but the extent of arbitrary killings by security forces reflects on government's lack of authority
12	Court sentences of corporal punishment	**YES**	Rights respected
13	Indefinite detention without charge	no	Particularly of rebels captured by government forces. Many incommunicado detentions – usually denied by the military and the police
14	Compulsory membership of state organizations or parties	**YES**	Rights respected
15	Compulsory religion or state ideology in schools	**YES**	Rights respected
16	Deliberate state policies to control artistic works	**YES**	Rights respected
17	Political censorship of press	yes	Communist papers illegal. Regarded as "voice" of insurgents. Other journals are targets for intimidation by the military and right-wing groups
18	Censorship of mail or telephone tapping	**YES**	Rights respected

	FREEDOM FOR OR RIGHTS TO:		COMMENTS
19	Peaceful political opposition	**YES**	Rights respected
20	Multiparty elections by secret and universal ballot	**YES**	Voting from age 18
21	Political and legal equality for women	yes	Women are beginning to play a significant role in politics and professions. The present president is a woman
22	Social and economic equality for women	yes	Traditional subordinate role. Situation improving in cities but worse in rural areas
23	Social and economic equality for ethnic minorities	**YES**	Rights respected

FREEDOM FOR OR RIGHTS TO:		COMMENTS
24	Independent newspapers	**yes**	Except when supporting insurgents. The deaths of many journalists, by both right-wing and leftist murder squads, create an ambience of risk and violence
25	Independent book publishing	**YES**	Rights respected
26	Independent radio and television networks	**yes**	Majority privately owned. Violence by insurgents and pressures from security forces may affect independence
27	All courts to total independence	**yes**	Independence, in practice, affected by threats, corruption, and political affiliations
28	Independent trade unions	**yes**	Wide emergency powers against strikes damaging the national welfare and security

LEGAL RIGHTS:		COMMENTS
29	From deprivation of nationality	**YES**	Rights respected
30	To be considered innocent until proved guilty	**no**	The extent of armed conflict and urban killings encourages major abuses by security forces. Forms of corruption include payment for release from false arrest
31	To free legal aid when necessary and counsel of own choice	**yes**	Uneven pattern over whole country. Large areas under insurgent rule. Legal aid most likely in metropolitan areas
32	From civilian trials in secret	**yes**	Pressures from military to retain the right to hold secret trials
33	To be brought promptly before a judge or court	**yes**	Within 36 hours of arrest but security forces may hold suspected insurgents for long periods (see question 8). Frequent police corruption to supplement low pay
34	From police searches of home without a warrant	**yes**	Many abuses as security forces search for suspected rebels. Particularly in remote areas
35	From arbitrary seizure of personal property	**yes**	Military action against Communist insurgents has meant displacement of local population. Confiscation of property and some looting by undisciplined military forces

PERSONAL RIGHTS:		COMMENTS
36	To interracial, interreligious, or civil marriage	**YES**	Marriage legal at 18 for both sexes. Trade in wives for foreigners ("mail-order wives") banned under new law

PERSONAL RIGHTS:		COMMENTS
37	Equality of sexes during marriage and for divorce proceedings	yes	Tradition favors husbands. Religious law among Muslims of the south prohibits intermarriage
38	To practice any religion	YES	Rights respected
39	To use contraceptive pills and devices	YES	Rights respected
40	To noninterference by state in strictly private affairs	yes	The extent of abuses, harassment, and intrusions, by ill-disciplined police and security forces reflects on the state's limited authority

Poland

Human rights rating: 83%

YES 24 yes 15 no 1 NO 0

Population: 38,400,000
Life expectancy: 71.8
Infant mortality (0–5 years)
  per 1,000 births: 18
United Nations covenants ratified:
Civil and Political Rights, Economic,
Social and Cultural Rights, Convention
on Equality for Women

Form of government: Multiparty
democracy
GNP per capita (US$): 1,860
% of GNP spent by state:
On health: 4.0
On military: 3.3
On education: 4.5

FACTORS AFFECTING HUMAN RIGHTS
Following the end of Communist rule, the country has progressed quickly toward a
multiparty democratic society. In 1990 the president was elected for the first time by a
free vote, a new constitution was adopted in 1991, and other reforms are planned.
Although human rights are now generally respected, the transition period to a free
market economy has meant increased unemployment and material shortages. With the
end of communism, the authority of the Roman Catholic church has increased.

	FREEDOM TO:		COMMENTS
1	Travel in own country	**YES**	Rights respected
2	Travel outside own country	**YES**	Rights respected
3	Peacefully associate and assemble	**YES**	Rights respected
4	Teach ideas and receive information	**YES**	Rights respected
5	Monitor human rights violations	**YES**	Rights respected
6	Publish and educate in ethnic language	**YES**	Rights respected

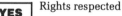

	FREEDOM FROM:		COMMENTS
7	Serfdom, slavery, forced or child labor	**YES**	Rights respected
8	Extrajudicial killings or "disappearances"	yes	Allegations of deaths despite disbanding of secret police. Government investigating a number of suspicious cases

FREEDOM FROM:		COMMENTS
9	Torture or coercion by the state	yes	Occasional beatings of prisoners, usually by guards retained in service from previous regime
10	Compulsory work permits or conscription of labor	yes	The law on conscripted labor of the previous Communist regime in process of being abolished
11	Capital punishment by the state	no	First abolished 1889. Reintroduced 1926. Early abolition anticipated
12	Court sentences of corporal punishment	YES	Rights respected
13	Indefinite detention without charge	YES	Rights respected
14	Compulsory membership of state organizations or parties	YES	Rights respected
15	Compulsory religion or state ideology in schools	yes	The influence of the dominant Roman Catholic church, following supportive Ministry of Education decree, extending compulsory religious instruction
16	Deliberate state policies to control artistic works	YES	Rights respected
17	Political censorship of press	YES	Rights respected
18	Censorship of mail or telephone tapping	YES	Rights respected

FREEDOM FOR OR RIGHTS TO:		COMMENTS
19	Peaceful political opposition	YES	A new democratic constitution (1991) codifies the sequence of changes and reforms since the end of the Communist regime
20	Multiparty elections by secret and universal ballot	YES	First democratically elected president in 1990, first parliamentary elections October 1991. Vote from age 18 years
21	Political and legal equality for women	yes	Position improving. Many women in senior government positions
22	Social and economic equality for women	yes	An economy in a transitional phase from communism to capitalism has meant increase in unemployment, affecting women most of all
23	Social and economic equality for ethnic minorities	YES	Rights respected
24	Independent newspapers	YES	A liquidation commission is selling or transferring many state-owned newspapers. Those retained by state publish without government interference

FREEDOM FOR OR RIGHTS TO:		COMMENTS
25	Independent book publishing	**YES**	Publishing of books adapting to the new "free market" situation
26	Independent radio and television networks	yes	Nearly all stations state controlled but generally reflect the democratic objectives of the changed government
27	All courts to total independence	yes	But many judges from previous Communist regime continue until retirement, possibly reflecting influences of that period
28	Independent trade unions	yes	A new trade union law (1991) is expected to correct certain anomalies

LEGAL RIGHTS:		COMMENTS
29	From deprivation of nationality	yes	Legally still permissible but unlikely to be practiced
30	To be considered innocent until proved guilty	**YES**	Rights respected
31	To free legal aid when necessary and counsel of own choice	yes	Free legal services but court may appoint counsel
32	From civilian trials in secret	**YES**	Rights respected
33	To be brought promptly before a judge or court	yes	Within 48 hours but continuing breaches until new legal guarantees become more established
34	From police searches of home without a warrant	**YES**	Rights respected
35	From arbitrary seizure of personal property	**YES**	Rights respected

PERSONAL RIGHTS:		COMMENTS
36	To interracial, interreligious, or civil marriage	yes	The increased authority of the dominant Roman Catholic church, with government concurrence, insists on its own form of marriage
37	Equality of sexes during marriage and for divorce proceedings	yes	The state-favored Catholic church, with its opposition to divorce, forms of birth control, and abortion, may introduce new factors into marriages contracted under the previous more secular system
38	To practice any religion	**YES**	Rights respected
39	To use contraceptive pills and devices	yes	Limited methods. Situation increasingly influenced by Catholic edicts
40	To noninterference by state in strictly private affairs	**YES**	Rights respected

Portugal

Human rights rating: 92%

YES 31 **yes** 9 **no** 0 NO 0

Population: 10,300,000
Life expectancy: 74.0
Infant mortality (0–5 years) per 1,000 births: 17
United Nations covenants ratified: Civil and Political Rights, Economic, Social and Cultural Rights, Convention on Equality for Women

Form of government: Parliamentary democracy
GNP per capita (US$): 3,650
% of GNP spent by state:
On health: 5.7
On military: 3.3
On education: 4.3

FACTORS AFFECTING HUMAN RIGHTS
The return to a democratic system was completed in two stages. A long-serving dictator was ousted in 1974, and the Council of the Revolution, which was dominated by the military, was finally followed by free, multiparty elections. The government, as with all others in the European Community, is now answerable to the people, there is a free press, and the previous monopoly of broadcasting has ended. Black immigrants from Portugal's previous African colonies complain of discrimination.

	FREEDOM TO:		COMMENTS
1	Travel in own country	YES	Rights respected
2	Travel outside own country	YES	Rights respected
3	Peacefully associate and assemble	YES	24-hour notice to authorities before public rallies
4	Teach ideas and receive information	YES	Rights respected
5	Monitor human rights violations	YES	Rights respected
6	Publish and educate in ethnic language	YES	Rights respected

	FREEDOM FROM:		COMMENTS
7	Serfdom, slavery, forced or child labor	yes	Degree of tolerated but unlawful child labor. Legal employment from age 14
8	Extrajudicial killings or "disappearances"	YES	Rights respected
9	Torture or coercion by the state	yes	Occasional police abuses, overzealous action during interrogation of suspects and brutality by police guards

	FREEDOM FROM:		COMMENTS
10	Compulsory work permits or conscription of labor	**YES**	Rights respected
11	Capital punishment by the state	**YES**	Abolished 1867 (first European country)
12	Court sentences of corporal punishment	**YES**	Rights respected
13	Indefinite detention without charge	**YES**	Rights respected
14	Compulsory membership of state organizations or parties	**YES**	Rights respected
15	Compulsory religion or state ideology in schools	**YES**	Rights respected
16	Deliberate state policies to control artistic works	**YES**	Rights respected
17	Political censorship of press	yes	Wide but seldom exercised rights to ban articles considered to defame civil and military authorities
18	Censorship of mail or telephone tapping	**YES**	Rights respected

	FREEDOM FOR OR RIGHTS TO:		COMMENTS
19	Peaceful political opposition	**YES**	Rights respected
20	Multiparty elections by secret and universal ballot	**YES**	Form of proportional representation. Presidential elections every 5 years
21	Political and legal equality for women	yes	Participation of women increasing though still underrepresented at senior levels
22	Social and economic equality for women	yes	Traditional inequalities in pay and employment lessening as modern democracy becomes more established
23	Social and economic equality for ethnic minorities	yes	The many black immigrants from Portugal's previous African colonies suffer minor disadvantages
24	Independent newspapers	**YES**	Rights respected
25	Independent book publishing	**YES**	Rights respected
26	Independent radio and television networks	yes	Privately owned networks recently authorized. Previously a government monopoly. Despite present guarantees of objectivity, still accused of bias
27	All courts to total independence	**YES**	Rights respected
28	Independent trade unions	**YES**	Rights respected

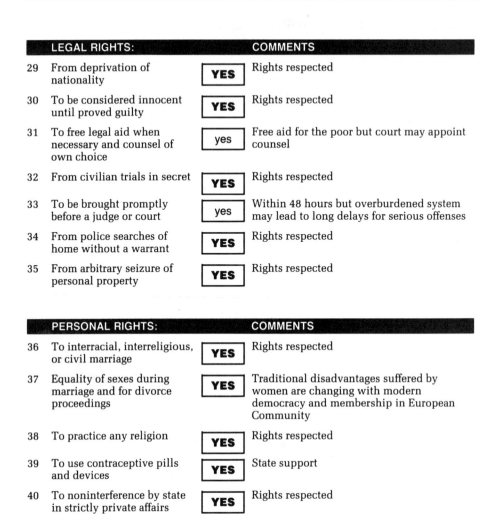

LEGAL RIGHTS:		COMMENTS
29	From deprivation of nationality	**YES**	Rights respected
30	To be considered innocent until proved guilty	**YES**	Rights respected
31	To free legal aid when necessary and counsel of own choice	yes	Free aid for the poor but court may appoint counsel
32	From civilian trials in secret	**YES**	Rights respected
33	To be brought promptly before a judge or court	yes	Within 48 hours but overburdened system may lead to long delays for serious offenses
34	From police searches of home without a warrant	**YES**	Rights respected
35	From arbitrary seizure of personal property	**YES**	Rights respected

PERSONAL RIGHTS:		COMMENTS
36	To interracial, interreligious, or civil marriage	**YES**	Rights respected
37	Equality of sexes during marriage and for divorce proceedings	**YES**	Traditional disadvantages suffered by women are changing with modern democracy and membership in European Community
38	To practice any religion	**YES**	Rights respected
39	To use contraceptive pills and devices	**YES**	State support
40	To noninterference by state in strictly private affairs	**YES**	Rights respected

Romania

Human rights rating: 82%

YES 18 **yes** 20 **no** 2 NO 0

Population: 23,300,000
Life expectancy: 70.8
**Infant mortality (0–5 years)
  per 1,000 births**: 28
United Nations covenants ratified:
Civil and Political Rights, Economic,
Social and Cultural Rights, Convention
on Equality for Women

Form of government: Multiparty
democracy
GNP per capita (US$): 2,560
% of GNP spent by state:
On health: 1.9
On military: 1.6
On education: 1.8

FACTORS AFFECTING HUMAN RIGHTS
The Communist regime which ended with the execution of the dictator Nicolae
Ceausescu has not been followed by a democracy as respectful of human rights as some
of the other transformed countries of Eastern Europe. The multiparty elections of 1990,
which resulted in a majority for a political party dominated by former Communists,
were marred by serious irregularities and intimidation. Although a new constitution
and other reforms are being introduced, the presence of some high officials and
members of the previous state security in important positions has caused concern. The
traditional discrimination against Gypsies continues and the large Hungarian minority
complain of inequalities. The trend, however, is for continuing improvement with
respect to human rights.

	FREEDOM TO:		COMMENTS
1	Travel in own country	**YES**	Rights respected
2	Travel outside own country	**YES**	Rights respected
3	Peacefully associate and assemble	yes	In the confrontations that followed the end of the Ceausescu dictatorship, many demonstrations bloodily suppressed by police and security forces. Peaceful meetings now permitted but some surveillance
4	Teach ideas and receive information	yes	A few post-Ceausescu problems have to be resolved in universities – such as continued employment of obdurate ex-Communists
5	Monitor human rights violations	**YES**	Since the end of the previous regime, both national and international human rights groups have been active
6	Publish and educate in ethnic language	yes	Despite the overthrow of the Ceausescu regime, the Hungarian minority (10% of population) and Gypsies are subject, in practice, to language and cultural discrimination

FREEDOM FROM:		COMMENTS
7	Serfdom, slavery, forced or child labor	**YES**	Rights respected
8	Extrajudicial killings or "disappearances"	yes	Security forces, involved in clashes between Romanians and the Hungarian minority, have been accused of indiscriminate shooting, resulting in numerous deaths
9	Torture or coercion by the state	**no**	The transition from an oppressive regime to one moving toward democracy has meant abuses by a minority of police and security forces still influenced by past practices
10	Compulsory work permits or conscription of labor	**YES**	Rights respected
11	Capital punishment by the state	**YES**	Abolished 1989
12	Court sentences of corporal punishment	**YES**	Rights respected
13	Indefinite detention without charge	**YES**	Rights respected
14	Compulsory membership of state organizations or parties	**YES**	Rights respected
15	Compulsory religion or state ideology in schools	**YES**	Rights respected
16	Deliberate state policies to control artistic works	**YES**	Rights respected
17	Political censorship of press	yes	Proposed new press law, intended to establish freedom, will still inhibit criticism of certain major institutions, including the president
18	Censorship of mail or telephone tapping	yes	Most of the previous surveillance practices have disappeared. Occasional abuses by former Securitate officials where still employed

FREEDOM FOR OR RIGHTS TO:		COMMENTS
19	Peaceful political opposition	yes	Many allegations of intimidation. Governing party accused of unfair practices. New constitution, if adopted, is expected to clarify situation
20	Multiparty elections by secret and universal ballot	yes	First multiparty elections in over 40 years. Many charges of intimidation and fraud, particularly by former Communist officials retaining some authority
21	Political and legal equality for women	yes	Underrepresented in politics and professions
22	Social and economic equality for women	yes	Despite equality in law, women used as cheap, unskilled labor

	FREEDOM FOR OR RIGHTS TO:		COMMENTS
23	Social and economic equality for ethnic minorities	yes	But Gypsies suffer discrimination and social hostility, the traditional behavior toward this minority. The position of the Hungarian minority has improved though complaints are still expressed
24	Independent newspapers	yes	Independent newspapers subject to violent attacks by political opponents. Supply of newsprint vulnerable to controls by government
25	Independent book publishing	yes	Former state-owned companies enjoy certain near-monopoly advantages not available to smaller publishers
26	Independent radio and television networks	yes	Although still state owned and controlled, independent stations are being established
27	All courts to total independence	yes	The judicial system is to be reformed under the new constitution
28	Independent trade unions	yes	In the interim period before new union legislation is introduced, affiliations with international bodies are being established – with guidance from the International Labor Organization

	LEGAL RIGHTS:		COMMENTS
29	From deprivation of nationality	YES	Rights respected
30	To be considered innocent until proved guilty	yes	The new judicial system not yet totally established, permitting abuses by a police force trained under the previous regime
31	To free legal aid when necessary and counsel of own choice	YES	Rights respected
32	From civilian trials in secret	yes	Government retains right to try civilians in secret
33	To be brought promptly before a judge or court	no	Previous Code of Criminal Procedure, permitting a 60-day preliminary investigative period, still retained
34	From police searches of home without a warrant	yes	Occasional abuses as police and possibly ex-Securitate officers adapt to the freer society
35	From arbitrary seizure of personal property	YES	Rights respected

	PERSONAL RIGHTS:		COMMENTS
36	To interracial, interreligious, or civil marriage	YES	Rights respected

PERSONAL RIGHTS:		COMMENTS
37	Equality of sexes during marriage and for divorce proceedings	**YES**	Rights respected
38	To practice any religion	**YES**	Rights respected
39	To use contraceptive pills and devices	**YES**	Position transformed. Contraception was banned under the Ceausescu regime
40	To noninterference by state in strictly private affairs	yes	With a democratic constitution currently being drafted, some abuses by officials – usually anonymous telephone threats by ex-Securitate officials still employed by the state

Rwanda

Human rights rating: 48%

YES 8 yes 13 no 13 NO 6

Population: 7,200,000
Life expectancy: 49.5
Infant mortality (0–5 years)
  per 1,000 births: 201
United Nations covenants ratified:
Civil and Political Rights, Economic,
Social and Cultural Rights, Convention
on Equality for Women

Form of government: One-party system
GNP per capita (US$): 320
% of GNP spent by state:
On health: 0.6
On military: 1.9
On education: 3.2

FACTORS AFFECTING HUMAN RIGHTS

A one-party state with a military leader who has been in power nearly 20 years. An ethnic minority, the Tutsi, 10% of the population, which has controlled the country in the past, rebelled in 1990 against the central government. The initial invasion came from Tutsi forces in neighboring Uganda. Its defeat after 2 months of fighting was followed by summary executions, torture, detentions, and increased surveillance by the Rwandan government. A June 1991 referendum voted for a multiparty system, and a political opposition is being formed. Apart from human rights violations in most areas, the Tutsi claim that most of them have fled the country and are banned from returning.

	FREEDOM TO:		COMMENTS
1	Travel in own country	no	Movement controlled by compulsory residence and work registration. Countrywide police checks
2	Travel outside own country	YES	Rights respected
3	Peacefully associate and assemble	NO	Nothing that conflicts with government policies. More stringent controls after the abortive invasion by Rwandan Tutsi insurgents. Periodic curfews. Jehovah's Witnesses imprisoned for holding meetings
4	Teach ideas and receive information	no	Criticism of government may mean imprisonment. Brief closure of university for political reasons. Tutsi minority, 10% of the population, subject to quotas in education
5	Monitor human rights violations	yes	International pressures have improved freedom to monitor
6	Publish and educate in ethnic language	yes	Government policies favor the Hutu majority, 90% of the population

FREEDOM FROM:		COMMENTS
7	Serfdom, slavery, forced or child labor	yes	Regional pattern of child labor, particularly in agriculture
8	Extrajudicial killings or "disappearances"	NO	Position worsened after the Tutsi insurgents attempted to overthrow the government. Some summary executions as well as an estimated 200 deaths in military operations
9	Torture or coercion by the state	NO	Beatings during interrogation of prisoners following the Tutsi insurgency. Some deaths from maltreatment. Torture increased with threat to government
10	Compulsory work permits or conscription of labor	yes	A national development program may call on citizens for specific projects
11	Capital punishment by the state	NO	By firing squad. Many crimes, including sorcery and rape. Until the 1991 Tutsi insurgency, the policy had been to commute death sentences
12	Court sentences of corporal punishment	YES	Rights respected
13	Indefinite detention without charge	NO	A small number of political detainees has been augmented by an estimated 2,000 prisoners from the Tutsi insurgency attempt
14	Compulsory membership of state organizations or parties	YES	With the change to a multiparty system, compulsory membership of the ruling party no longer applies
15	Compulsory religion or state ideology in schools	YES	Rights respected
16	Deliberate state policies to control artistic works	yes	Public exhibitions of creative works reflect a prudent self-censorship
17	Political censorship of press	no	The constitutional freedom of the press becomes, in practice, strict censorship. Some arrests of journalists
18	Censorship of mail or telephone tapping	no	Surveillance increased during and after the Tutsi insurgency

FREEDOM FOR OR RIGHTS TO:		COMMENTS
19	Peaceful political opposition	yes	A June 1991 referendum voted for a multiparty system. Political opposition being organized
20	Multiparty elections by secret and universal ballot	yes	As question 19. Free elections being planned
21	Political and legal equality for women	no	Discrimination and traditional inequalities in all areas and at all levels
22	Social and economic equality for women	no	Major differences in pay and employment, except in agricultural labor, which is dominated by women

	FREEDOM FOR OR RIGHTS TO:		COMMENTS
23	Social and economic equality for ethnic minorities	**YES**	Rights respected but the Tutsi minority complain of discrimination in government services – though enjoying dominance in commerce
24	Independent newspapers	no	No daily newspaper. The major weekly is state owned
25	Independent book publishing	no	Limited book publishing affected by censorship and the need for prudence
26	Independent radio and television networks	**NO**	State owned and controlled
27	All courts to total independence	no	Government pressures in political and subversion cases. Outcome frequently prejudged
28	Independent trade unions	yes	The new national political charter permits freedom for a union movement still at the stage of development

	LEGAL RIGHTS:		COMMENTS
29	From deprivation of nationality	no	Situation confused because of the large numbers of Tutsis taking refuge in Uganda. Exiles may be targets for government-sanctioned assassination
30	To be considered innocent until proved guilty	yes	In view of inadequate prison system and close control of movements, courts grant provisional release to most defendants awaiting trial or sentence
31	To free legal aid when necessary and counsel of own choice	no	System underfunded and inadequate. Situation varies from court to court
32	From civilian trials in secret	yes	Periodic states of emergency permit rare secret trials
33	To be brought promptly before a judge or court	yes	Within 48 hours but lengthy delays if suspected crime is regarded as "political"
34	From police searches of home without a warrant	no	Despite legal requirements, arbitrary searches. Justified by "security" and "document checking"
35	From arbitrary seizure of personal property	yes	Some unlawful seizures by police, particularly from Tutsi homes during periods of ethnic conflicts

	PERSONAL RIGHTS:		COMMENTS
36	To interracial, interreligious, or civil marriage	**YES**	Rights respected. Males may marry at 18 years, females at 15

PERSONAL RIGHTS:		COMMENTS
37	Equality of sexes during marriage and for divorce proceedings	**no**	The new family code, protecting the rights of wives, unlikely in practice to change traditional inequalities – all to the advantage of husbands
38	To practice any religion	yes	Restrictions on Jehovah's Witnesses, sometimes seen as "subversive" by the government
39	To use contraceptive pills and devices	**YES**	Rights respected
40	To noninterference by state in strictly private affairs	**YES**	Rights respected

Saudi Arabia

Human rights rating: 29%

YES 5 yes 5 no 8 NO 22

Population: 14,100,000
Life expectancy: 64.5
Infant mortality (0–5 years)
  per 1,000 births: 95
United Nations covenants ratified:
None

Form of government: Islamic kingdom
GNP per capita (US$): 6,200
% of GNP spent by state:
On health: 4.0
On military: 22.7
On education: 10.6

FACTORS AFFECTING HUMAN RIGHTS

The king is an absolute ruler. In the absence of a written constitution, elected assemblies, or secular law, the country scrupulously adheres to the Koran and Shari'a law. Although a member of the United Nations, the country regards the human rights instruments of the organization as conflicting with its beliefs and contrary to the absolute correctness of the Islamic faith. The recent gulf war, with the presence of a large Western military force to help defend the world's major oil-producing country, temporarily introduced attitudes and values alien to the more orthodox Muslims. With the departure of the Western presence, conflicting trends are evident. On the one hand there is a reaction to reassert conservative values, on the other a desire – necessarily limited – for a degree of democracy.

FREEDOM TO:		COMMENTS
1 Travel in own country	yes	Women forbidden to drive cars and to travel alone
2 Travel outside own country	no	Exit visa required. A woman must have husband's or male guardian's permission before applying for a visa
3 Peacefully associate and assemble	NO	Public criticism of the royal family, religion, or the authorities may incur prison sentence
4 Teach ideas and receive information	NO	All ideas hostile to Islam forbidden. Wide network of informers. The religious police pursue all those guilty of un-Islamic behavior. No Freud, Marx, etc.
5 Monitor human rights violations	NO	No Saudi organizations. Requests from international monitors are occasionally acknowledged but never agreed to
6 Publish and educate in ethnic language	YES	Rights respected

FREEDOM FROM:		**COMMENTS**
7	Serfdom, slavery, forced or child labor	yes	Traditional customs permit child labor. Domestic servants, by their status, have few legal rights
8	Extrajudicial killings or "disappearances"	**YES**	Rights respected
9	Torture or coercion by the state	no	Many abuses by overzealous police. Beatings, general maltreatment, etc.
10	Compulsory work permits or conscription of labor	no	Strict control of labor, most of which is by foreign immigrants
11	Capital punishment by the state	**NO**	Beheading. Sometimes in public. Stoning for sexual crimes. Approximately 20 in 1990
12	Court sentences of corporal punishment	**NO**	Amputations of hand or foot for theft. Strict Shari'a law. Flogging must not exceed 60 lashes per session. Those receiving 1,000 to 1,500 lashes have them spread over a period
13	Indefinite detention without charge	**NO**	Many suffer years of detention, which may only be discovered when accused's family persist with inquiries
14	Compulsory membership of state organizations or parties	**NO**	All Saudis must be Muslim. Compulsion to be faithful to the religion of the state. Apostasy a capital offense
15	Compulsory religion or state ideology in schools	**NO**	Strict Islamic instruction
16	Deliberate state policies to control artistic works	**NO**	Shari'a law limits artistic expression to "inoffensive" subjects. "No part of the female body must ever be displayed." Public cinemas forbidden
17	Political censorship of press	no	Understood guidelines. No criticism of rulers or religion, no mention of taboo subjects such as alcohol, pork, etc.
18	Censorship of mail or telephone tapping	**NO**	Surveillance by Ministry of Islam

FREEDOM FOR OR RIGHTS TO:		**COMMENTS**
19	Peaceful political opposition	**NO**	No political parties permitted to exist
20	Multiparty elections by secret and universal ballot	**NO**	Elections would mean an exercise of free will and free choice and would therefore violate Koranic certainty
21	Political and legal equality for women	**NO**	The slight liberalizing effect of the presence of Western armies in the country during the gulf war has not continued. No place for women in government

FREEDOM FOR OR RIGHTS TO: COMMENTS

22	Social and economic equality for women	**NO**	Women at disadvantage in most areas. The word of two women equals that of one man in Shari'a court. May not drive vehicles; must enter public transport by special entrances, etc.
23	Social and economic equality for ethnic minorities	no	Position worsened by gulf war. Discrimination against those from countries favoring Iraq, although illegal, practiced at all levels. But most obviously against Yemenis
24	Independent newspapers	yes	Independence demands religious, social, and most other areas of conformity
25	Independent book publishing	no	As question 24. Many taboos
26	Independent radio and television networks	**NO**	Owned and controlled by the state. The more comprehensive news service during the gulf war is being reversed
27	All courts to total independence	yes	But the Ministry of Justice can change judges and impose its authority
28	Independent trade unions	**NO**	Illegal but traditions permit worker/employer negotiations

LEGAL RIGHTS: COMMENTS

29	From deprivation of nationality	**YES**	Rights respected
30	To be considered innocent until proved guilty	**YES**	Rights respected
31	To free legal aid when necessary and counsel of own choice	yes	Limited concession of defending counsel for the needy, though within Shari'a rulings
32	From civilian trials in secret	no	Many in secret. Frequent arbitrary powers for both civil and military courts
33	To be brought promptly before a judge or court	**NO**	Long delays, incommunicado detentions, before appearance in Shari'a courts
34	From police searches of home without a warrant	**NO**	Grounds for suspicion permit police searches without warrant. Religious police particularly intrusive
35	From arbitrary seizure of personal property	**YES**	Rights respected

PERSONAL RIGHTS: COMMENTS

36	To interracial, interreligious, or civil marriage	**NO**	Saudis cannot marry foreigners without special permission. No civil or interreligious marriage. Population 99% Muslim

PERSONAL RIGHTS:		COMMENTS
37	Equality of sexes during marriage and for divorce proceedings		Husband may divorce "at will." Under Shari'a law women are subordinate including their inheritance rights
38	To practice any religion		No non-Islamic places of worship. Other religions can only pray in private. Atheism a possible capital offense
39	To use contraceptive pills and devices	no	Little public support. Availability limited. A ban in 1975 not strictly enforced
40	To noninterference by state in strictly private affairs	NO	Raids by the religious police. Law permits execution of adulterers, homosexuals, and apostates, though sentence seldom carried out

Senegal

Human rights rating: 71%

Population: 7,300,000
Life expectancy: 48.3
Infant mortality (0–5 years)
per 1,000 births: 189
United Nations covenants ratified:
Civil and Political Rights, Economic,
Social and Cultural Rights, Convention
on Equality for Women

Form of government: Multiparty
system
GNP per capita (US$): 650
% of GNP spent by state:
On health: 1.1
On military: 2.3
On education: 4.6

FACTORS AFFECTING HUMAN RIGHTS
The multiparty elections of recent years have been marked by allegations of fraud and
irregularities but many of the faults have been corrected by changes to the electoral
code. An independence movement in the Casamance region has involved the military
forces in major operations, and human rights throughout the country have been affected
by the fighting. Among the violations are the destruction of villages suspected of
helping the insurgents, detentions – including journalists and polticians – and
increased torture of prisoners during interrogation.

FREEDOM TO: COMMENTS

1	Travel in own country	**YES**	Rights respected
2	Travel outside own country	**YES**	Rights respected
3	Peacefully associate and assemble	yes	Occasional bannings under pretext of threat to public order may be politically motivated
4	Teach ideas and receive information	**YES**	Rights respected
5	Monitor human rights violations	yes	Less government cooperation with international groups since current Casamance independence unrest began
6	Publish and educate in ethnic language	**YES**	Rights respected

FREEDOM FROM: COMMENTS

7	Serfdom, slavery, forced or child labor	yes	Widespread traditional child labor
8	Extrajudicial killings or "disappearances"		Increasing separatist violence in Casamance area has resulted in a small number of civilian deaths, some by summary executions by the military

FREEDOM FROM: COMMENTS

9	Torture or coercion by the state	no	Many instances of torture of captured Casamance rebels
10	Compulsory work permits or conscription of labor	**YES**	Rights respected
11	Capital punishment by the state	no	Legal but no recent executions
12	Court sentences of corporal punishment	yes	Unruly prison inmates may be whipped as "swift punishment"
13	Indefinite detention without charge	no	Confused situation with regard to large numbers of imprisoned Casamance suspects. Also detentions of journalists and political opponents
14	Compulsory membership of state organizations or parties	**YES**	Rights respected
15	Compulsory religion or state ideology in schools	yes	Some government support of compulsory religious instruction in Muslim schools
16	Deliberate state policies to control artistic works	**YES**	Rights respected
17	Political censorship of press	yes	Laws against criticizing certain state functions and public officials. Support for regionalist movements may lead to imprisonment
18	Censorship of mail or telephone tapping	**YES**	Rights respected

FREEDOM FOR OR RIGHTS TO: COMMENTS

19	Peaceful political opposition	**YES**	Rights respected
20	Multiparty elections by secret and universal ballot	yes	Evidence of electoral fraud and minor infringements favoring governing Socialist party
21	Political and legal equality for women	no	The 80% Muslim majority follow traditional beliefs but women are increasingly involved in politics, particularly students
22	Social and economic equality for women	no	Traditional customs continue to deny equality. Female circumcision and infibulation are not illegal and surveys place practices as high as 20% of women
23	Social and economic equality for ethnic minorities	**YES**	Discrimination, if practiced, is seldom on tribal or racial grounds
24	Independent newspapers	yes	Occasional charges against publishers criticizing military forces, particularly during current Casamance unrest
25	Independent book publishing	**YES**	Rights respected

FREEDOM FOR OR RIGHTS TO: COMMENTS

26	Independent radio and television networks	no	State run and owned. Selective programming but greater readiness to broadcast opposition views
27	All courts to total independence	**YES**	Rights respected
28	Independent trade unions	yes	Major unions are closely associated with governing party

LEGAL RIGHTS: COMMENTS

29	From deprivation of nationality	**YES**	Rights respected
30	To be considered innocent until proved guilty	**YES**	Rights respected
31	To free legal aid when necessary and counsel of own choice	yes	Counsel appointed by court; usually because of illiteracy of defendant
32	From civilian trials in secret	**YES**	Rights respected
33	To be brought promptly before a judge or court	yes	Within 72 hours but instances of prolonged periods in custody
34	From police searches of home without a warrant	**YES**	Rights respected
35	From arbitrary seizure of personal property	**YES**	Rights respected

PERSONAL RIGHTS: COMMENTS

36	To interracial, interreligious, or civil marriage	**YES**	Younger generation choosing civil marriage though older people adhere to Shari'a law. Legal age 20 for males, 16 for females
37	Equality of sexes during marriage and for divorce proceedings	no	Traditional and religious influences still predominate – to husband's advantage. But less compliance by younger generation
38	To practice any religion	**YES**	Rights respected
39	To use contraceptive pills and devices	**YES**	Government support
40	To noninterference by state in strictly private affairs	**YES**	Rights respected

Sierra Leone

Human rights rating: 67%

Population: 4,200,000
Life expectancy: 42.0
Infant mortality (0–5 years)
    per 1,000 births: 261
United Nations covenants ratified:
Convention on Equality for Women

Form of government: One-party system
GNP per capita (US$): 300
% of GNP spent by state:
On health: 0.7
On military: 1.2
On education: 3.0

FACTORS AFFECTING HUMAN RIGHTS

The country has had a one-party system since 1978 although current changes throughout Africa have forced the government to permit a vote for a multiparty system. A political opposition is now being formed. The country is one of the poorest in the world, though it has rich resources that remain undeveloped or corruptly managed, and there are frequent violent demonstrations against poverty and social misery. The need for an improvement in the standard of living, rather than government respect for civil and political rights, is regarded by the people as the most urgent priority.

FREEDOM TO: COMMENTS

1	Travel in own country	**YES**	Rights respected
2	Travel outside own country	**YES**	Rights respected
3	Peacefully associate and assemble	no	Only demonstrations supporting the governing party are permitted
4	Teach ideas and receive information	yes	Relative freedom for academics and students but degree of prudence advisable. Student riots in 1990 resulted in brief closures of universities
5	Monitor human rights violations	**YES**	Rights respected
6	Publish and educate in ethnic language	**YES**	Rights respected

---

FREEDOM FROM: COMMENTS

7	Serfdom, slavery, forced or child labor	yes	Regional pattern of child labor
8	Extrajudicial killings or "disappearances"	yes	Overzealous police action against unarmed demonstrators resulted in deaths. Also deaths in prison from maltreatment
9	Torture or coercion by the state	yes	Local abuses. Police and prison wardens frequently ill-disciplined

FREEDOM FROM:		COMMENTS
10	Compulsory work permits or conscription of labor	**YES**	Rights respected
11	Capital punishment by the state	**NO**	Death sentences continue to be imposed but no information on executions
12	Court sentences of corporal punishment	**YES**	Rights respected
13	Indefinite detention without charge	no	President has authority to order indefinite detention under the Public Emergency Act
14	Compulsory membership of state organizations or parties	**YES**	Rights respected
15	Compulsory religion or state ideology in schools	**YES**	Rights respected
16	Deliberate state policies to control artistic works	**YES**	Rights respected
17	Political censorship of press	yes	No direct censorship. Occasional imprisonment of editors and journalists. In one instance judge stated sentence was "a lesson to others"
18	Censorship of mail or telephone tapping	**YES**	Rights respected

FREEDOM FOR OR RIGHTS TO:		COMMENTS
19	Peaceful political opposition	no	The vote for a multiparty constitution in August 1991 may be followed by peaceful political opposition
20	Multiparty elections by secret and universal ballot	**NO**	If multiparty elections proceed from the intention to change the constitution, 13 years of absolute power by the president and the All People's party will come to an end
21	Political and legal equality for women	no	Increasing influence, particularly in tribal life. But little equality at senior government level
22	Social and economic equality for women	no	Economy male dominated. Most apparent in indigenous tribal areas. Circumcision of girls a common practice – and legal
23	Social and economic equality for ethnic minorities	yes	Smaller tribes complain of discrimination favoring the Mende and Temne tribes which form 60% of population
24	Independent newspapers	yes	A number of small independent newspapers but understood guidelines followed, particularly in their respect for the president. Occasional seizures
25	Independent book publishing	yes	Must avoid controversial matters. Usually only commercial considerations

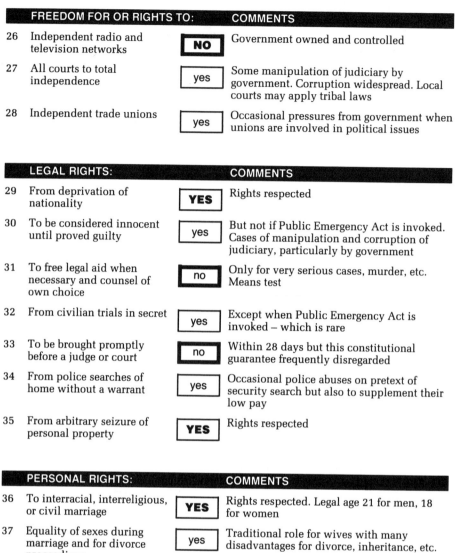

FREEDOM FOR OR RIGHTS TO: COMMENTS

26 Independent radio and   **NO**   Government owned and controlled
    television networks

27 All courts to total   yes   Some manipulation of judiciary by
    independence   government. Corruption widespread. Local
                                           courts may apply tribal laws

28 Independent trade unions   yes   Occasional pressures from government when
                                           unions are involved in political issues

LEGAL RIGHTS: COMMENTS

29 From deprivation of   **YES**   Rights respected
    nationality

30 To be considered innocent   yes   But not if Public Emergency Act is invoked.
    until proved guilty   Cases of manipulation and corruption of
                                           judiciary, particularly by government

31 To free legal aid when   no   Only for very serious cases, murder, etc.
    necessary and counsel of   Means test
    own choice

32 From civilian trials in secret yes   Except when Public Emergency Act is
                                           invoked – which is rare

33 To be brought promptly   no   Within 28 days but this constitutional
    before a judge or court   guarantee frequently disregarded

34 From police searches of   yes   Occasional police abuses on pretext of
    home without a warrant   security search but also to supplement their
                                           low pay

35 From arbitrary seizure of   **YES**   Rights respected
    personal property

PERSONAL RIGHTS: COMMENTS

36 To interracial, interreligious, **YES** Rights respected. Legal age 21 for men, 18
    or civil marriage   for women

37 Equality of sexes during   yes   Traditional role for wives with many
    marriage and for divorce   disadvantages for divorce, inheritance, etc.
    proceedings

38 To practice any religion   **YES**   Rights respected

39 To use contraceptive pills   **YES**   Limited government support
    and devices

40 To noninterference by state yes   Rights respected but occasional security
    in strictly private affairs   searches may be unjustified – except for
                                           corrupt motives

Singapore

Human rights rating: 60%

YES 9 **yes** 17 **no** 11 NO 3

Population: 2,700,000
Life expectancy: 74.0
Infant mortality (0–5 years)
  per 1,000 births: 12
United Nations covenants ratified: None

Form of government: Parliamentary system (nominal)
GNP per capita (US$): 9,070
% of GNP spent by state:
On health: 1.3
On military: 5.5
On education: 5.0

FACTORS AFFECTING HUMAN RIGHTS

The People's Action party has maintained power since the end of British colonial rule in 1959. An authoritarian prime minister relinquished his position after 25 years but retains the title of senior minister. The administration justifies its firm management of society and the economy, enforced by a comprehensive Internal Security Act, by the fact that Singapore is one of the most prosperous countries in Asia, there is little social unrest, and elections confirm the people's choice of a managed system. The party holds over 95% of parliamentary seats, judges are indirectly appointed by the prime minister, under the Internal Security Act political opponents are sometimes detained without charge, and the press practices "self-censorship."

	FREEDOM TO:		COMMENTS
1	Travel in own country	**YES**	Change of address to be reported
2	Travel outside own country	yes	Occasional refusal of passports to former detainees. Government has wide discretion
3	Peacefully associate and assemble	no	Permits denied "in the public interest" for many protest demonstrations
4	Teach ideas and receive information	yes	Academics follow understood guidelines if they seek career advantages. Students require "suitability" certificates for university enrollment, a form of "loyalty" test
5	Monitor human rights violations	no	Some clandestine – or prudent – forms of monitoring but government opposes them as a challenge to its authority. Visits by international bodies obstructed by "visa" requirements
6	Publish and educate in ethnic language	**YES**	Rights respected

FREEDOM FROM:		COMMENTS
7	Serfdom, slavery, forced or child labor	**YES**	Rights respected
8	Extrajudicial killings or "disappearances"	**YES**	Rights respected
9	Torture or coercion by the state	yes	Well-documented cases of mistreatment and psychological coercion under the Internal Security Act
10	Compulsory work permits or conscription of labor	**YES**	Rights respected
11	Capital punishment by the state	**NO**	Many. By hanging – and not always reported. Usually for drug trafficking
12	Court sentences of corporal punishment	no	Caning. Maximum 40 strokes. Frequently administered with prison sentence
13	Indefinite detention without charge	no	President may order 2 years' detention under the Internal Security Act and have it renewed indefinitely. Usually for security or subversion offences
14	Compulsory membership of state organizations or parties	**YES**	Rights respected
15	Compulsory religion or state ideology in schools	yes	Forms of pro-government indoctrination in schools. This includes traditional Confucianism
16	Deliberate state policies to control artistic works	yes	Less puritan policy toward art forms which promote the offenses of "permissiveness," disrespect for authority, and overt sexuality
17	Political censorship of press	no	Self-censorship on understood sensitive issues. Foreign newspapers frequently restricted. Position worsened during 1990
18	Censorship of mail or telephone tapping	no	Wide surveillance, particularly of outspoken opponents of the government and ruling party

FREEDOM FOR OR RIGHTS TO:		COMMENTS
19	Peaceful political opposition	yes	The government party regularly wins at least 95% of seats in Parliament. The presence of two or three other members has provoked threats and a variety of disadvantages
20	Multiparty elections by secret and universal ballot	yes	In reality, after 39 years of unbroken power, the government considers elections to be periodic confirmation of national support
21	Political and legal equality for women	yes	Women's presence increasing in professions though limited representation in politics
22	Social and economic equality for women	yes	Minor pay and employment inequalities

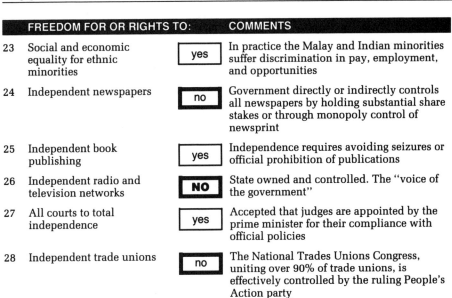

	FREEDOM FOR OR RIGHTS TO:		COMMENTS
23	Social and economic equality for ethnic minorities	yes	In practice the Malay and Indian minorities suffer discrimination in pay, employment, and opportunities
24	Independent newspapers	no	Government directly or indirectly controls all newspapers by holding substantial share stakes or through monopoly control of newsprint
25	Independent book publishing	yes	Independence requires avoiding seizures or official prohibition of publications
26	Independent radio and television networks	**NO**	State owned and controlled. The "voice of the government"
27	All courts to total independence	yes	Accepted that judges are appointed by the prime minister for their compliance with official policies
28	Independent trade unions	no	The National Trades Unions Congress, uniting over 90% of trade unions, is effectively controlled by the ruling People's Action party

	LEGAL RIGHTS:		COMMENTS
29	From deprivation of nationality	yes	Rare instances of government opponents residing overseas losing citizenship rights – a form of exile
30	To be considered innocent until proved guilty	no	As Internal Security Act detentions do not come under judicial review, arbitrary arrests encouraged
31	To free legal aid when necessary and counsel of own choice	no	Free aid only for capital offenses. No personal choice of counsel
32	From civilian trials in secret	**NO**	Those detained under the Internal Security Act may be tried in secret. Wide definition of security charges, which may be outside judicial review
33	To be brought promptly before a judge or court	yes	Normally within 48 hours but delays may occur at the discretion of the authorities
34	From police searches of home without a warrant	yes	Permitted under criminal law but usually in drives against narcotics dealers
35	From arbitrary seizure of personal property	**YES**	Rights respected

	PERSONAL RIGHTS:		COMMENTS
36	To interracial, interreligious, or civil marriage	**YES**	Rights respected

PERSONAL RIGHTS:		COMMENTS
37	Equality of sexes during marriage and for divorce proceedings	yes	Confucian ideals, encouraged by the government, perpetuate a traditional role of female subservience. Mostly of the older generation
38	To practice any religion	yes	1990 Religious Harmony Law provides for 2 years' imprisonment or fines up to US$10,000 for causing racial tensions, including opposition to "secular government policies"
39	To use contraceptive pills and devices	YES	From time to time population programs introduced to improve "quality" of future generations
40	To noninterference by state in strictly private affairs	no	Government policies include compulsory location of some families and intrusion into family planning. The degree of social conformity demanded by the state may be seen as a form of interference in private affairs

South Africa

Human rights rating: 50%

YES 8 **yes** 13 **no** 15 NO 4

Population: 35,300,000
Life expectancy: 61.7
Infant mortality (0–5 years)
  per 1,000 births: 91
United Nations covenants ratified:
None
Form of government: Minority
parliamentary republic

GNP per capita (US$): 2,290
% of GNP spent by state:
On health: 0.6
On military: 3.9
On education: 4.6
(Some of the statistics above do not
convey the major differences between
the white and black sections of the
population)

FACTORS AFFECTING HUMAN RIGHTS
The changes at present taking place in South Africa are of historic and universal
significance. The white minority, which has enjoyed social and economic supremacy
since South Africa was first settled by Europeans, has accepted the inevitability of
greater political equality for the black majority (75% of the population) and eventual
government by that majority. The present negotiations are protracted but making
progress, previously banned political parties are no longer restricted, and most of the
acts perpetuating racial discrimination have been repealed. Violence, however, remains
widespread, not only between security forces notorious for past human rights crimes
and blacks, but between the major black opposition party and the mainly Zulu Inkatha
movement.

	FREEDOM TO:		COMMENTS
1	Travel in own country	yes	Minor restrictions affect citizens of black "homelands" who may regard themselves as South African
2	Travel outside own country	yes	Previous travel restrictions of both black and white opponents of the government now lifted. But passports may still be refused or revoked
3	Peacefully associate and assemble	no	Violent clashes, sometimes between opposing groups of armed demonstrators, occur frequently, usually in or near black townships. Police may overreact, adding to the high number of deaths. Situation affects granting of permits
4	Teach ideas and receive information	no	Blacks limited in the education and information they receive because of the disproportionately small share of the national expenditure

FREEDOM TO:		COMMENTS
5	Monitor human rights violations	yes	Situation improved. Monitors report government cooperation. However, the extent of periodic social and racial violence may invoke emergency acts which limit "on-the-spot" monitoring
6	Publish and educate in ethnic language	**YES**	Rights respected

FREEDOM FROM:		COMMENTS
7	Serfdom, slavery, forced or child labor	no	Regional pattern of child labor. Parole system for convicts provides cheap labor for white farmers. Continuing reports of Mozambican refugees accepting "slave" conditions on neighboring South African farms
8	Extrajudicial killings or "disappearances"	no	Government commission confirms killings and bombings by security forces and their "hit squads" over a number of years. But official measures to eliminate killings have not prevented the occasional "unexplained" death
9	Torture or coercion by the state	no	Despite official attempts to reduce torture practices, and bring culprits to trial, state violence remains a reality during interrogations
10	Compulsory work permits or conscription of labor	**YES**	Rights respected
11	Capital punishment by the state	**NO**	Abolition under consideration. No early decision expected. Many still awaiting execution. In "homelands," 2 recent executions
12	Court sentences of corporal punishment	no	15 lashes (legal limit). A frequent punishment
13	Indefinite detention without charge	no	Amendment to Internal Security Act in June 1991 permits indefinite detention for interrogation. In 10-day renewable periods – and incommunicado
14	Compulsory membership of state organizations or parties	**YES**	Rights respected
15	Compulsory religion or state ideology in schools	**YES**	Rights respected
16	Deliberate state policies to control artistic works	yes	The continuing social and economic disadvantages suffered by blacks limit, to the point of controlling, the form and the volume of their artistic works

FREEDOM FROM:		COMMENTS
17	Political censorship of press — **yes**	Despite the lifting of many restrictive laws, others remain under various acts. Occasional prosecutions. Journalists harassed at local level by police and political extremists. Sometimes violently
18	Censorship of mail or telephone tapping — **no**	Surveillance – though in a less-conspicuous manner now that the press has greater freedom

FREEDOM FOR OR RIGHTS TO:		COMMENTS
19	Peaceful political opposition — **no**	The process of involving blacks – the great majority of the population – in the political life of the country is taking a slow and erratic course. And characterized by white violence against white, white against black, and black against black
20	Multiparty elections by secret and universal ballot — **NO**	The more enlightened policies of the current white government have gone as far as discussing the involvement of blacks in government. Previously banned parties now permitted
21	Political and legal equality for women — **no**	Inequalities cross divisions of race. African women also have to contend with traditional and customary discrimination
22	Social and economic equality for women — **no**	As question 21
23	Social and economic equality for ethnic minorities — **NO**	Whatever the recent advances in the government's attempts to create greater equality, the reality remains endemic disadvantages for all nonwhite citizens
24	Independent newspapers — **YES**	Restrictions withdrawn 1990
25	Independent book publishing — **yes**	The Publications Board, an appeal body, is following the more liberal policies of the government and those being adopted socially
26	Independent radio and television networks — **NO**	State owned and controlled. Tentative progress toward a more balanced presentation of news programs. Task force established to study restructuring
27	All courts to total independence — **yes**	Court's independence may be influenced by the ethnic, social, and political character of the judiciary
28	Independent trade unions — **yes**	Black unions increasing membership now that most restrictions lifted. Public employees forbidden to strike. Government relations with International Labor Organization improving

LEGAL RIGHTS: COMMENTS

| 29 | From deprivation of nationality | no | Situation complex due to policy toward the designated "homelands," which may mean forced "emigration" of South African blacks |

| 30 | To be considered innocent until proved guilty | no | Many arrests in designated "unrest areas" where security forces may abuse their arbitrary powers |

| 31 | To free legal aid when necessary and counsel of own choice | no | Counsel appointed for most serious cases (murder and similar). Most defendants are black and denied legal help for lesser offenses |

| 32 | From civilian trials in secret | yes | Reserve powers for secret trials – if in the interests of "security" |

| 33 | To be brought promptly before a judge or court | no | The extent of security trials and ordinary criminal proceedings makes delays and postponements unavoidable |

| 34 | From police searches of home without a warrant | yes | The repeal in 1990 of the Separate Amenities Act, which entrenched "white privilege" throughout the country, has removed the pretext for frequent police "swoops." Nevertheless, black homes may still suffer intimidating searches |

| 35 | From arbitrary seizure of personal property | yes | But summary destruction of "shantytowns" as part of government resettlement plans may, in practice, mean deprivation of material possessions |

PERSONAL RIGHTS: COMMENTS

| 36 | To interracial, interreligious, or civil marriage | YES | Rights respected |

| 37 | Equality of sexes during marriage and for divorce proceedings | yes | Multiracial women's organizations are actively seeking to correct inequalities – some of which affect wives of all races |

| 38 | To practice any religion | YES | Rights respected |

| 39 | To use contraceptive pills and devices | YES | All races |

| 40 | To noninterference by state in strictly private affairs | yes | The ending of the "apartheid" policy has not meant the end of minor official harassments and social threats which affect private lives. Particularly in traditionally Afrikaans areas |

Spain

Human rights rating: 87%

YES 27 yes 13 no 0 NO 0

Population: 39,000,000
Life expectancy: 77.0
Infant mortality (0–5 years)
   per 1,000 births: 12
United Nations covenants ratified:
Civil and Political Rights, Economic,
Social and Cultural Rights, Convention
on Equality for Women

Form of government: Parliamentary
monarchy
GNP per capita (US$): 7,740
% of GNP spent by state:
On health: 4.3
On military: 2.3
On education: 3.2

FACTORS AFFECTING HUMAN RIGHTS

Since the end of Fascist rule in 1975, Spain has taken its place with the multiparty democratic countries of Western Europe. The government is answerable to the people, the media are relatively free and independent, and membership in the European Community has markedly increased the country's prosperity. The major area of human rights concern relates to the ETA terrorist movement which seeks a separate Basque fatherland. In response to assassinations and bombings, nearly 100 in 1990, the security forces have been active and vigilant. They have also been accused of "planned killings" of suspected terrorists and comparable arbitrary actions. A few military officers have been charged with these offenses.

	FREEDOM TO:		COMMENTS
1	Travel in own country	YES	Rights respected
2	Travel outside own country	YES	Rights respected
3	Peacefully associate and assemble	YES	Rights respected
4	Teach ideas and receive information	YES	No pro-terrorist sentiments to be propagated
5	Monitor human rights violations	YES	Rights respected
6	Publish and educate in ethnic language	YES	Rights respected

	FREEDOM FROM:		COMMENTS
7	Serfdom, slavery, forced or child labor	yes	Child labor in remote rural areas

FREEDOM FROM:		COMMENTS
8	Extrajudicial killings or "disappearances"	yes	The continuing threat from violent terrorist groups – ETA and GRAPO – has resulted in many deaths, including those of security forces. But many allegations of "planned killings" by security forces
9	Torture or coercion by the state	yes	Serious local abuses of suspected terrorists by overzealous police and military
10	Compulsory work permits or conscription of labor	YES	Rights respected
11	Capital punishment by the state	YES	Abolished 1978
12	Court sentences of corporal punishment	YES	Rights respected
13	Indefinite detention without charge	yes	2 years' detention for terrorist suspects being investigated may be renewed for similar period
14	Compulsory membership of state organizations or parties	YES	Rights respected
15	Compulsory religion or state ideology in schools	yes	Strict Roman Catholic instruction in the many church schools
16	Deliberate state policies to control artistic works	YES	Rights respected
17	Political censorship of press	yes	Allegations of harassment of journalists regarded as biased in reporting of terrorism; no "insulting the king"
18	Censorship of mail or telephone tapping	yes	Surveillance widespread because of the threat from terrorist factions

FREEDOM FOR OR RIGHTS TO:		COMMENTS
19	Peaceful political opposition	YES	Rights respected
20	Multiparty elections by secret and universal ballot	YES	Rights respected
21	Political and legal equality for women	yes	Women's representation at highest levels improving but still limited
22	Social and economic equality for women	yes	Many instances of pay, employment, and opportunity inequalities
23	Social and economic equality for ethnic minorities	yes	But Gypsies, 3% of population, despite constitutional guarantees, suffer discrimination in all areas
24	Independent newspapers	YES	Rights respected
25	Independent book publishing	yes	Independence does not include the right to show "disrespect" toward the king, church, judiciary, and military

FREEDOM FOR OR RIGHTS TO:		COMMENTS
26	Independent radio and television networks	**YES**	Both government and private stations. Occasional official "guidance" on security and terrorist matters
27	All courts to total independence	**YES**	Rights respected
28	Independent trade unions	**YES**	Rights respected

LEGAL RIGHTS:		COMMENTS
29	From deprivation of nationality	**YES**	Rights respected
30	To be considered innocent until proved guilty	yes	Frequent allegations of abuses by overzealous security forces in pursuit of terrorists
31	To free legal aid when necessary and counsel of own choice	**YES**	Rights respected
32	From civilian trials in secret	**YES**	Rights respected
33	To be brought promptly before a judge or court	**YES**	Within 72 hours except for captured terrorists
34	From police searches of home without a warrant	**YES**	Except in pursuit of terrorists, usually in Basque region
35	From arbitrary seizure of personal property	**YES**	Rights respected

PERSONAL RIGHTS:		COMMENTS
36	To interracial, interreligious, or civil marriage	**YES**	Marriage legal at age 18
37	Equality of sexes during marriage and for divorce proceedings	yes	Despite constitutional guarantees, in practice the wife's role is frequently influenced by tradition and customs
38	To practice any religion	**YES**	Rights respected
39	To use contraceptive pills and devices	**YES**	Government support
40	To noninterference by state in strictly private affairs	**YES**	Rights respected

| Sri Lanka | Human rights rating: | 47% |

YES 10 yes 12 no 10 NO 8

Population: 17,200,000
Life expectancy: 70.9
Infant mortality (0–5 years) per 1,000 births: 36
United Nations covenants ratified: Civil and Political Rights, Economic, Social and Cultural Rights, Convention on Equality for Women

Form of government: Multiparty with executive presidency
GNP per capita (US$): 420
% of GNP spent by state:
On health: 1.7
On military: 5.7
On education: 2.4

FACTORS AFFECTING HUMAN RIGHTS

Attempts to mediate in the civil war between the Sinhalese majority (74%) and the Tamil minority (17%) have failed to bring to an end a confrontation that started in the mid-1980s and has resulted in many thousands of deaths each year, not only of the armed factions but of civilians of all ethnic communities. The forces of the multiparty government have failed to reoccupy the northeast peninsula of the island, which remains under the control of the most militant of the Tamil separatists. Further killings, numbering many thousands, have been committed by an armed Sinhalese nationalist group. In the circumstances, respect for human rights is not a government priority, and various emergency regulations give it virtually unlimited powers.

	FREEDOM TO:		COMMENTS
1	Travel in own country	no	The degree of control of the northern peninsula by Tamil separatist insurgents prevents travel in that area. Local curfews, travel permits, searches, etc.
2	Travel outside own country	YES	Rights usually respected but many Tamils choose to reach India by clandestine routes
3	Peacefully associate and assemble	no	The extent of violence and constant ethnic confrontation limits public meetings, usually under emergency regulations
4	Teach ideas and receive information	yes	Subject to restrictions on activities "harmful to the national interest"
5	Monitor human rights violations	yes	Groups active but face restrictions on a few sensitive issues. Nevertheless, considering the extent of violence and lawlessness, the degree of permitted monitoring is commendable
6	Publish and educate in ethnic language	yes	Sinhalese and Tamil official languages with English used as a lingua franca. A few disadvantages for Tamil, 17% of the population

FREEDOM FROM:		COMMENTS
7	Serfdom, slavery, forced or child labor	yes	Certain child labor abuses
8	Extrajudicial killings or "disappearances"	**NO**	A virtual war situation exists between the government and Tamil separatist insurgents with thousands of military deaths in action, and of civilians, human rights activists, and individual protesters. Plus thousands of "disappearances" from all sections
9	Torture or coercion by the state	**NO**	Constant. Usually during interrogation and by all parties. Complaints too numerous for official investigation
10	Compulsory work permits or conscription of labor	**YES**	Rights respected
11	Capital punishment by the state	**NO**	By hanging
12	Court sentences of corporal punishment	no	Caning
13	Indefinite detention without charge	**NO**	Estimates of approximately 15,000. Mostly Tamils. Equally, many thousands of Sinhalese and minority Tamils held in Jaffna peninsula by the ruling insurgents
14	Compulsory membership of state organizations or parties	**YES**	Rights respected
15	Compulsory religion or state ideology in schools	**NO**	Compulsory. Buddhism the religion of the Sinhalese majority, Hinduism that of the Tamils
16	Deliberate state policies to control artistic works	**YES**	Rights respected
17	Political censorship of press	no	Apart from controls under emergency regulations, the government owns major newspapers. Assassinations of editors, journalists, and distributors also inhibit a free press
18	Censorship of mail or telephone tapping	**NO**	Wide surveillance under emergency regulations

FREEDOM FOR OR RIGHTS TO:		COMMENTS
19	Peaceful political opposition	**YES**	Local elections of May 1991, with a 75–80% vote, passed peacefully, though the separatist-held Jaffna area was excluded
20	Multiparty elections by secret and universal ballot	yes	Periodic bannings of small parties accused of dividing the community
21	Political and legal equality for women	yes	Although a woman has served in the past as prime minister, few reach senior government positions. Inequalities in the professions

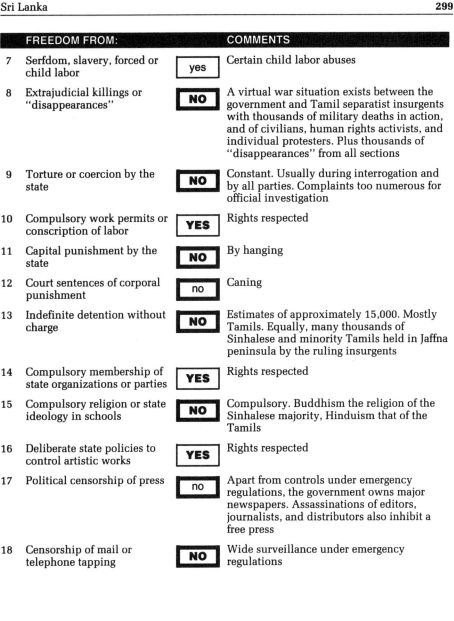

FREEDOM FOR OR RIGHTS TO:		COMMENTS
22	Social and economic equality for women	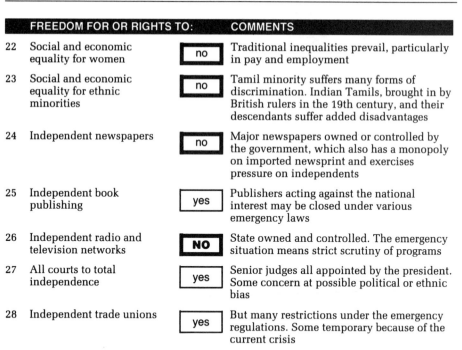 no	Traditional inequalities prevail, particularly in pay and employment
23	Social and economic equality for ethnic minorities	no	Tamil minority suffers many forms of discrimination. Indian Tamils, brought in by British rulers in the 19th century, and their descendants suffer added disadvantages
24	Independent newspapers	no	Major newspapers owned or controlled by the government, which also has a monopoly on imported newsprint and exercises pressure on independents
25	Independent book publishing	yes	Publishers acting against the national interest may be closed under various emergency laws
26	Independent radio and television networks	**NO**	State owned and controlled. The emergency situation means strict scrutiny of programs
27	All courts to total independence	yes	Senior judges all appointed by the president. Some concern at possible political or ethnic bias
28	Independent trade unions	yes	But many restrictions under the emergency regulations. Some temporary because of the current crisis

LEGAL RIGHTS:		COMMENTS
29	From deprivation of nationality	**YES**	Rights respected
30	To be considered innocent until proved guilty	no	Arbitrary actions by all security forces made worse by the extent of violence, suspicions, and ethnic hostility
31	To free legal aid when necessary and counsel of own choice	yes	Means test. Court appoints counsel
32	From civilian trials in secret	no	Permitted under Prevention of Terrorism Act
33	To be brought promptly before a judge or court	no	Normally within 24 hours but possible 90-day delays because of the emergency. Such delays have led to indefinite detentions
34	From police searches of home without a warrant	**NO**	Under emergency regulations, particularly during periods of armed violence, police and military have arbitrary powers
35	From arbitrary seizure of personal property	yes	Situation confused. Despite constitutional guarantees, frequent breakdown of discipline among government forces leads to looting, destruction of property, etc.

PERSONAL RIGHTS:		COMMENTS
36	To interracial, interreligious, or civil marriage	**YES**	Rights respected
37	Equality of sexes during marriage and for divorce proceedings	yes	Traditional and religious factors affect women's equality in both marriage and divorce – to their disadvantage
38	To practice any religion	**YES**	Rights respected
39	To use contraceptive pills and devices	**YES**	Rights respected
40	To noninterference by state in strictly private affairs	**YES**	Rights respected

Sudan

Human rights rating: 18%

YES 1 yes 3 no 15 NO 21

Population: 25,200,000
Life expectancy: 50.8
Infant mortality (0–5 years)
per 1,000 births: 175
United Nations covenants ratified:
Civil and Political Rights, Economic,
Social and Cultural Rights

Form of government: Military council
GNP per capita (US$): 480
% of GNP spent by state:
On health: 0.2
On military: 5.9
On education: 4.0

FACTORS AFFECTING HUMAN RIGHTS

The country suffers most of the misfortunes of mankind. A military council was formed in 1989 after a relatively democratic government was overthrown in a bloodless coup, and a state of emergency, effective for most of the period that has followed, permits widespread killings, torture, detentions, and a judiciary system dominated by Islamic fundamentalists. This regime is challenged by the rebel movement in the south of the country, largely composed of African tribes following Christian or animist beliefs. As well as the civil war in the south, however, the area is frequently affected by famine, responsible for the deaths of tens of thousands, homelessness, and official neglect. The country is a member of the United Nations.

	FREEDOM TO:		COMMENTS
1	Travel in own country	no	The south of country controlled by a rebel liberation movement. Travel also affected by martial law regulations, which apply to many areas
2	Travel outside own country	no	Exit visas – strictly controlled. Denied to political opponents of regime
3	Peacefully associate and assemble	NO	Protest rallies forbidden after the military coup of 1989. No political demonstrations
4	Teach ideas and receive information	NO	Restrictions on academics and students. Hostility toward government may lead to dismissal. University life increasingly dominated by fundamentalists
5	Monitor human rights violations	no	International groups occasionally visit, though programs strictly controlled. Local groups usually operate secretly
6	Publish and educate in ethnic language	yes	New government seeking to enforce wider use of Arabic, particularly in universities

FREEDOM FROM:		COMMENTS
7	Serfdom, slavery, forced or child labor	**NO**	Serfdom and degrees of slavery practiced. Children sold; and to a neighboring Islamic country. Also concubinage
8	Extrajudicial killings or "disappearances"	**NO**	Countercoups have been followed by summary executions, but worst excesses committed by government tribal militia forces against communities, rebels, and tribal enemies in the south
9	Torture or coercion by the state	**NO**	Beatings, electric shocks, semisuffocation, broken bones, extraction of teeth and nails, sleep and food deprivation
10	Compulsory work permits or conscription of labor	yes	Random mobilizing of labor in both north and south of country
11	Capital punishment by the state	**NO**	Hanging, shooting, stoning, and widespread summary executions
12	Court sentences of corporal punishment	**NO**	Many punishments under Shari'a law. Of women as well as men. Courts sentence freely to public floggings
13	Indefinite detention without charge	**NO**	Permitted under State of Emergency Law. No reasons given. No appeal procedures. Estimates range to 1,000 detainees
14	Compulsory membership of state organizations or parties	**YES**	Rights respected
15	Compulsory religion or state ideology in schools	no	The military coup has been followed by stricter Islamic instruction in schools
16	Deliberate state policies to control artistic works	**NO**	With the adoption of form of Islamic fundamentalism by the military council, free expression, the female figure, and many other subjects forbidden
17	Political censorship of press	no	Many newspapers banned after 1989 military coup. Those permitted to appear conform to government and religious directives. Journalists imprisoned
18	Censorship of mail or telephone tapping	**NO**	Constant surveillance, particularly of subversives and the southern war zone

FREEDOM FOR OR RIGHTS TO:		COMMENTS
19	Peaceful political opposition	**NO**	Absolute rule by the Revolution Command Council
20	Multiparty elections by secret and universal ballot	**NO**	No elections. The military regime is supported by an increasingly authoritarian Islamic fundamentalist movement
21	Political and legal equality for women	no	Inequalities at all political and professional levels. Position worsening as government moves toward Islamic fundamentalism

FREEDOM FOR OR RIGHTS TO:		COMMENTS
22	Social and economic equality for women	no	Wide areas of oppression. Major work inequalities, inferior status in many southern tribes, female circumcision and infibulation affecting majority of women, etc.
23	Social and economic equality for ethnic minorities	NO	South of country, an area neglected by the richer north, populated by Africans. Africans and other dark-skinned minorities suffer countrywide discrimination
24	Independent newspapers	NO	Either government controlled or strongly influenced by Islamic fundamentalists
25	Independent book publishing	no	Nothing to be published that conflicts with the policies of the military council or the beliefs of the fundamentalists
26	Independent radio and television networks	NO	State owned and controlled. Recently by Islamic fundamentalist elements within the network
27	All courts to total independence	NO	Wave of Islamic fundamentalism influencing courts' independence. Senior Christian judges dismissed. No southerners on the Supreme Court
28	Independent trade unions	no	New union laws being drafted. Restoration of limited rights after temporary total abolition in 1989. No strikes

LEGAL RIGHTS:		COMMENTS
29	From deprivation of nationality	no	Known opponents living abroad may be refused entry or be denied passports
30	To be considered innocent until proved guilty	NO	Purges of those suspected of supporting southern rebels and of non-Muslims, with or without reasons for suspicion
31	To free legal aid when necessary and counsel of own choice	no	Limited appointment of "friend of the court" as defense counsel. System arbitrary
32	From civilian trials in secret	NO	Of many categories of crime as well as security or military cases
33	To be brought promptly before a judge or court	no	Within 48 hours but with wide disregard of legal rights under emergency or martial laws
34	From police searches of home without a warrant	NO	Arbitrary actions by security or military forces. Becoming worse as Islamic fundamentalism is extended
35	From arbitrary seizure of personal property	NO	Worst of many violations are by tribal militia formed to fight southern Liberation Army. Looting, pillaging, abductions, etc.

PERSONAL RIGHTS: **COMMENTS**

36 To interracial, interreligious,   Over 70% of population Muslim.
    or civil marriage   Intermarriage rare and widely opposed

37 Equality of sexes during   Under country's Islamic law, disadvantages
    marriage and for divorce   for wives include those of inheritance,
    proceedings   polygamy, divorce, husband's permission to
                                                               travel abroad, etc.

38 To practice any religion   Islam, religion of most of the north, receives
                                                               many advantages. Christianity and others
                                                               frequently under government surveillance.
                                                               Limits on places of worship. Some
                                                               harassment

39 To use contraceptive pills   yes   Minimum state involvement. Method may
    and devices   be determined by religion

40 To noninterference by state   no   New psychological measures by government
    in strictly private affairs   to induce obedience into family life. Local
                                                               pressures such as forbidding women to be
                                                               on streets after 7 pm and taxi drivers
                                                               refusing service to women

Sweden

Human rights rating: 98%

YES 37 yes 3 no 0 NO 0

Population: 8,400,000
Life expectancy: 77.4
Infant mortality (0–5 years) per 1,000 births: 7
United Nations covenants ratified: Civil and Political Rights, Economic, Social and Cultural Rights, Convention on Equality for Women

Form of government: Parliamentary monarchy
GNP per capita (US$): 19,300
% of GNP spent by state:
On health: 8.0
On military: 2.9
On education: 7.6

FACTORS AFFECTING HUMAN RIGHTS
The individual respected. Government answerable to the people. A prosperous economy, democratic traditions, a free press, and a cohesive society. The country has also avoided involvement in the European wars of this century.

FREEDOM TO: / COMMENTS

#	Freedom To		Comments
1	Travel in own country	YES	Rights respected
2	Travel outside own country	YES	Rights respected
3	Peacefully associate and assemble	YES	Rights respected
4	Teach ideas and receive information	YES	Rights respected
5	Monitor human rights violations	YES	Rights respected
6	Publish and educate in ethnic language	YES	Rights respected

FREEDOM FROM: / COMMENTS

#	Freedom From		Comments
7	Serfdom, slavery, forced or child labor	YES	Rights respected
8	Extrajudicial killings or "disappearances"	YES	Rights respected
9	Torture or coercion by the state	YES	Rights respected
10	Compulsory work permits or conscription of labor	YES	Rights respected

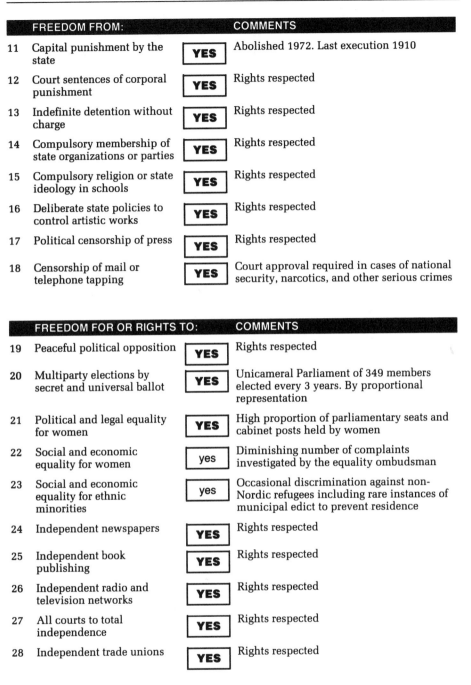

	FREEDOM FROM:		COMMENTS
11	Capital punishment by the state	YES	Abolished 1972. Last execution 1910
12	Court sentences of corporal punishment	YES	Rights respected
13	Indefinite detention without charge	YES	Rights respected
14	Compulsory membership of state organizations or parties	YES	Rights respected
15	Compulsory religion or state ideology in schools	YES	Rights respected
16	Deliberate state policies to control artistic works	YES	Rights respected
17	Political censorship of press	YES	Rights respected
18	Censorship of mail or telephone tapping	YES	Court approval required in cases of national security, narcotics, and other serious crimes

	FREEDOM FOR OR RIGHTS TO:		COMMENTS
19	Peaceful political opposition	YES	Rights respected
20	Multiparty elections by secret and universal ballot	YES	Unicameral Parliament of 349 members elected every 3 years. By proportional representation
21	Political and legal equality for women	YES	High proportion of parliamentary seats and cabinet posts held by women
22	Social and economic equality for women	yes	Diminishing number of complaints investigated by the equality ombudsman
23	Social and economic equality for ethnic minorities	yes	Occasional discrimination against non-Nordic refugees including rare instances of municipal edict to prevent residence
24	Independent newspapers	YES	Rights respected
25	Independent book publishing	YES	Rights respected
26	Independent radio and television networks	YES	Rights respected
27	All courts to total independence	YES	Rights respected
28	Independent trade unions	YES	Rights respected

	LEGAL RIGHTS:		COMMENTS
29	From deprivation of nationality	YES	Rights respected

LEGAL RIGHTS:		COMMENTS
30	To be considered innocent until proved guilty	**YES**	Rights respected
31	To free legal aid when necessary and counsel of own choice	yes	Means test. Counsel of own choice under means test only when prison sentence could be over 6 months
32	From civilian trials in secret	**YES**	Rights respected
33	To be brought promptly before a judge or court	**YES**	To be held for only 12 hours without charge
34	From police searches of home without a warrant	**YES**	Rights respected. And warrants issued only when there is evidence of serious crimes
35	From arbitrary seizure of personal property	**YES**	Rights respected

PERSONAL RIGHTS:		COMMENTS
36	To interracial, interreligious, or civil marriage	**YES**	Marriage legal at age 18 for both sexes
37	Equality of sexes during marriage and for divorce proceedings	**YES**	Rights respected
38	To practice any religion	**YES**	Rights respected
39	To use contraceptive pills and devices	**YES**	Rights respected
40	To noninterference by state in strictly private affairs	**YES**	Rights respected

Switzerland

Human rights rating: 96%

YES 33 yes 7 no 0 NO 0

Population: 6,600,000
Life expectancy: 77.4
Infant mortality (0–5 years)
 per 1,000 births: 8
United Nations covenants ratified:
Not a member of the United Nations

Form of government: Democratic federation
GNP per capita (US$): 27,500
% of GNP spent by state:
On health: 6.8
On military: 1.9
On education: 4.8

FACTORS AFFECTING HUMAN RIGHTS
The individual respected. Government answerable to the people. Many administrative powers widely dispersed to cantons and local communes. A prosperous economy, a free press, and, because of the country's traditional neutrality, a favorite location for international organizations. The three major languages are German, French, and Italian, and foreign workers and their families make up nearly 15% of the population. There is some concern about the extent of foreign immigration and police surveillance has been increased.

	FREEDOM TO:		COMMENTS
1	Travel in own country	YES	Rights respected
2	Travel outside own country	YES	Rights respected
3	Peacefully associate and assemble	YES	Police permits required but only refused when a threat of violence
4	Teach ideas and receive information	YES	Rights respected
5	Monitor human rights violations	YES	Rights respected
6	Publish and educate in ethnic language	YES	Rights respected. Government support for minority languages in a multilingual society

	FREEDOM FROM:		COMMENTS
7	Serfdom, slavery, forced or child labor	YES	Rights respected
8	Extrajudicial killings or "disappearances"	YES	Rights respected
9	Torture or coercion by the state	YES	Rights respected

FREEDOM FROM:		COMMENTS
10	Compulsory work permits or conscription of labor	**YES**	Rights respected
11	Capital punishment by the state	**YES**	Rights respected
12	Court sentences of corporal punishment	**YES**	Rights respected
13	Indefinite detention without charge	**YES**	Rights respected
14	Compulsory membership of state organizations or parties	**YES**	Rights respected
15	Compulsory religion or state ideology in schools	**YES**	Rights respected
16	Deliberate state policies to control artistic works	**YES**	Rights respected
17	Political censorship of press	**YES**	Over 100 independent daily newspapers
18	Censorship of mail or telephone tapping	yes	Some surveillance by police and military on pretext of national security. Files on many citizens kept secret by the Military Department

FREEDOM FOR OR RIGHTS TO:		COMMENTS
19	Peaceful political opposition	**YES**	Voting at age 20
20	Multiparty elections by secret and universal ballot	**YES**	National Council of 200 members. Elections every 4 years. Referenda also held to decide certain issues of major public or local concern
21	Political and legal equality for women	yes	Advances being made but there is still discrimination at senior government level and areas of traditional prejudice against women in politics
22	Social and economic equality for women	yes	Pay and employment differences, some as high as 30% in pay
23	Social and economic equality for ethnic minorities	yes	Position aggravated by large proportion of immigrant labor. Controls on residence, work, and social security conditions
24	Independent newspapers	**YES**	Rights respected
25	Independent book publishing	**YES**	Rights respected
26	Independent radio and television networks	**YES**	Most are state owned but legal safeguards prevent political or official interference
27	All courts to total independence	**YES**	Rights respected

FREEDOM FOR OR RIGHTS TO:	COMMENTS

28 Independent trade unions   **YES**   Rights respected

LEGAL RIGHTS:	COMMENTS

29 From deprivation of nationality   yes   Rare cases of deprivation of double nationals if they are considered to prejudice Swiss neutrality

30 To be considered innocent until proved guilty   **YES**   Rights respected

31 To free legal aid when necessary and counsel of own choice   **YES**   26 cantons with different court procedures but usually free legal aid on application and counsel of own choice

32 From civilian trials in secret   **YES**   Rights respected

33 To be brought promptly before a judge or court   **YES**   Within 24 hours but time-consuming cantonal procedures have to be respected

34 From police searches of home without a warrant   yes   Occasional cases of abuse at cantonal level

35 From arbitrary seizure of personal property   **YES**   Rights respected

PERSONAL RIGHTS:	COMMENTS

36 To interracial, interreligious, or civil marriage   **YES**   Rights respected

37 Equality of sexes during marriage and for divorce proceedings   yes   Women may suffer minor inequalities during divorce

38 To practice any religion   **YES**   Rights respected

39 To use contraceptive pills and devices   **YES**   Rights respected

40 To noninterference by state in strictly private affairs   **YES**   Rights respected

Syria

Human rights rating: 30%

YES 7 yes 5 no 10 NO 18

Population: 12,500,000
Life expectancy: 66.1
Infant mortality (0–5 years)
 per 1,000 births: 62
United Nations covenants ratified:
Civil and Political Rights, Economic,
Social and Cultural Rights

Form of government: One-party state
GNP per capita (US$): 1,680
% of GNP spent by state:
On health: 0.4
On military: 14.7
On education: 2.9

FACTORS AFFECTING HUMAN RIGHTS
The president is an absolute ruler, there has been a state of emergency since 1963, no
political opposition is allowed to the governing Ba'ath party, approximately 7,500
dissidents are under detention, torture is widely practiced with official approval, and
some opponents have "disappeared." Among the excuses given for this human rights
record is the continuing state of war with Israel. Syria gave military support to the
Western powers in the recent gulf war and for this it has been rewarded with many
material concessions and greater political respectability. The country has ratified the
major covenants of the United Nations.

	FREEDOM TO:		COMMENTS
1	Travel in own country	**YES**	Rights respected
2	Travel outside own country	no	Exit permits required. Movement of dissidents or suspects therefore controlled
3	Peacefully associate and assemble	**NO**	Only with police permit. No rallies or protests hostile to the government
4	Teach ideas and receive information	**NO**	Seats of learning either directly or indirectly supervised. Network of informers
5	Monitor human rights violations	**NO**	Severe restrictions prohibit all but the clandestine. Activists seized
6	Publish and educate in ethnic language	**YES**	Rights respected

	FREEDOM FROM:		COMMENTS
7	Serfdom, slavery, forced or child labor	yes	Regional pattern of child labor
8	Extrajudicial killings or "disappearances"	**NO**	Opponents of the president or system occasionally "eliminated"

FREEDOM FROM: COMMENTS

9	Torture or coercion by the state		Widely practiced with official approval. Beatings, scorchings, electric shocks, bone breaking, etc.
10	Compulsory work permits or conscription of labor		Rights respected
11	Capital punishment by the state		By hanging or shooting. Sometimes in public. For treason, murder, robbery, or rape. Also for outspoken political dissidence
12	Court sentences of corporal punishment		The degree of torture with official approval must be equated with corporal punishment by the state
13	Indefinite detention without charge		Approximately 7,500 political detainees. Some sources give a higher figure
14	Compulsory membership of state organizations or parties		Advantages for those belonging to the ruling Ba'ath party. Including work and civil service priorities
15	Compulsory religion or state ideology in schools		Wide compulsory indoctrination of children in favor of president and party
16	Deliberate state policies to control artistic works		Severe restrictions on erotica and Western vulgarity. New art trends may come under scrutiny
17	Political censorship of press	NO	No meaningful criticism of president. Severe penalties
18	Censorship of mail or telephone tapping	NO	Constant surveillance. Arbitrary arrests may follow evidence of political dissidence and other "offenses"

FREEDOM FOR OR RIGHTS TO: COMMENTS

19	Peaceful political opposition	NO	The president is an absolute ruler. No opposition permitted. Reinforced by almost continuous state of emergency
20	Multiparty elections by secret and universal ballot	NO	Only the ruling Ba'ath party. Presidential elections every 7 years. Vote favoring incumbent 99%
21	Political and legal equality for women	yes	Women's status in both political life and professions continues to improve
22	Social and economic equality for women	no	Koranic law and traditional factors reinforce pay and employment discrimination
23	Social and economic equality for ethnic minorities	no	Kurds, Druze, and Palestinian minorities, since the end of the gulf war, suffer discrimination
24	Independent newspapers	NO	Mostly state owned and controlled. Remainder adhere to well-established guidelines

FREEDOM FOR OR RIGHTS TO:		COMMENTS
25	Independent book publishing	**NO**	Survival depends on compliance with understood guidelines. The favored Union of Arab Writers has power to censor books
26	Independent radio and television networks	**NO**	Wholly owned and controlled by the state
27	All courts to total independence	**NO**	Under emergency powers, almost continuous for 30 years, military may enforce transfer of cases to their own courts
28	Independent trade unions	no	In practice, the local form of unions is controlled and coerced by government

LEGAL RIGHTS:		COMMENTS
29	From deprivation of nationality	**YES**	Rights respected
30	To be considered innocent until proved guilty	no	No redress for false arrest. Many on wide interpretation of "suspicion"
31	To free legal aid when necessary and counsel of own choice	yes	Means test, though counsel nominated by court may refuse to represent since a successful defense may mean conflict with the authorities
32	From civilian trials in secret	**NO**	Permitted and practiced under state of emergency
33	To be brought promptly before a judge or court	**NO**	Separate courts for civil, military, religious, and other categories of crime. Prompt court appearance only for minor offenses
34	From police searches of home without a warrant	**NO**	Homes of dissidents raided frequently and arbitrarily
35	From arbitrary seizure of personal property	**YES**	Rights respected

PERSONAL RIGHTS:		COMMENTS
36	To interracial, interreligious, or civil marriage	no	With 88% of population Muslim, in practice civil law rights seldom apply
37	Equality of sexes during marriage and for divorce proceedings	no	The government does not interfere with traditional Muslim practices and beliefs, which are usually to husband's advantage
38	To practice any religion	**YES**	Rights respected. No state religion but the president must be Muslim
39	To use contraceptive pills and devices	**YES**	Rights respected
40	To noninterference by state in strictly private affairs	no	The harassment of surveillance and a network of government "informers" create fear in private life and inhibit nonconformism in private behavior

Tanzania

Human rights rating: 41%

YES 6 **yes** 9 **no** 17 NO 8

Population: 27,300,000
Life expectancy: 54
Infant mortality (0–5 years)
  per 1,000 births: 173
United Nations covenants ratified:
Civil and Political Rights, Economic,
Social and Cultural Rights, Convention
on Equality for Women

Form of government: One-party state
GNP per capita (US$): 160
% of GNP spent by state:
On health: 1.2
On military: 3.3
On education: 1.7

FACTORS AFFECTING HUMAN RIGHTS

A single-party regime has governed Tanzania since its formation in 1964. An attempt to
introduce a form of "African socialism" was abandoned after one of the poorest
countries in the world became even poorer. There is a system of national surveillance of
the population by "10-family cells," and the security forces are supplemented by
semilegal vigilante units. Political and social opposition is not tolerated but there are
increasing displays of public dissatisfaction. The degree of detentions, instances of
torture, and the occasional extrajudicial killing vary with the perceived threat to the
government.

FREEDOM TO:		COMMENTS
1 Travel in own country	yes	Migrations into urban areas may be restricted. Position improving
2 Travel outside own country	yes	Many regulations limit travel abroad, including currency requirements, tax clearance, etc.
3 Peacefully associate and assemble	NO	Nothing that conflicts with the ruling party. Stricter on the offshore islands of Zanzibar and Pemba
4 Teach ideas and receive information	no	Periodic closing of university following excessive criticism of president. Academics follow understood guidelines. Dissident students arrested
5 Monitor human rights violations	no	International monitors unwelcome. Police occasionally harass investigations by local groups
6 Publish and educate in ethnic language	YES	Rights respected

	FREEDOM FROM:		COMMENTS
7	Serfdom, slavery, forced or child labor	no	Limited forced resettlement still occurs. Child labor a traditional practice
8	Extrajudicial killings or "disappearances"	no	Central government tends to overlook the occasional killing by local militia groups. Instances of "mob justice" deaths
9	Torture or coercion by the state	no	Beatings, harsh interrogation, coercive semi-starvation, etc.
10	Compulsory work permits or conscription of labor	no	The system of conscripting labor to collective farms has been modified but has not been discontinued
11	Capital punishment by the state	NO	By hanging. Treason, murder, economic sabotage, etc.
12	Court sentences of corporal punishment	yes	But caning still in the penal code
13	Indefinite detention without charge	no	Instances of long detention on direct orders of president. Numbers vary with perceived threats to the government
14	Compulsory membership of state organizations or parties	yes	But membership of ruling party an advantage in social and professional life
15	Compulsory religion or state ideology in schools	YES	Rights respected
16	Deliberate state policies to control artistic works	YES	Rights respected
17	Political censorship of press	no	Understood guidelines. Occasional "unofficial" warnings from authorities
18	Censorship of mail or telephone tapping	no	Widely practiced. A network of 10-cell neighborhood informants reinforces the security system

	FREEDOM FOR OR RIGHTS TO:		COMMENTS
19	Peaceful political opposition	NO	One-party system. Resistance most evident on Zanzibar and Pemba
20	Multiparty elections by secret and universal ballot	NO	Single-party elections every 5 years, with all candidates belonging to the ruling party. Subject of permitting opposition a topic for debate – but only as a maneuver
21	Political and legal equality for women	no	Progress toward introducing women into higher ranks of the party but traditional resistance remains strong
22	Social and economic equality for women	no	Wide discrimination, particularly in rural areas, where they provide most of the agricultural labor. Some female circumcision
23	Social and economic equality for ethnic minorities	no	Occasional harassment. Over 100 ethnic groups. Some persecution of Zanzibaris

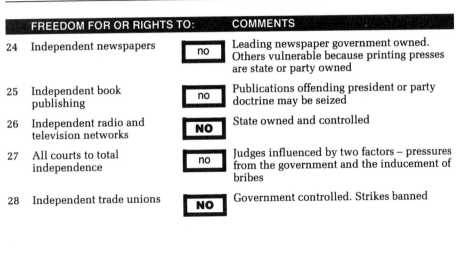

FREEDOM FOR OR RIGHTS TO:		COMMENTS
24	Independent newspapers	no	Leading newspaper government owned. Others vulnerable because printing presses are state or party owned
25	Independent book publishing	no	Publications offending president or party doctrine may be seized
26	Independent radio and television networks	**NO**	State owned and controlled
27	All courts to total independence	no	Judges influenced by two factors – pressures from the government and the inducement of bribes
28	Independent trade unions	**NO**	Government controlled. Strikes banned

LEGAL RIGHTS:		COMMENTS
29	From deprivation of nationality	yes	Legally permitted but no recent withdrawals of passports
30	To be considered innocent until proved guilty	yes	But police have wide powers and may arrest without justification. Targets include "layabouts" and "loiterers"
31	To free legal aid when necessary and counsel of own choice	yes	For serious crimes. Means test
32	From civilian trials in secret	yes	Rare secret trials in security cases
33	To be brought promptly before a judge or court	yes	24 hours. Usually honored except for cases affecting security or political dissidence
34	From police searches of home without a warrant	**NO**	Police raids with or without warrants. In practice, arbitrary powers
35	From arbitrary seizure of personal property	**NO**	Native lands and property of Barabaig herdspeople seized by government. Other seizures to implement collective land schemes

PERSONAL RIGHTS:		COMMENTS
36	To interracial, interreligious, or civil marriage	**YES**	Rights respected
37	Equality of sexes during marriage and for divorce proceedings	no	Wide variations among ethnic peoples. Subservience of wives, disadvantages in inheritance and divorce
38	To practice any religion	**YES**	Rights respected
39	To use contraceptive pills and devices	**YES**	Government support

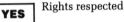

PERSONAL RIGHTS: COMMENTS

40   To noninterference by state   Private affairs suffer under an intrusive
     in strictly private affairs   system, particularly from the 10-cell
                                                              neighborhood surveillance units

Thailand

Human rights rating: 62%

YES 13 yes 15 no 10 NO 2

Population: 55,700,000
Life expectancy: 66.1
Infant mortality (0–5 years)
  per 1,000 births: 35
United Nations covenants ratified:
None

Form of government: Constitutional monarchy
GNP per capita (US$): 1,000
% of GNP spent by state:
On health: 1.0
On military: 4.0
On education: 3.2

FACTORS AFFECTING HUMAN RIGHTS
Early in 1991 the military seized power from a civilian government but promised multiparty elections in 1992. Despite periods of civilian rule and a constitutional monarch, the military decide national policy and effectively control the country. The justification for human rights violations is claimed to be the dangers from Communist guerrillas, militant Muslims, and criminal gangs, particularly in rural areas. The situation on the Cambodian frontier is complicated by the presence of considerable Cambodian rebel forces and refugee camps, some of which are, by agreement, on Thai territory.

	FREEDOM TO:		COMMENTS
1	Travel in own country	yes	Certain minorities from neighboring countries – Cambodian, Chinese, Vietnamese, and other refugees – are restricted in their movements
2	Travel outside own country	YES	Rights respected but checks to prevent the trade in "women prostitutes for export." Also when single women apply for passports
3	Peacefully associate and assemble	yes	The pattern of bloodless military coups, accompanied by brief periods of martial law, may result in arbitrary bannings. Particularly of workers and students
4	Teach ideas and receive information	yes	Some surveillance of left-wing groups in universities. Criticism of monarchy a forbidden practice
5	Monitor human rights violations	YES	Rights respected
6	Publish and educate in ethnic language	YES	Rights respected

	FREEDOM FROM:		COMMENTS
7	Serfdom, slavery, forced or child labor	no	Child labor estimated at 5% of population. "Sex service industry" employs children from age 11 or 12
8	Extrajudicial killings or "disappearances"	yes	A few well-documented excess killings by the military against Muslim and Communist rebel groups. Also by police to control a high level of serious crime
9	Torture or coercion by the state	no	Beating and maltreatment by police and in prisons. Little supervision by government
10	Compulsory work permits or conscription of labor	YES	Rights respected
11	Capital punishment by the state	NO	For murder, treason, and drug offenses. By machine gun
12	Court sentences of corporal punishment	yes	Floggings for misdemeanors in prison
13	Indefinite detention without charge	yes	Reserve powers under periodic declarations of martial law. Under Anti-Communist Act, 480 days without charge but few detentions at present
14	Compulsory membership of state organizations or parties	YES	Rights respected
15	Compulsory religion or state ideology in schools	YES	Rights respected
16	Deliberate state policies to control artistic works	YES	Rights respected
17	Political censorship of press	no	Self-censorship practiced and varies with the severity of military or civilian government, whichever is in power. Comments limited on monarchy, religion, national security, etc.
18	Censorship of mail or telephone tapping	yes	Some abuses of law during martial law periods – usually against left-wing groups

	FREEDOM FOR OR RIGHTS TO:		COMMENTS
19	Peaceful political opposition	yes	The military seize power from time to time but opposition parties are able to propagate policies
20	Multiparty elections by secret and universal ballot	no	Military have agreed a general election in 1992 after they ousted a five-party coalition in February 1991
21	Political and legal equality for women	no	A male-dominated society. Advances by women "token" despite relative constitutional equality
22	Social and economic equality for women	no	Discrimination throughout society. Pay, civil service, and educational inequalities

FREEDOM FOR OR RIGHTS TO:		COMMENTS
23	Social and economic equality for ethnic minorities	**yes**	Inequalities most obvious as the large numbers of refugees from neighboring countries seek employment, shelter, and social rights. Source of cheap labor
24	Independent newspapers	**yes**	Controls and bannings most evident during periods of martial law
25	Independent book publishing	**yes**	As question 24
26	Independent radio and television networks	**no**	Major networks owned and run by the state, including one by the army. Some private stations now on the air
27	All courts to total independence	**yes**	Pressures from the military during occasional periods of martial law. Some corruption of courts
28	Independent trade unions	**no**	Unions in state owned organizations abolished in 1991. Less than 15% in unions. No strikes throughout public sector

LEGAL RIGHTS:		COMMENTS
29	From deprivation of nationality	**YES**	Rights respected
30	To be considered innocent until proved guilty	**YES**	Rights respected
31	To free legal aid when necessary and counsel of own choice	**NO**	Aid comes from voluntary system run by legal associations
32	From civilian trials in secret	**no**	Only when considered to be of national importance – including defamation and disrespect toward king, queen, and others of the royal family
33	To be brought promptly before a judge or court	**yes**	Delays are usually the result of an overburdened system
34	From police searches of home without a warrant	**yes**	Occasional abuses in frequent actions against Communists and suspected subversives
35	From arbitrary seizure of personal property	**YES**	Rights respected

PERSONAL RIGHTS:		COMMENTS
36	To interracial, interreligious, or civil marriage	**YES**	Legal marriage age 17 for both sexes
37	Equality of sexes during marriage and for divorce proceedings	**no**	Husbands enjoy advantages during divorce. Traditional pattern of male dominance in marriage

PERSONAL RIGHTS: | COMMENTS

38	To practice any religion	**YES**	Minor restrictions on a few cult religions
39	To use contraceptive pills and devices	**YES**	Government support
40	To noninterference by state in strictly private affairs	yes	The failure of the government to exercise law or reduce poverty has created a "sex trade" estimated to involve 1% of the population, so distorting normal family life

| Togo | Human rights rating: | 48% |

YES 7 yes 9 no 18 NO 6

Population: 3,500,000
Life expectancy: 54.0
Infant mortality (0–5 years)
   per 1,000 births: 150
United Nations covenants ratified:
Civil and Political Rights, Economic,
Social and Cultural Rights, Convention
on Equality for Women

Form of government: One-party system
GNP per capita (US$): 370
% of GNP spent by state:
On health: 1.7
On military: 3.2
On education: 6.5

FACTORS AFFECTING HUMAN RIGHTS
Demonstrations protesting against an oppressive single-party system and calling for a
multiparty democracy have forced the president to agree to the introduction of a new
constitution and a referendum on the issue of free elections. The president has ruled the
country since 1967, maintaining his authority through the army and security forces, but
in August 1991, his powers were restricted. An interim administration was deposed by
the armed forces and a "government of national unity" formed. The situation is
unpredictable.

FREEDOM TO: COMMENTS

1	Travel in own country	**YES**	Rights respected
2	Travel outside own country	yes	But political opponents may be refused passports. Exit visas necessary
3	Peacefully associate and assemble	no	No protests or rallies against the ruling party or the president. Recent peaceful demonstration opposed by police, resulting in a number of deaths. But see Factors
4	Teach ideas and receive information	no	Government surveillance. A network of informers check on loyalty of academics and student agitators
5	Monitor human rights violations	yes	Local human rights groups now permitted. Their effectiveness has still to be tested. Monitors proceed with caution
6	Publish and educate in ethnic language	**YES**	Rights respected

FREEDOM FROM: COMMENTS

7	Serfdom, slavery, forced or child labor	yes	Wide pattern of child labor – particularly in rural areas

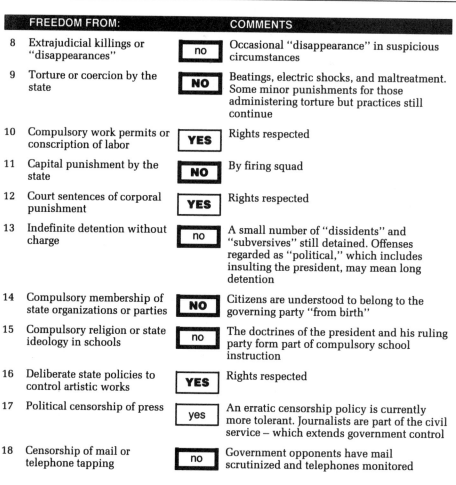

FREEDOM FROM: **COMMENTS**

8   Extrajudicial killings or   `no`   Occasional "disappearance" in suspicious
    "disappearances"   circumstances

9   Torture or coercion by the   **NO**   Beatings, electric shocks, and maltreatment.
    state   Some minor punishments for those
                                              administering torture but practices still
                                              continue

10 Compulsory work permits or   **YES**   Rights respected
    conscription of labor

11 Capital punishment by the   **NO**   By firing squad
    state

12 Court sentences of corporal   **YES**   Rights respected
    punishment

13 Indefinite detention without   `no`   A small number of "dissidents" and
    charge   "subversives" still detained. Offenses
                                              regarded as "political," which includes
                                              insulting the president, may mean long
                                              detention

14 Compulsory membership of   **NO**   Citizens are understood to belong to the
    state organizations or parties   governing party "from birth"

15 Compulsory religion or state   `no`   The doctrines of the president and his ruling
    ideology in schools   party form part of compulsory school
                                              instruction

16 Deliberate state policies to   **YES**   Rights respected
    control artistic works

17 Political censorship of press   yes   An erratic censorship policy is currently
                                              more tolerant. Journalists are part of the civil
                                              service – which extends government control

18 Censorship of mail or   `no`   Government opponents have mail
    telephone tapping   scrutinized and telephones monitored

FREEDOM FOR OR RIGHTS TO: **COMMENTS**

19 Peaceful political opposition   `no`   The single-party system is beginning to
                                               permit limited discussion on the alternative
                                               of multiparty democracy. But see Factors

20 Multiparty elections by   **NO**   After 23 years of absolute power, the
    secret and universal ballot   president announced the holding of a
                                               referendum at the end of 1991 on the issue
                                               of a multiparty system. But see Factors

21 Political and legal equality   `no`   Traditional inequality in politics and
    for women   professions, pension rights, and government
                                               service

22 Social and economic   `no`   Despite equality in law, subservience
    equality for women   socially. Women have a significant presence
                                               in urban business but, less profitably, also as
                                               cheap agricultural labor

FREEDOM FOR OR RIGHTS TO:		COMMENTS
23	Social and economic equality for ethnic minorities	yes	Complaints by different ethnic groups of discrimination by government. The largest, the Ewe, 20–25% of the population
24	Independent newspapers	no	The only daily newspaper is government controlled and has a circulation of approximately 10,000
25	Independent book publishing	yes	Prudence necessary for survival. Small potential market. Book imports from France, though some seized
26	Independent radio and television networks	NO	State owned and controlled
27	All courts to total independence	no	Courts subject to presidential decrees. Tribal courts follow customary system
28	Independent trade unions	NO	The national trade union movement is controlled by the governing political party. Strikes provoke violent action by the security forces

LEGAL RIGHTS:		COMMENTS
29	From deprivation of nationality	yes	Many political opponents remain in exile. No deprivation of nationality but possible imprisonment or death on their return
30	To be considered innocent until proved guilty	no	The system does not guarantee assumption of innocence. Vague interpretation of the offense of "insulting" public servants. 5 years' prison for public utterances of such insults
31	To free legal aid when necessary and counsel of own choice	no	Little provision for the needy. Except when sentence may be capital punishment
32	From civilian trials in secret	yes	Rare. But president has virtually absolute powers. See Factors
33	To be brought promptly before a judge or court	no	Within 48 hours by law but many "political crimes" can mean indefinite delays. Regional governors also have delaying powers
34	From police searches of home without a warrant	no	Many abuses. The wide interpretation of "security" searches excuses arbitrary entry
35	From arbitrary seizure of personal property	no	Abuses by ill-disciplined security forces. Forcible resettlement of villagers, inadequate compensation for seizures. Some looting

PERSONAL RIGHTS:		COMMENTS
36	To interracial, interreligious, or civil marriage	YES	Rights respected

PERSONAL RIGHTS:		COMMENTS

37 Equality of sexes during marriage and for divorce proceedings — **no** — Tribal or traditional law, affecting the majority, favors husbands. On divorce he acquires all property

38 To practice any religion — **yes** — 50% follow animism. Limits on a number of religious associations, and Jehovah's Witnesses may not practice in public

39 To use contraceptive pills and devices — **YES** — Rights respected

40 To noninterference by state in strictly private affairs — **no** — Laws such as compulsion to belong to the governing party, forbidding "European" first names, network of neighborhood informers intrude into private affairs

Trinidad

Human rights rating: 84%

Population: 1,300,000
Life expectancy: 71.6
Infant mortality (0–5 years) per 1,000 births: 27
United Nations covenants ratified: Civil and Political Rights, Economic, Social and Cultural Rights, Convention on Equality for Women

Form of government: Parliamentary democracy
GNP per capita (US$): 3,350
% of GNP spent by state:
On health: 3.0
On military: 1.0
On education: 5.8

FACTORS AFFECTING HUMAN RIGHTS
The government of this multiparty democracy was briefly challenged in July 1990 by an armed band of Muslim radicals. Parliament was captured, the television station seized, and there was widespread damage to public buildings. The defeat of the rebels was followed by a state of emergency, which ended after 4 months. Human rights are generally honored and the country has the highest income per head in the Caribbean. The population of Africans, 43%, and Asian Indians, 40%, coexist with relatively little tension.

	FREEDOM TO:		COMMENTS
1	Travel in own country	YES	For a brief period, during the July 1990 state of emergency, limited travel restrictions when Muslim radicals attempted a coup
2	Travel outside own country	YES	Rights respected
3	Peacefully associate and assemble	YES	Rights respected
4	Teach ideas and receive information	YES	Rights respected
5	Monitor human rights violations	YES	Rights respected
6	Publish and educate in ethnic language	YES	Rights respected

	FREEDOM FROM:		COMMENTS
7	Serfdom, slavery, forced or child labor	yes	Regional pattern of child labor

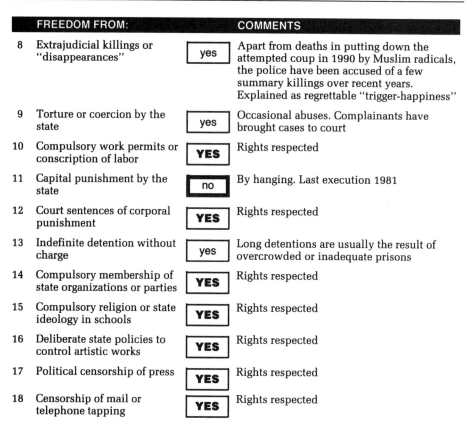

FREEDOM FROM:		COMMENTS
8	Extrajudicial killings or "disappearances"	**yes**	Apart from deaths in putting down the attempted coup in 1990 by Muslim radicals, the police have been accused of a few summary killings over recent years. Explained as regrettable "trigger-happiness"
9	Torture or coercion by the state	**yes**	Occasional abuses. Complainants have brought cases to court
10	Compulsory work permits or conscription of labor	**YES**	Rights respected
11	Capital punishment by the state	**no**	By hanging. Last execution 1981
12	Court sentences of corporal punishment	**YES**	Rights respected
13	Indefinite detention without charge	**yes**	Long detentions are usually the result of overcrowded or inadequate prisons
14	Compulsory membership of state organizations or parties	**YES**	Rights respected
15	Compulsory religion or state ideology in schools	**YES**	Rights respected
16	Deliberate state policies to control artistic works	**YES**	Rights respected
17	Political censorship of press	**YES**	Rights respected
18	Censorship of mail or telephone tapping	**YES**	Rights respected

FREEDOM FOR OR RIGHTS TO:		COMMENTS
19	Peaceful political opposition	**YES**	Rights respected
20	Multiparty elections by secret and universal ballot	**YES**	Rights respected
21	Political and legal equality for women	**yes**	Inequalities persist despite some progress in all areas
22	Social and economic equality for women	**yes**	Traditional attitudes of male dominance perpetuate minor inequalities in pay and employment
23	Social and economic equality for ethnic minorities	**YES**	Rights respected but there is still discrimination against those with darker skins
24	Independent newspapers	**YES**	Rights respected
25	Independent book publishing	**YES**	Rights respected

FREEDOM FOR OR RIGHTS TO:		COMMENTS
26	Independent radio and television networks	yes	Privately owned stations lisenced from 1990. Government-owned stations accused of occasional political and social bias
27	All courts to total independence	**YES**	Rights respected
28	Independent trade unions	**YES**	Rights respected

LEGAL RIGHTS:		COMMENTS
29	From deprivation of nationality	**YES**	Rights respected
30	To be considered innocent until proved guilty	**YES**	Rights respected
31	To free legal aid when necessary and counsel of own choice	yes	At discretion of court
32	From civilian trials in secret	**YES**	Rights respected
33	To be brought promptly before a judge or court	yes	Within 48 hours but some long delays, particularly in treatment of recent Muslim coup attempt
34	From police searches of home without a warrant	yes	Cases of unofficial police abuses before and after the state of emergency (briefly introduced at the time of coup attempt)
35	From arbitrary seizure of personal property	**YES**	Rights respected

PERSONAL RIGHTS:		COMMENTS
36	To interracial, interreligious, or civil marriage	**YES**	Rights respected. No minimum age for marriage
37	Equality of sexes during marriage and for divorce proceedings	yes	Minor property inequalities to wives' disadvantage
38	To practice any religion	**YES**	Rights respected
39	To use contraceptive pills and devices	**YES**	Government support
40	To noninterference by state in strictly private affairs	yes	Law permits deportation of visiting homosexuals – if apprehended

Tunisia

Human rights rating: 60%

Population: 8,200,000
Life expectancy: 66.7
Infant mortality (0–5 years) per 1,000 births: 66
United Nations covenants ratified: Civil and Political Rights, Economic, Social and Cultural Rights, Convention on Equality for Women

Form of government: Multiparty system
GNP per capita (US$): 1,230
% of GNP spent by state:
On health: 2.2
On military: 6.2
On education: 5.4

FACTORS AFFECTING HUMAN RIGHTS

A parliamentary democracy with the opposition boycotting the elections has meant that all seats in the Chamber of Deputies were won by the ruling president's party. The opposition claimed that there had been electoral irregularities, intimidation, and an unfair system. Much of the police surveillance and harassment is directed at Muslim activists, who are regarded as the major threat to the Islamic state. The press and broadcasting are either restricted or controlled by the government, and academic life follows understood guidelines.

FREEDOM TO: | COMMENTS

#	Freedom		Comments
1	Travel in own country	yes	Identity cards must be carried. Complaints of harassment and brief detentions at checkpoints in open country and near frontiers
2	Travel outside own country	yes	Restrictions on passports of a few political opponents
3	Peacefully associate and assemble	no	Many clashes with police during student demonstrations. Banned political parties denied permits
4	Teach ideas and receive information	yes	Academics practice prudence on sensitive subjects. Student militancy has increased government surveillance
5	Monitor human rights violations	yes	Activities of monitors expanding. Some government cooperation
6	Publish and educate in ethnic language	YES	Rights respected

FREEDOM FROM: | COMMENTS

#	Freedom		Comments
7	Serfdom, slavery, forced or child labor	yes	Child labor follows regional pattern

FREEDOM FROM:		COMMENTS
8	Extrajudicial killings or "disappearances"	yes	A number of unexplained deaths of prisoners while in custody
9	Torture or coercion by the state	no	Many instances. Beatings, electric shocks, etc. Mostly local abuses by ill-disciplined police and prison wardens
10	Compulsory work permits or conscription of labor	**YES**	Rights respected
11	Capital punishment by the state	**NO**	By hanging. For murder, treason, arson, rape of minors, looting, etc.
12	Court sentences of corporal punishment	**YES**	Rights respected
13	Indefinite detention without charge	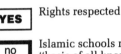 no	Long periods without charge. Usually of political and trade union opponents. Some held incommunicado
14	Compulsory membership of state organizations or parties	**YES**	Rights respected
15	Compulsory religion or state ideology in schools	 no	Islamic schools regard their faith as the "basis of all knowledge and education"
16	Deliberate state policies to control artistic works	yes	But nothing that insults or satirizes religion or the president
17	Political censorship of press	no	Considerable censorship. Severe defamation laws. Wide interpretation of subversion charges
18	Censorship of mail or telephone tapping	no	Degree of surveillance depends on threat to government

FREEDOM FOR OR RIGHTS TO:		COMMENTS
19	Peaceful political opposition	no	Limitations. Major opposition party still banned
20	Multiparty elections by secret and universal ballot	yes	Elections boycotted by opposition. The governing party therefore gained all seats in the Chamber of Deputies
21	Political and legal equality for women	yes	Progress toward equality in the professions is more evident than prominence in politics. Islamic traditions of woman's role still prevail
22	Social and economic equality for women	no	Many inequalities. Worst in rural areas
23	Social and economic equality for ethnic minorities	**YES**	Rights respected
24	Independent newspapers	no	Both publishers and printers subject to restrictions. Occasional seizures, though they must have court approval. Newsprint virtually a government monopoly

	FREEDOM FOR OR RIGHTS TO:		COMMENTS
25	Independent book publishing	no	Understood guidelines. Self-censorship of contents advisable. A number of important books banned in 1990
26	Independent radio and television networks	no	State owned and controlled. But recent limited concessions to the broadcasting of opposition views
27	All courts to total independence	no	Ministry of Justice appoints all judges and senior court officials. Wide degree of government control when their authority threatened
28	Independent trade unions	yes	Limitations on freedom to strike and on union meetings held for political reasons

	LEGAL RIGHTS:		COMMENTS
29	From deprivation of nationality	YES	Rights respected
30	To be considered innocent until proved guilty	yes	Constitutional protection but wide use of "arrests on suspicion"
31	To free legal aid when necessary and counsel of own choice	yes	Free aid for the needy but court appoints counsel
32	From civilian trials in secret	yes	A military tribunal may insist on secret trials of a wide interpretation of "security" cases
33	To be brought promptly before a judge or court	yes	10-day limit. Many violations
34	From police searches of home without a warrant	no	Islamic militants considered "a threat to state security." A pretext for arbitrary searches of homes
35	From arbitrary seizure of personal property	YES	Rights respected

	PERSONAL RIGHTS:		COMMENTS
36	To interracial, interreligious, or civil marriage	no	No interreligious marriage for practicing Muslims (99% of population) without government approval. Males may legally marry at 20 years, females at 17 years
37	Equality of sexes during marriage and for divorce proceedings	no	Wives denied equality. Islamic traditions of husband as head of the family. Polygamy abolished 1957
38	To practice any religion	yes	Baha'i religion considered heretical and banned. Islam the official state religion. Proselytizing forbidden
39	To use contraceptive pills and devices	YES	Encouraged by the state, which permits abortions on demand

PERSONAL RIGHTS:	COMMENTS

40 To noninterference by state
in strictly private affairs   **YES**   Rights respected

Turkey

Human rights rating: 44%

YES 6 yes 11 no 15 NO 8

Population: 55,900,000
Life expectancy: 65.1
Infant mortality (0–5 years) per 1,000 births: 90
United Nations covenants ratified:
Convention on Equality for Women

Form of government: Multiparty system
GNP per capita (US$): 1,280
% of GNP spent by state:
On health: 0.5
On military: 4.9
On education: 2.8

FACTORS AFFECTING HUMAN RIGHTS

Despite government claims that the human rights situation has improved, and an intention to apply for membership in the European Community, the situation in reality is little changed. Multiparty elections are held and usually honored but these have not created a democratic society. The major threat of the PKK (Kurdish) separatist movement, which operates mostly in a part of the country under a state of emergency, is given as the reason for comparable killings by both the security forces and the terrorists, but police and military atrocities extend to other opponents of the government. These may suffer systematic torture and incommunicado detention. Use of the Kurdish language in Turkey is severely restricted and all printed matter in the language is illegal.

	FREEDOM TO:		COMMENTS
1	Travel in own country	YES	But subject to the security situation in the 10 provinces under emergency laws
2	Travel outside own country	no	Passports denied to security risks and a few political dissidents
3	Peacefully associate and assemble	no	Permission required. Purpose of meeting closely scrutinized for antigovernment activities.
4	Teach ideas and receive information	no	No student demonstrations. Teachers dismissed on political grounds
5	Monitor human rights violations	no	Official restrictions and harassment of human rights groups justified by security considerations. Some arrests. A new parliamentary human rights commission has been formed
6	Publish and educate in ethnic language	no	Certain ethnic languages restricted by law, including Kurdish, the language of one-tenth of the country's population. No printed works in Kurdish

FREEDOM FROM:		COMMENTS
7 Serfdom, slavery, forced or child labor	yes	Child labor in rural areas
8 Extrajudicial killings or "disappearances"	no	Well-documented reports of deaths in detention, described as "suicides" by officials
9 Torture or coercion by the state	**NO**	Beatings, electric shocks, force-feeding of human excrement, breaking of bones. When perpetrators brought to court, sentences usually light. Under-16s also tortured
10 Compulsory work permits or conscription of labor	**YES**	Rights respected
11 Capital punishment by the state	**NO**	Hanging. For murder, antistate activities, treason, etc. Some previous crimes incurring death penalty now rescinded
12 Court sentences of corporal punishment	**YES**	Rights respected
13 Indefinite detention without charge	**NO**	Thousands of opponents, including militant Kurds, held incommunicado for long periods. Particularly under emergency laws or the new Anti-Terrorism Law
14 Compulsory membership of state organizations or parties	**YES**	Rights respected
15 Compulsory religion or state ideology in schools	no	98% Muslim population. Periods of compulsory religion in schools
16 Deliberate state policies to control artistic works	no	Kurdish dancing, singers, and songs forbidden. Some arrests. Censorship of many avant-garde works
17 Political censorship of press	**NO**	Journalists imprisoned, harassed, and tortured for favoring separatism, communism, "insulting government," criticizing Turkey while abroad, etc. Longest known sentence 700 years
18 Censorship of mail or telephone tapping	**NO**	Surveillance widespread, particularly in martial law provinces

FREEDOM FOR OR RIGHTS TO:		COMMENTS
19 Peaceful political opposition	yes	Political debate influenced by strict media controls and subjects that may not be freely discussed
20 Multiparty elections by secret and universal ballot	yes	But certain smaller parties remain banned, including the Communist party. Form of proportional representation in practice
21 Political and legal equality for women	no	A little progress toward meaningful representation but tradition generally still prevails

FREEDOM FOR OR RIGHTS TO: COMMENTS

22	Social and economic equality for women	no	Wide discrimination in pay and employment. Women have an "understood role"
23	Social and economic equality for ethnic minorities	yes	Kurds suffer from widespread official discrimination. Sometimes referred to as "mountain people"
24	Independent newspapers	no	Proprietors and editors frequently imprisoned or fined for comments or criticisms relating to many taboo subjects
25	Independent book publishing	no	Publishers serve long sentences if they risk issuing books hostile or insulting to president, military, etc.
26	Independent radio and television networks	**NO**	State owned and controlled
27	All courts to total independence	yes	Regional governors may assume control in 10 provinces under emergency laws – overruling courts
28	Independent trade unions	yes	But police surveillance to check on political activities. Various restrictions on strikes, including long "delaying" tactics

LEGAL RIGHTS: COMMENTS

29	From deprivation of nationality	no	Usually for "unpatriotic" activities while abroad. Passports withdrawn
30	To be considered innocent until proved guilty	no	The extent of arrests of dissidents, journalists, Kurds seeking separatism, and other groups under surveillance results in many arbitrary and groundless arrests
31	To free legal aid when necessary and counsel of own choice	yes	For nonpolitical cases. Counsel appointed by court
32	From civilian trials in secret	**NO**	Permitted in provinces covered by emergency or antiterrorist measures
33	To be brought promptly before a judge or court	no	Maximum 8 hours but exceptions include political dissidents. Among them are students parading "no to war" placards
34	From police searches of home without a warrant	**NO**	The continuing military action against PKK (Kurdish) terrorists is a pretext for many violations of lawfully permitted searches
35	From arbitrary seizure of personal property	**YES**	Rights respected

PERSONAL RIGHTS:		COMMENTS
36	To interracial, interreligious, or civil marriage	yes	Little intermarriage in practice. Strict social and religious pressures to conform to Islamic marriage laws. Males may marry at 17 years, females at 15
37	Equality of sexes during marriage and for divorce proceedings	yes	Position improving but husband's consent frequently required by wives (for passport, to start a business, etc.)
38	To practice any religion	yes	Wide surveillance of minority religions suspected of political activities. Also of Islamic fundamentalists
39	To use contraceptive pills and devices	**YES**	Rights respected
40	To noninterference by state in strictly private affairs	yes	The extent of security surveillance and the social harassment of significant minorities create fear and tension in private lives

Uganda

Human rights rating: 46%

Population: 18,800,000
Life expectancy: 52.0
Infant mortality (0–5 years)
  per 1,000 births: 167
United Nations covenants ratified:
Economic, Social and Cultural Rights,
Convention on Equality for Women

Form of government: Military council
GNP per capita (US$): 280
% of GNP spent by state:
On health: 0.2
On military: 4.2
On education: 1.5

FACTORS AFFECTING HUMAN RIGHTS

A 5-year guerrilla war which brought to power in 1986 the present government has not seen the end of widespread killings and destruction in one of the poorest countries in the world. Government forces are still fighting a few rebel units including a Holy Spirit Movement of fanatical extremists which practices summary killings, kidnappings, and brutal atrocities. The president, a military general, promised that his government would be an interim one until elections could take place, but his period of rule has now been extended to 1995. Intertribal feuding adds to the high level of violence and atrocities. Unlimited detention under a security act is lawful and over 1,000 are being currently held.

	FREEDOM TO:		COMMENTS
1	Travel in own country	yes	In practice free travel denied by military operations against insurgents and armed bands
2	Travel outside own country	yes	Travel abroad is restricted by lack of foreign currency and passport delays
3	Peacefully associate and assemble	**NO**	All political rallies, protests, or associations which might challenge government authority forbidden
4	Teach ideas and receive information	no	Nothing that conflicts with official policies and programs
5	Monitor human rights violations	yes	Position improving. The government is displaying greater tolerance toward national and international monitoring
6	Publish and educate in ethnic language	**YES**	Rights respected

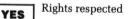

	FREEDOM FROM:		COMMENTS
7	Serfdom, slavery, forced or child labor	yes	Regional pattern of child labor

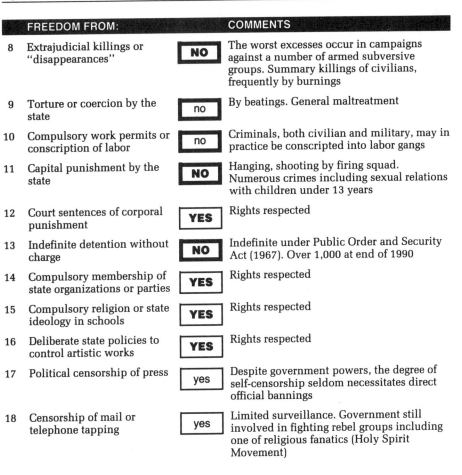

FREEDOM FROM: COMMENTS

8	Extrajudicial killings or "disappearances"	**NO**	The worst excesses occur in campaigns against a number of armed subversive groups. Summary killings of civilians, frequently by burnings
9	Torture or coercion by the state	no	By beatings. General maltreatment
10	Compulsory work permits or conscription of labor	no	Criminals, both civilian and military, may in practice be conscripted into labor gangs
11	Capital punishment by the state	**NO**	Hanging, shooting by firing squad. Numerous crimes including sexual relations with children under 13 years
12	Court sentences of corporal punishment	**YES**	Rights respected
13	Indefinite detention without charge	**NO**	Indefinite under Public Order and Security Act (1967). Over 1,000 at end of 1990
14	Compulsory membership of state organizations or parties	**YES**	Rights respected
15	Compulsory religion or state ideology in schools	**YES**	Rights respected
16	Deliberate state policies to control artistic works	**YES**	Rights respected
17	Political censorship of press	yes	Despite government powers, the degree of self-censorship seldom necessitates direct official bannings
18	Censorship of mail or telephone tapping	yes	Limited surveillance. Government still involved in fighting rebel groups including one of religious fanatics (Holy Spirit Movement)

FREEDOM FOR OR RIGHTS TO: COMMENTS

19	Peaceful political opposition	**NO**	In 1990 government banned political activity until 1995
20	Multiparty elections by secret and universal ballot	**NO**	No national elections or formation of political parties before 1995. President has stated: "There is no reason why a single political party cannot be democratic"
21	Political and legal equality for women	yes	Government improving status of women and encouraging participation at higher levels. District council seats reserved for women
22	Social and economic equality for women	no	Traditional discrimination against women. In inheritance matters, divorce, pay, and employment
23	Social and economic equality for ethnic minorities	no	The country is divided by violence between ethnic groups and the government, by intertribal feuding, etc. Wide discrimination

FREEDOM FOR OR RIGHTS TO: COMMENTS

24   Independent newspapers   **yes**   The high cost of newsprint, with supplies
                                                    controlled by government, affects
                                                    independence. The official newspaper, *The
                                                    New Vision*, enjoys many advantages

25   Independent book   **yes**   The government publishing house remains
     publishing   the only significant one

26   Independent radio and   **NO**   State owned and controlled
     television networks

27   All courts to total   **no**   President's authority extends to dismissing
     independence   or transferring judges for political or security
                                                    cases

28   Independent trade unions   **YES**   Rights respected. The union movement is
                                                    limited by its small membership – 10% of
                                                    the population

LEGAL RIGHTS: COMMENTS

29   From deprivation of   **YES**   Rights respected
     nationality

30   To be considered innocent   **no**   Many long detentions with prisoners
     until proved guilty   unaware of their offenses. Arrests on
                                                    inadequate evidence

31   To free legal aid when   **no**   Situation arbitrary. Means available to court
     necessary and counsel of   very limited
     own choice

32   From civilian trials in secret **NO**   For treason trials. Attempts to overthrow
                                                    government by violence, subversion, etc.

33   To be brought promptly   **no**   Within 24 hours by law but many long
     before a judge or court   delays, some caused by inadequate system
                                                    and undermanned courts

34   From police searches of   **NO**   Violations usually occur in areas of
     home without a warrant   operations against insurgents. Also arbitrary
                                                    searches in major towns

35   From arbitrary seizure of   **NO**   Continuous depredations in insurgency
     personal property   areas with looting, burnings, etc., by all
                                                    factions

PERSONAL RIGHTS: COMMENTS

36   To interracial, interreligious, **YES**   Males may marry at 18 years, females at 16
     or civil marriage

37   Equality of sexes during   **no**   Customary laws usually prevail. To
     marriage and for divorce   husband's advantage. Wives given legal
     proceedings   advice. Official leaflet informs citizens:
                                                    "Wife beating is against the law"

38   To practice any religion   **YES**   Rights respected

PERSONAL RIGHTS:		COMMENTS
39	To use contraceptive pills and devices	yes	Some limitations. Government influenced by ecclesiastical pressures
40	To noninterference by state in strictly private affairs	**YES**	Rights respected

USSR

Human rights rating: 54%

(at dissolution of 1917 Union, August 1991)

YES 7 yes 19 no 13 NO 1

Population: 288,600,000
Life expectancy: 70.6
Infant mortality (0–5 years)
per 1,000 births: 32
United Nations covenants ratified:
Civil and Political Rights, Economic,
Social and Cultural Rights, Convention
on Equality for Women

Form of government: One-party
Communist state (at dissolution)
GNP per capita (US$): 4,550
% of GNP spent by state:
On health: 3.2
On military: 11.5
On education: 5.2

FACTORS AFFECTING HUMAN RIGHTS

One of the most remarkable events of the 20th century has been the swift disintegration as a single state and superpower of the Union of Soviet Socialist Republics. After the failed coup attempt of August 1991, many of the republics declared their independence, later to form the Commonwealth of Independent States, and those remaining within the Union have still to agree on the form or the terms of their association. The newly independent states such as the Baltic republics have not yet established the political or constitutional structure of their countries, and it is therefore impossible to complete meaningful assessments of human rights. The questionnaire below provides a historical record for the benefit of human rights archives.

	FREEDOM TO:		COMMENTS
1	Travel in own country	yes	Obligation of internal passport usually disregarded but place of residence must still be registered
2	Travel outside own country	no	New, more liberal, law promised. Meanwhile exit visas continue. 1991 cost of passport equals 3 months' average wage
3	Peacefully associate and assemble	yes	The new freedom may be limited by frequent local states of emergency with "old guard" officials making arbitrary decisions. Position remains unpredictable
4	Teach ideas and receive information	yes	Despite new liberalization, certain restrictions continue on academic life. A 1990 decree is intended to protect "the honor and dignity of the president"
5	Monitor human rights violations	yes	Monitoring gradually being accepted though groups complain of lack of cooperation by "old guard" officials
6	Publish and educate in ethnic language	yes	The current tensions between the republics and the central government have resulted in greater "Russification" at senior levels and among officials

FREEDOM FROM:		COMMENTS
7	Serfdom, slavery, forced or child labor	**no**	Labor camps (internal exile) still in existence, though numbers being reduced with the expansion of more liberal rule
8	Extrajudicial killings or "disappearances"	**no**	Considerable reduction in unauthorized killings by security forces and military. Many killings in suppressing civil unrest and nationalistic militancy in a number of the republics
9	Torture or coercion by the state	**no**	Fewer abuses by police at local level. Compulsory psychiatric treatment of dissidents still in evidence where the "old guard" survive in power
10	Compulsory work permits or conscription of labor	**no**	Position improving but law not repealed. Convicts most affected by forced labor practices
11	Capital punishment by the state	**NO**	Approximately 300 per year. Draft legislation intended to reduce number of capital offenses from 18 to 6
12	Court sentences of corporal punishment	**YES**	Rights respected
13	Indefinite detention without charge	**no**	Some imprisonment under recent laws but long detentions still enforced, particularly of political dissidents in regions remote from main urban areas
14	Compulsory membership of state organizations or parties	**YES**	Rights respected
15	Compulsory religion or state ideology in schools	**YES**	Rights respected
16	Deliberate state policies to control artistic works	**YES**	Rights respected
17	Political censorship of press	**yes**	The previous severe censorship has been generally discontinued but there are still a few restrictions. Also most printing presses and paper supplies under government control
18	Censorship of mail or telephone tapping	**no**	The new law to limit surveillance not respected where "old guard" officials enjoy virtual independence

FREEDOM FOR OR RIGHTS TO:		COMMENTS
19	Peaceful political opposition	**yes**	Political freedom transformed by recent changes but the immediate transitional period remains unpredictable. The president is not subject to a popular vote before 1995

FREEDOM FOR OR RIGHTS TO: COMMENTS

20 Multiparty elections by secret and universal ballot — **yes**
"Leading role of the Communist party" abolished in 1990. Local and national elections now being conducted but widespread protests where Communists still retain the power to manipulate vote

21 Political and legal equality for women — **yes**
Most senior posts held by men though women approach equal numbers at lower government levels

22 Social and economic equality for women — **yes**
Traditional discrimination in some male-dominated areas, particularly in Muslim regions (12% of population)

23 Social and economic equality for ethnic minorities — **no**
"Russification" of senior government posts. Minorities suffer discrimination in security and military forces. Excessive suppression by military of nationalist movements in Armenia, Azerbaijan, and other smaller republics

24 Independent newspapers — **yes**
Relative freedom under new system of registration but major newspapers are state controlled and have many advantages

25 Independent book publishing — **no**
Independent publishing, despite the new freedom, may suffer bannings by local "old guard" officials. Still dependent on the state for material supplies

26 Independent radio and television networks — **no**
Almost totally owned and controlled by the state but liberalization of programs making significant progress

27 All courts to total independence — **yes**
"Old guard" Communist officials remain in some positions of authority and may influence the nomination of judges

28 Independent trade unions — **yes**
Independent trade unions now permitted and are being formed. Official unions enjoy many privileges – including leisure and holiday facilities

LEGAL RIGHTS: COMMENTS

29 From deprivation of nationality — **no**
Still legal but citizenship restored to a number of distinguished exiles

30 To be considered innocent until proved guilty — **yes**
Cases before appeal judges frequently dismissed to protect the credibility of the lower court. Ideological bias no longer a factor

31 To free legal aid when necessary and counsel of own choice — **yes**
New legal procedures improve previous limited facilities and an uncaring system

32 From civilian trials in secret — **yes**
Situation changing under amended law. Secret trials "unconstitutional"

LEGAL RIGHTS:		COMMENTS
33	To be brought promptly before a judge or court	no	Within 4 hours but many violations, particularly in remote areas where courts may still be dominated by "old guard" officials
34	From police searches of home without a warrant	no	Wide discretion for security forces. The new liberalization has not prevented arbitrary searches
35	From arbitrary seizure of personal property	YES	Rights respected.

PERSONAL RIGHTS:		COMMENTS
36	To interracial, interreligious, or civil marriage	YES	Rights respected. Legal marriage at 18 years for both sexes
37	Equality of sexes during marriage and for divorce proceedings	YES	Rights respected
38	To practice any religion	yes	The law on religious freedom requires registration of all faiths. Procedure controlled by the separate republics, with local variations
39	To use contraceptive pills and devices	yes	Supplies a major problem. Unavailability of pills and devices has unforeseen consequences
40	To noninterference by state in strictly private affairs	yes	But some unauthorized police surveillance of foreign visitors. Residential restrictions still continue

United Kingdom   Human rights rating: 93%

YES 33 yes 7 no 0 NO 0

Population: 57,200,000
Life expectancy: 75.7
Infant mortality (0–5 years)
  per 1,000 births: 11
United Nations covenants ratified:
Civil and Political Rights, Economic,
Social and Cultural Rights, Convention
on Equality for Women

Form of government: Parliamentary
monarchy
GNP per capita (US$): 12,810
% of GNP spent by state:
On health: 5.3
On military: 5.0
On education: 5.3

FACTORS AFFECTING HUMAN RIGHTS
The individual respected. A government answerable to the people. A generally
adequate standard of living, liberal democratic traditions, a free press, and broadcasting
channels which are renowned for their objective news programs. An "unwritten"
constitution means that there is no formal guarantee of human rights, and attempts to
pass a Bill of Rights through Parliament have failed. The major problems of Northern
Ireland remain unresolved despite repeated political efforts to end the many terrorist
killings. The country's diminished international role is being accompanied by the need
to adapt its traditional economic policies and commonwealth ties and to form a closer
association with Europe. This is seen as a change of status that mortifies many in
government and among the public.

	FREEDOM TO:		COMMENTS
1	Travel in own country	YES	But persons suspected of terrorism relating to Northern Ireland can be excluded from any part of the UK
2	Travel outside own country	YES	No legal obligation by government to issue passport but it is rarely refused. No appeal, however, against withdrawal
3	Peacefully associate and assemble	YES	Banned only when law and order threatened. No legal right to assemble on public property
4	Teach ideas and receive information	yes	Rights respected for academic freedom but government may – and does – deny nonsecurity information to the public under the Official Secrets Act
5	Monitor human rights violations	YES	Rights respected
6	Publish and educate in ethnic language	YES	Rights respected

FREEDOM FROM:		COMMENTS
7	Serfdom, slavery, forced or child labor	**YES**	Minor instances of new immigrants continuing their traditions of child labor
8	Extrajudicial killings or "disappearances"	yes	Inquests into killings in Northern Ireland by security force personnel frequently postponed. Instances of "trigger-happy" young soldiers lacking restraint
9	Torture or coercion by the state	yes	Abuses by security forces against Northern Ireland terrorists. Frequent police harassment of nonwhites and occasional charges of overzealous police behavior when controlling factious rallies
10	Compulsory work permits or conscription of labor	**YES**	Rights respected
11	Capital punishment by the state	**YES**	Theoretically for treason and arson in the Royal Dockyards. Last hanging 1964
12	Court sentences of corporal punishment	**YES**	Rights respected
13	Indefinite detention without charge	**YES**	Rights respected
14	Compulsory membership of state organizations or parties	**YES**	Rights respected
15	Compulsory religion or state ideology in schools	**YES**	Rights respected
16	Deliberate state policies to control artistic works	**YES**	Rights respected
17	Political censorship of press	**YES**	Editors may be inhibited by the Official Secrets Act, libel laws, and the Contempt of Court Act, which may oblige journalists to disclose sources of information "in the interests of justice," etc.
18	Censorship of mail or telephone tapping	yes	Telephone tapping. In excess of 500 a year. Greater official activity in the control of computer "hacking." Minor abuses of privacy

FREEDOM FOR OR RIGHTS TO:		COMMENTS
19	Peaceful political opposition	**YES**	Rights respected
20	Multiparty elections by secret and universal ballot	**YES**	Rights respected
21	Political and legal equality for women	yes	Women specifically underrepresented in government and Parliament despite recent prime minister being a women
22	Social and economic equality for women	yes	Inequalities persist, despite legal requirements, in pay, employment, and opportunities

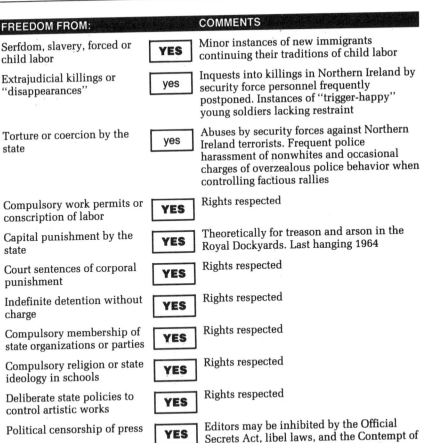

FREEDOM FOR OR RIGHTS TO:		COMMENTS
23	Social and economic equality for ethnic minorities	**YES**	But social and economic evidence indicates that certain ethnic groups suffer higher unemployment, fewer opportunities for work promotion, etc.
24	Independent newspapers	**YES**	Rights respected but in reality a few wealthy individuals control most of the national newspapers
25	Independent book publishing	**YES**	Rights respected. But publishers need prudence on blasphemy matters and security subjects that may conflict with official rulings
26	Independent radio and television networks	**YES**	Rights respected but government may veto programs on security grounds. Some public concern at ministerial sensitivity to criticisms
27	All courts to total independence	**YES**	Occasional charges of "class bias" of judges and magistrates, indicating insensitivity toward unfamiliar social realities
28	Independent trade unions	**YES**	Rights respected

LEGAL RIGHTS:		COMMENTS
29	From deprivation of nationality	**YES**	A few cases concerning deported Asian nonresidents have been disputed. The complex position of Hong Kong citizens and their "British" status still causes dissatisfaction among Chinese British subjects
30	To be considered innocent until proved guilty	**YES**	Rights respected
31	To free legal aid when necessary and counsel of own choice	**YES**	Rights respected but the system is becoming increasingly underfunded by government economies
32	From civilian trials in secret	**YES**	Rights respected
33	To be brought promptly before a judge or court	yes	Within 24–48 hours, extending to 4 days (7 days for suspected terrorists). Long remand periods. The Northern Ireland situation has forced constant review of judicial procedures
34	From police searches of home without a warrant	**YES**	Rights respected but the degree of violence in Northern Ireland permits entry without warrant on "reasonable grounds of suspicion"
35	From arbitrary seizure of personal property	**YES**	Rights respected

PERSONAL RIGHTS:		COMMENTS
36	To interracial, interreligious, or civil marriage	**YES**	Subject to immigration controls on certain restricted categories of would-be spouses
37	Equality of sexes during marriage and for divorce proceedings	**YES**	Rights respected
38	To practice any religion	**YES**	Rights respected
39	To use contraceptive pills and devices	**YES**	From age 16
40	To noninterference by state in strictly private affairs	**YES**	Rights respected

United States of America

Human rights rating: 90%

YES 31 yes 8 no 1 NO 0

Population: 249,200,000
Life expectancy: 75.9
Infant mortality (0–5 years) per 1,000 births: 13
United Nations covenants ratified: None

Form of government: Federal democracy
GNP per capita (US$): 19,840
% of GNP spent by state:
On health: 4.5
On military: 6.7
On education: 5.3

FACTORS AFFECTING HUMAN RIGHTS

The individual respected. Government answerable to the people. The country enjoys genuine democratic institutions, a high average level of prosperity, and a free press. During 1991, following its military triumph in the gulf war and after the disintegration of the USSR, the USA became the dominant world power and is taking initiatives to improve human rights in many oppressed countries and to bring peace to warring areas. At home, however, the human rights position varies marginally between 50 different states, and at the end of 1990 there were over 2,000 people under sentence of death in 34 of these states. Certain ethnic groups are at a disadvantage both socially and economically and this causes tensions in heavily populated cities. The USA has not ratified the human rights covenants of the United Nations.

	FREEDOM TO:		COMMENTS
1	Travel in own country	**YES**	Rights respected
2	Travel outside own country	**YES**	Rights respected
3	Peacefully associate and assemble	**YES**	Rights respected
4	Teach ideas and receive information	yes	Surveillance of groups hostile to aspects of US foreign policy. Seizures of computer equipment and censorship by US security service are increasing with attempts to monitor advanced systems of communication
5	Monitor human rights violations	**YES**	A major monitor of human rights in most countries
6	Publish and educate in ethnic language	**YES**	Rights respected

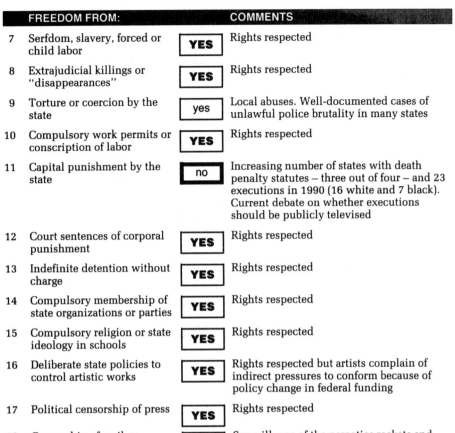

FREEDOM FROM:		COMMENTS
7	Serfdom, slavery, forced or child labor	**YES**	Rights respected
8	Extrajudicial killings or "disappearances"	**YES**	Rights respected
9	Torture or coercion by the state	yes	Local abuses. Well-documented cases of unlawful police brutality in many states
10	Compulsory work permits or conscription of labor	**YES**	Rights respected
11	Capital punishment by the state	no	Increasing number of states with death penalty statutes – three out of four – and 23 executions in 1990 (16 white and 7 black). Current debate on whether executions should be publicly televised
12	Court sentences of corporal punishment	**YES**	Rights respected
13	Indefinite detention without charge	**YES**	Rights respected
14	Compulsory membership of state organizations or parties	**YES**	Rights respected
15	Compulsory religion or state ideology in schools	**YES**	Rights respected
16	Deliberate state policies to control artistic works	**YES**	Rights respected but artists complain of indirect pressures to conform because of policy change in federal funding
17	Political censorship of press	**YES**	Rights respected
18	Censorship of mail or telephone tapping	yes	Surveillance of the narcotics rackets and gangster operations. The wider increase in telephone tapping includes investigations of innocent computer operators as well as criminal "hackers." Also the monitoring of electronic mail

FREEDOM FOR OR RIGHTS TO:		COMMENTS
19	Peaceful political opposition	**YES**	Rights respected
20	Multiparty elections by secret and universal ballot	**YES**	Rights respected but in a few states convicts are disenfranchised during their sentences
21	Political and legal equality for women	yes	Despite constitution, women remain underrepresented in politics and in senior government posts. 2% of US senators are women
22	Social and economic equality for women	yes	Average income for women calculated at 70% of men's pay. No female justice yet elected to Supreme Court

FREEDOM FOR OR RIGHTS TO: COMMENTS

23	Social and economic equality for ethnic minorities	yes	Wide inequalities, despite constitution, in many areas for blacks and Hispanics. A subsociety is largely populated by these deprived groups
24	Independent newspapers	**YES**	Rights respected
25	Independent book publishing	**YES**	Rights respected but increased commercialism is influencing publishers' preference for "safe" books for a "popular" market
26	Independent radio and television networks	**YES**	Indirect social and official pressures toward conformism. Advertisers on 8,000–9,000 commercial radio and television stations tending toward narrow diversity of programs
27	All courts to total independence	**YES**	Rights respected
28	Independent trade unions	**YES**	Rights respected

LEGAL RIGHTS: COMMENTS

29	From deprivation of nationality	**YES**	Rights respected
30	To be considered innocent until proved guilty	**YES**	Rights respected
31	To free legal aid when necessary and counsel of own choice	yes	Limited and short of funding. "Public defender" usually appointed by court
32	From civilian trials in secret	**YES**	Rights respected
33	To be brought promptly before a judge or court	**YES**	Increased to 48 hours from 24 hours
34	From police searches of home without a warrant	**YES**	Only when in "hot pursuit"
35	From arbitrary seizure of personal property	**YES**	Rights respected but see question 4

PERSONAL RIGHTS: COMMENTS

36	To interracial, interreligious, or civil marriage	**YES**	Rights respected
37	Equality of sexes during marriage and for divorce proceedings	**YES**	Rights respected
38	To practice any religion	**YES**	Rights respected

PERSONAL RIGHTS:		COMMENTS
39	To use contraceptive pills and devices	**YES**	The urgency to control AIDS epidemic has taken precedence over recent attempts to reduce juvenile pregnancies by discouraging use of contraceptives by the young, and so deterring intercourse
40	To noninterference by state in strictly private affairs	yes	As question 18. Foundation proposing extension to Bill of Rights to protect right of privacy, etc. State laws vary in tolerance of homosexuality and sexual deviants. Pressures to curtail present abortion rights causing widespread protests

Uruguay

YES 31 yes 9 no 0 NO 0

Population: 3,100,000
Life expectancy: 72.2
Infant mortality (0–5 years)
per 1,000 births: 27
United Nations covenants ratified:
Civil and Political Rights, Economic,
Social and Cultural Rights, Convention
on Equality for Women

Form of government: Multiparty
democracy
GNP per capita (US$): 2,470
% of GNP spent by state:
On health: 2.7
On military: 1.6
On education: 6.6

FACTORS AFFECTING HUMAN RIGHTS

After 12 years of military rule, the civilian government was restored and the country
enjoys a multiparty democratic system. It is a leading champion of human rights in
Latin America and has a free press and responsible political institutions. A few
investigations of human rights crimes by the previous military government continue
and action against guilty officials remains a public issue. There is also some unease
about a few abuses by the police, many of whom served under a regime that was
described as "the torture chamber of South America."

	FREEDOM TO:		COMMENTS
1	Travel in own country	YES	Rights respected
2	Travel outside own country	YES	Rights respected
3	Peacefully associate and assemble	YES	Rights respected
4	Teach ideas and receive information	YES	Rights respected
5	Monitor human rights violations	YES	Rights respected
6	Publish and educate in ethnic language	YES	Rights respected

	FREEDOM FROM:		COMMENTS
7	Serfdom, slavery, forced or child labor	yes	Limited rural child labor
8	Extrajudicial killings or "disappearances"	yes	Several suspicious deaths while in police custody. Investigations continuing
9	Torture or coercion by the state	yes	Some local abuses by overzealous police and prison wardens

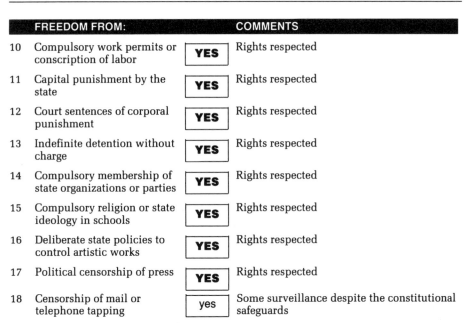

	FREEDOM FROM:		COMMENTS
10	Compulsory work permits or conscription of labor	**YES**	Rights respected
11	Capital punishment by the state	**YES**	Rights respected
12	Court sentences of corporal punishment	**YES**	Rights respected
13	Indefinite detention without charge	**YES**	Rights respected
14	Compulsory membership of state organizations or parties	**YES**	Rights respected
15	Compulsory religion or state ideology in schools	**YES**	Rights respected
16	Deliberate state policies to control artistic works	**YES**	Rights respected
17	Political censorship of press	**YES**	Rights respected
18	Censorship of mail or telephone tapping	yes	Some surveillance despite the constitutional safeguards

	FREEDOM FOR OR RIGHTS TO:		COMMENTS
19	Peaceful political opposition	**YES**	Rights respected
20	Multiparty elections by secret and universal ballot	**YES**	Rights respected. Elections every 5 years
21	Political and legal equality for women	yes	Following the restoration of democracy in 1984, equality being achieved in most areas
22	Social and economic equality for women	yes	Traditional and regional attitude toward women perpetuates pay and employment inequalities
23	Social and economic equality for ethnic minorities	yes	Black minority, 4% of population, suffer disadvantages and have limited opportunities in civil service and education
24	Independent newspapers	**YES**	Rights respected
25	Independent book publishing	**YES**	Rights respected
26	Independent radio and television networks	**YES**	Most stations privately owned and free of controls
27	All courts to total independence	**YES**	Rights respected
28	Independent trade unions	**YES**	Rights respected. A new labor law being introduced to reduce the numerous strikes

LEGAL RIGHTS:		COMMENTS
29	From deprivation of nationality	**YES**	Rights respected
30	To be considered innocent until proved guilty	**YES**	Rights respected
31	To free legal aid when necessary and counsel of own choice	yes	Means test. Lawyer appointed ("public defender")
32	From civilian trials in secret	**YES**	Rights respected
33	To be brought promptly before a judge or court	**YES**	Within 24 hours
34	From police searches of home without a warrant	yes	Occasional violations despite legal safeguards
35	From arbitrary seizure of personal property	**YES**	Rights respected

PERSONAL RIGHTS:		COMMENTS
36	To interracial, interreligious, or civil marriage	**YES**	Rights respected
37	Equality of sexes during marriage and for divorce proceedings	**YES**	Marriage legal at 21 for both sexes
38	To practice any religion	**YES**	Rights respected
39	To use contraceptive pills and devices	**YES**	Rights respected
40	To noninterference by state in strictly private affairs	**YES**	Rights respected

Venezuela

Human rights rating: 75%

YES 23 yes 12 no 3 NO 2

Population: 19,700,000
Life expectancy: 70
Infant mortality (0–5 years) per 1,000 births: 44
United Nations covenants ratified: Civil and Political Rights, Economic, Social and Cultural Rights, Convention on Equality for Women

Form of government: Multiparty democracy
GNP per capita (US$): 3,250
% of GNP spent by state:
On health: 2.2
On military: 1.6
On education: 4.3

FACTORS AFFECTING HUMAN RIGHTS

A democratic system with a free press, multiparty elections that have continued for over 30 years, and an oil-based economy providing the highest average income in South America. The human rights situation has, however, worsened over the last 5 years, principally because of arbitrary actions by the police and security forces. Extrajudicial killings, "disappearances," and well-documented instances of torture occur but little official action is taken against the perpetrators. Some of these excesses are against indigenous peoples in remote areas of the country. Corruption at all levels of the government and judiciary is widespread.

	FREEDOM TO:		COMMENTS
1	Travel in own country	YES	Rights respected
2	Travel outside own country	YES	Rights respected
3	Peacefully associate and assemble	YES	Rights respected
4	Teach ideas and receive information	YES	Rights respected
5	Monitor human rights violations	yes	Some unauthorized police harassment of monitors investigating unlawful killings, torture, and detentions
6	Publish and educate in ethnic language	YES	Rights respected

	FREEDOM FROM:		COMMENTS
7	Serfdom, slavery, forced or child labor	yes	Regional pattern of rural child labor

FREEDOM FROM:		COMMENTS
8	Extrajudicial killings or "disappearances"	**NO**	Many instances of unlawful killings and "disappearances" by security and police forces but few charges against the perpetrators
9	Torture or coercion by the state	**NO**	Well-documented reports, electric shocks, burning, beatings, rape, etc., by police and prison guards. Miscreants appear to go unpunished
10	Compulsory work permits or conscription of labor	**YES**	Rights respected
11	Capital punishment by the state	**YES**	Abolished 1863
12	Court sentences of corporal punishment	**YES**	Rights respected
13	Indefinite detention without charge	yes	A few long detentions are the result of an inadequate system, indifference, and corruption
14	Compulsory membership of state organizations or parties	**YES**	Rights respected
15	Compulsory religion or state ideology in schools	**YES**	Rights respected
16	Deliberate state policies to control artistic works	**YES**	Rights respected
17	Political censorship of press	**YES**	Rights respected
18	Censorship of mail or telephone tapping	yes	Some unauthorized telephone tapping

FREEDOM FOR OR RIGHTS TO:		COMMENTS
19	Peaceful political opposition	**YES**	Rights respected
20	Multiparty elections by secret and universal ballot	**YES**	Rights respected
21	Political and legal equality for women	yes	Traditional male attitudes limit women's promotion to senior posts – and in professions
22	Social and economic equality for women	yes	Pay and employment inequalities. Government campaign to inform women of their rights
23	Social and economic equality for ethnic minorities	yes	Well-documented evidence of discrimination against indigenous peoples. Also of police brutality in remote areas
24	Independent newspapers	**YES**	Rights respected
25	Independent book publishing	**YES**	Rights respected

FREEDOM FOR OR RIGHTS TO:		COMMENTS
26	Independent radio and television networks	**YES**	Both state owned and private stations – with little evidence of political bias
27	All courts to total independence	yes	The judicial process is influenced by both political factors and corruption
28	Independent trade unions	**YES**	Rights respected

LEGAL RIGHTS:		COMMENTS
29	From deprivation of nationality	**YES**	Rights respected
30	To be considered innocent until proved guilty	no	In many instances bribery of officials is the most effective way of proving innocence
31	To free legal aid when necessary and counsel of own choice	yes	Means test but court chooses counsel or "public defender"
32	From civilian trials in secret	no	Civilians drawn into wide definition of security cases, which may legally be tried in secret
33	To be brought promptly before a judge or court	no	Long delays. First court appearance should be within 8 days. Situation sometimes influenced by bribes
34	From police searches of home without a warrant	yes	Some unauthorized searches by security police – often, in fact, a search for gain
35	From arbitrary seizure of personal property	yes	In remote areas, security forces and National Guard occasionally loot and dispossess indigenous tribes

PERSONAL RIGHTS:		COMMENTS
36	To interracial, interreligious, or civil marriage	**YES**	Rights respected. Males may marry at 14, females at 12 years
37	Equality of sexes during marriage and for divorce proceedings	yes	Some inequalities suffered by women during divorce despite official efforts to correct these. Traditional attitudes a factor
38	To practice any religion	**YES**	Rights respected
39	To use contraceptive pills and devices	**YES**	Limited government support
40	To noninterference by state in strictly private affairs	**YES**	Rights respected

Vietnam

Human rights rating: 27%

YES 5 **yes** 3 **no** 14 NO 18

Population: 66,700,000
Life expectancy: 62.7
Infant mortality (0–5 years) per 1,000 births: 84
United Nations covenants ratified: Civil and Political Rights, Economic, Social and Cultural Rights, Convention on Equality for Women

Form of government: One-party Communist state
GNP per capita (US$): 220
% of GNP spent by state:
On health: n/a
On military: n/a
On education: n/a

FACTORS AFFECTING HUMAN RIGHTS

Since the country was unified in 1975, after a long civil war, the Communist party has imposed its ideology and authority on all areas of government and public life. Its failure to improve the economy significantly led to the *doi moi* or "renovation policy" but this has had only limited success. The end of communism in the Soviet Union and the resultant reduction or loss of trade and aid have, however, meant a reversal of the more liberal trend and an improvement of human rights. The continuing problems of the economy are, most recently, dominating government policy and, following the peace agreement in neighboring Cambodia, Vietnam is attempting to increase foreign trade and improve international relations.

	FREEDOM TO:		COMMENTS
1	Travel in own country	yes	But permits required for change of residence
2	Travel outside own country	no	A limited relaxation has been caused by the many departures of emigrants and clandestine refugees and the government's wish to retain the loyalties of the increasing numbers of overseas Vietnamese
3	Peacefully associate and assemble	NO	Only in support of government and the Communist party
4	Teach ideas and receive information	NO	All education conforms to party ideology. Curricula controlled
5	Monitor human rights violations	no	Limited and supervised visits by international human rights investigators
6	Publish and educate in ethnic language	YES	Rights respected

	FREEDOM FROM:		COMMENTS
7	Serfdom, slavery, forced or child labor	no	Reeducation camps, which are currently being revived, are sources of forced labor

FREEDOM FROM:		COMMENTS
8	Extrajudicial killings or "disappearances"	no	Usually the result of torture, hard labor, neglect, and semistarvation of prisoners
9	Torture or coercion by the state	no	Many beatings during interrogation and as punishment in reeducation centers
10	Compulsory work permits or conscription of labor	**NO**	The national program, with many workers drafted to New Economic Zones, relies on conscripted labor
11	Capital punishment by the state	**NO**	By shooting. The major program of eliminating most dangerous enemies from the previously independent south has been completed
12	Court sentences of corporal punishment	no	The degree of officially condoned corporal punishment in prisons and camps must be seen as a state crime
13	Indefinite detention without charge	**NO**	Estimates suggest 2,000 in detention. Commonest in south of country. Duration of confinement in reeducation camps varies with the "ability to learn well"
14	Compulsory membership of state organizations or parties	**YES**	Rights respected
15	Compulsory religion or state ideology in schools	**NO**	Political and ideological indoctrination of children
16	Deliberate state policies to control artistic works	**NO**	Surveillance by regional "council of arts." Writers and artists rearrested after policy change by party and government
17	Political censorship of press	**NO**	Newspapers should be submitted for scrutiny before circulation. Tighter censorship in 1991
18	Censorship of mail or telephone tapping	**NO**	Wide surveillance. After a brief relaxation, controls more vigorously applied

FREEDOM FOR OR RIGHTS TO:		COMMENTS
19	Peaceful political opposition	**NO**	Nonexistent. Opposition not tolerated
20	Multiparty elections by secret and universal ballot	**NO**	A National Assembly of approximately 500 members gives formal approval to party policies
21	Political and legal equality for women	yes	Senior posts usually held by men. Traditional Confucian ideas of woman's role still a strong influence
22	Social and economic equality for women	no	The higher status enjoyed by women during the north-south war is reverting to the traditional inequalities

FREEDOM FOR OR RIGHTS TO:		COMMENTS
23	Social and economic equality for ethnic minorities	no	Chinese, Indian, and other ethnic groups suffer discrimination. Suspicion persists toward nationals of the previously independent south
24	Independent newspapers	NO	1989 Press Law states that all media represent the Communist party. No privately owned newspapers
25	Independent book publishing	NO	All control rests with the ruling party
26	Independent radio and television networks	NO	State owned and controlled. Virtually the "voice of the party"
27	All courts to total independence	NO	Courts and local tribunals all subject to party directives
28	Independent trade unions	NO	Unions are an extension of the Communist party. Strikes "unpatriotic" and "antisocial"

LEGAL RIGHTS:		COMMENTS
29	From deprivation of nationality	no	Vietnamese who have fled their country may lose their nationality
30	To be considered innocent until proved guilty	NO	With the state exercising absolute control, convincing proof may be manufactured
31	To free legal aid when necessary and counsel of own choice	no	Trial is in the hands of the state, with counsel an obedient government employee
32	From civilian trials in secret	NO	Sometimes *in camera*. For many categories of crime
33	To be brought promptly before a judge or court	no	New procedures during the *doi moi* (renovation) period introduced greater safeguards but long delays still usual
34	From police searches of home without a warrant	yes	Some abuses. Particularly against the population of the previously independent south of the country
35	From arbitrary seizure of personal property	no	Property of political prisoners and of those who have fled the country may be seized

PERSONAL RIGHTS:		COMMENTS
36	To interracial, interreligious, or civil marriage	YES	Males may marry at 20 years, females at 18
37	Equality of sexes during marriage and for divorce proceedings	YES	Rights respected
38	To practice any religion	no	Arrests of priests and monks, particularly Buddhists, charged with unlawful religious and political activities

PERSONAL RIGHTS:		COMMENTS
39	To use contraceptive pills and devices	**YES**	State support
40	To noninterference by state in strictly private affairs	no	Constant surveillance by neighborhood informants watching for political or social unorthodoxy creates fear and form of intrusion into private life

Yemen

Human rights rating: 49%

YES 8 yes 8 no 20 NO 4

Population: 11,700,000
Life expectancy: 51.5
Infant mortality (0–5 years) per 1,000 births: n/a
United Nations covenants ratified:
Civil and Political Rights, Economic,
Social and Cultural Rights, Convention
on Equality for Women

Form of government: Presidential
council (transitional)
GNP per capita (US$): 595
% of GNP spent by state:
On health: 1.2
On military: 9.1
On education: 5.6

FACTORS AFFECTING HUMAN RIGHTS
In 1990 the previous Marxist state of south Yemen and the orthodox Islamic state of the
northern Yemen Arab Republic united to form a single country. There have been
significant human rights improvements, including the release of political prisoners of
both of the previous governments. The transitional Presidential Council, which will
remain in power during a 30-month period, is planning multiparty elections and has
already introduced a new constitution. It has also ratified the two major human rights
covenants of the United Nations, which had not been done before.

	FREEDOM TO:		COMMENTS
1	Travel in own country	yes	Identity checks. Some arbitrary behavior by police at checkpoints, usually to extract bribes
2	Travel outside own country	yes	The country has a large emigrant labor force though exit visas necessary
3	Peacefully associate and assemble	yes	Rallies and protests being treated more indulgently by the security forces of the unified country but demonstrators follow understood guidelines
4	Teach ideas and receive information	no	Situation improving following Unity Agreement but self-censorship advisable. The country's composition of 99% Muslims compels conformism on many subjects
5	Monitor human rights violations	yes	Position transformed following the Unity Agreement between the two previously independent states. Monitoring now possible
6	Publish and educate in ethnic language	YES	Rights respected

	FREEDOM FROM:		COMMENTS
7	Serfdom, slavery, forced or child labor	yes	Regional pattern of child labor though officially not under 12 years of age
8	Extrajudicial killings or "disappearances"	no	A number of "disappearances" in the immediate (1989–90) preunification period remain unexplained. Situation unpredictable as two previously oppressive security forces are merged
9	Torture or coercion by the state	no	Fewer cases of torture as country adapts to unification. Practice of keeping prisoners in shackles to be discontinued
10	Compulsory work permits or conscription of labor	YES	Rights respected
11	Capital punishment by the state	no	No sentences carried out since national unification in May 1990. Abolition may be under review
12	Court sentences of corporal punishment	no	Flogging a possible punishment of cases under Shari'a law
13	Indefinite detention without charge	no	Many releases of prisoners under the previous regimes of north and south Yemen. Situation improving though not totally reassuring
14	Compulsory membership of state organizations or parties	YES	Rights respected
15	Compulsory religion or state ideology in schools	NO	Islamic instruction compulsory in most schools. Changes in previous Marxist schools of the former People's Democratic Republic. Situation in transition
16	Deliberate state policies to control artistic works	no	The previous limitations of an orthodox Islamic state and a Marxist state have not disappeared with a declaration of national unity. Western values and "vulgarity" not favored
17	Political censorship of press	no	Situation improving with the adoption of a democratic constitution but self-censorship, plus restrictions in practice, related to the previous authoritarian state limit press freedom
18	Censorship of mail or telephone tapping	no	The merged security services of the two previously independent countries continue wide surveillance

	FREEDOM FOR OR RIGHTS TO:		COMMENTS
19	Peaceful political opposition	yes	The unified country is moving toward a democratic system with little disruption, though return of Yemenis expelled during and after the gulf war, and equaling 7% of population, may affect situation

FREEDOM FOR OR RIGHTS TO:		COMMENTS
20	Multiparty elections by secret and universal ballot	**no**	After a 30-month period of unification of north and south Yemen, multiparty elections are planned for late 1992. Numerous parties now being formed
21	Political and legal equality for women	**NO**	New constitution does not admit equality of women, who continue in their traditional role. Women discouraged in the north to become involved in politics
22	Social and economic equality for women	**no**	Greater equality in the previous Marxist part of the country but, overall, pay and employment disadvantages, particularly in the more traditional north
23	Social and economic equality for ethnic minorities	**YES**	Rights respected
24	Independent newspapers	**no**	Although the Unity Agreement has been followed by many independent newspapers, many editions seized for "subversive" reports and features. Period of transition to greater democracy
25	Independent book publishing	**yes**	Circumspection required. The unification of a Marxist state with a traditional Islamic state requires publishing prudence in the interim period
26	Independent radio and television networks	**NO**	State owned and controlled
27	All courts to total independence	**no**	The new constitution still under consideration. Meanwhile, previous judicial systems, including that of Islamic Shari'a law, continue – with limited independence in practice
28	Independent trade unions	**no**	New union laws expected during the 30-month transitional unification period but the old codes still apply – though less rigorously

LEGAL RIGHTS:		COMMENTS
29	From deprivation of nationality	**YES**	Rights respected following country's unification
30	To be considered innocent until proved guilty	**no**	Arbitrary arrests may be resolved by bribes and similar forms of "negotiations"
31	To free legal aid when necessary and counsel of own choice	**yes**	Traditional practices with assistance for defendant depending on the particular court system and "tribal ties"
32	From civilian trials in secret	**YES**	Rights respected

LEGAL RIGHTS: COMMENTS

33 To be brought promptly
    before a judge or court   **no**   The new constitution states within 24 hours
                                           but this conflicts with long delays in
                                           practice. Bribery of officials at all levels

34 From police searches of   **no**   The system of arbitrary searches in the
    home without a warrant   previous parts of the newly united Yemen
                                           may be moderating but still continues

35 From arbitrary seizure of   **no**   The extent of unlawful seizures by police
    personal property   and security forces of the merged north and
                                           south Yemen may be lessening. Mostly in
                                           remote areas

PERSONAL RIGHTS: COMMENTS

36 To interracial, interreligious, **no**   Country 99% Muslim. Intermarriage an
    or civil marriage   offence against the divine will and under
                                              Shari'a law could invoke a mandatory death
                                              penalty

37 Equality of sexes during   **NO**   Despite equality under the law of the
    marriage and for divorce   previous Marxist regime of south Yemen,
    proceedings   traditional Islamic law permits polygamy,
                                          dowry marriages, child brides, etc.
                                          Husband's permission needed for travel,
                                          taking a job, etc.

38 To practice any religion   **no**   Islam is the state religion and therefore
                                          apostasy can invoke the death penalty. No
                                          proselytizing. Situation confused as the
                                          previous Marxist half of country merges
                                          with the orthodox Islamic half

39 To use contraceptive pills   **YES**   Rights respected
    and devices

40 To noninterference by state **YES**   Rights respected
    in strictly private affairs

Yugoslavia (at dissolution of 1945 Communist republic, mid-1991)

Human rights rating: 55%

YES 9 **yes** 18 **no** 10 NO 3

Population: 23,800,000
Life expectancy: 72.6
Infant mortality (0–5 years)
  per 1,000 births: 28
United Nations covenants ratified:
Civil and Political Rights, Economic,
Social and Cultural Rights, Convention
on Equality for Women

Form of government: One-party
socialist state
GNP per capita (US$): 2,520
% of GNP spent by state:
On health: 4.3
On military: 4.0
On education: 3.8

FACTORS AFFECTING HUMAN RIGHTS
The political, social, and economic changes in Eastern Europe during 1990–91 hastened
the crisis in Yugoslavia which led to one of the constituent republics, Slovenia,
becoming independent and another, Croatia, involved in a civil war with the federal
national army (dominated by Serbia) after secession from Yugoslavia was refused. There
is further unrest in other republics. Historically, the Balkan area has been one of
repeated wars and ethnic conflicts. The information below relates to the period
immediately before the first of the breakaway republics became independent.

FREEDOM TO:		COMMENTS
1 Travel in own country	**YES**	Rights respected
2 Travel outside own country	yes	Restrictions on a small number of known dissidents. Ethnic Albanians and Gypsies occasionally refused passports
3 Peacefully associate and assemble	no	Demonstrations in favor of various separatist causes, free elections at federal level, etc., violently suppressed by police. Formations of associations strictly scrutinized
4 Teach ideas and receive information	yes	Academics and professors displaying separatist sympathies may risk career advantages. Situation worst in Serbia
5 Monitor human rights violations	yes	Occasional interference with visiting international monitors. Local organizations increasingly active
6 Publish and educate in ethnic language	no	Major discrimination in Serbia against Albanians in Kosovo province. Of language, literature, academic life,, etc.

FREEDOM FROM:		COMMENTS
7	Serfdom, slavery, forced or child labor	**YES**	Rights respected
8	Extrajudicial killings or "disappearances"	**NO**	Many arbitrary killings in the period before full-scale war between the republic of Croatia and the Serbian-dominated federal army. Particularly of protesting ethnic Albanians
9	Torture or coercion by the state	**NO**	Usually in Kosovo province where the ruling Serbian police conduct campaign of suppression against Albanian minority
10	Compulsory work permits or conscription of labor	no	In Kosovo province the Serbian authorities may take action against those not holding a labor permit. This may take the form of direction of labor to specific employment
11	Capital punishment by the state	**NO**	By firing squad
12	Court sentences of corporal punishment	**YES**	Rights respected
13	Indefinite detention without charge	no	Many political prisoners, usually dissident Albanians in parts of Serbia. Thousands held for the permitted 3-month period of investigatory detention
14	Compulsory membership of state organizations or parties	**YES**	Rights respected
15	Compulsory religion or state ideology in schools	no	Only in those republics that continue to teach Marxism-Leninism
16	Deliberate state policies to control artistic works	yes	The Communist government in Serbia, unlike the other republics, continues surveillance of cultural groups. Funding, concessions, etc., may be limited or withdrawn
17	Political censorship of press	yes	Despite more press freedom in recent years, local political leaders may intimidate newspapers. Two copies of every issue must be sent to the public prosecutor
18	Censorship of mail or telephone tapping	no	Surveillance, particularly of those supporting separatist "nationalism." Most extensive in Serbia

FREEDOM FOR OR RIGHTS TO:		COMMENTS
19	Peaceful political opposition	no	Not at federal level. No general agreement by republics on a more democratic (national) constitution. Ethnic conflicts during federal elections

FREEDOM FOR OR RIGHTS TO:		COMMENTS
20	Multiparty elections by secret and universal ballot	no	Federal government maintains one-party rule but in 1990 the separate republics introduced their own multiparty elections. Complaints in Serbia of violations by the former Communist party
21	Political and legal equality for women	yes	Increasing presence at senior government level and in the professions though these are still dominated by men
22	Social and economic equality for women	yes	Local traditions in south of country continue and perpetuate the inferior status of women. 9% of population Muslim
23	Social and economic equality for ethnic minorities	yes	Albanian minority suffers discrimination in Serbia and Gypsies throughout the country. The ethnic admixture in most republics also creates tensions
24	Independent newspapers	no	Most restrictions in Serbia where major Albanian-language newspaper closed. The Serbian domination of federal security forces occasionally affects press freedoms elsewhere
25	Independent book publishing	yes	Apart from situation in Serbia, book banning rare
26	Independent radio and television networks	yes	Independent channels have begun broadcasting in most republics but control persists in Serbia
27	All courts to total independence	no	Constitutional changes improving independence of courts at both federal and republic levels. Political pressures on judiciary still, however, a factor, and ethnic bias may influence trials
28	Independent trade unions	yes	Independent unions now established throughout country after the ending of previous Communist domination though some restrictions in Serbia

LEGAL RIGHTS:		COMMENTS
29	From deprivation of nationality	YES	Rights respected
30	To be considered innocent until proved guilty	yes	Arbitrary arrests of Albanians in Serbia. Constant surveillance for "nationalist" activities. Guilt frequently assumed
31	To free legal aid when necessary and counsel of own choice	yes	Legal aid but court may appoint defense lawyer
32	From civilian trials in secret	yes	Recent liberalization in most republics makes such trials unlikely but still legal for "antistate" activities

LEGAL RIGHTS:		COMMENTS
33	To be brought promptly before a judge or court	yes	Criminal codes vary in the different republics. Many being revised but 3-month investigatory detentions may delay court appearance
34	From police searches of home without a warrant	yes	Some exceptions by security forces in Kosovo province in operations against Albanian separatists
35	From arbitrary seizure of personal property	yes	Usually affecting the Albanian minority in Serbia where seizures may be described as confiscations of property intended for criminal purposes

PERSONAL RIGHTS:		COMMENTS
36	To interracial, interreligious, or civil marriage	YES	Rights respected
37	Equality of sexes during marriage and for divorce proceedings	YES	Rights respected
38	To practice any religion	YES	Rights respected but the differing ethnic groups, with their own religions, may clash at local level over religious issues
39	To use contraceptive pills and devices	YES	Rights respected
40	To noninterference by state in strictly private affairs	yes	The only area of significant interference is in Kosovo where Serbia attempts to dominate all aspects of the lives of Albanians affect private affairs

Zaire

Human rights rating: 40%

YES 4 yes 12 no 17 NO 7

Population: 35,000,000
Life expectancy: 53
Infant mortality (0–5 years)
 per 1,000 births: 32
United Nations covenants ratified:
Civil and Political Rights, Economic,
Social and Cultural Rights, Convention
on Equality for Women

Form of government: One-party state
GNP per capita (US$): 170
% of GNP spent by state:
On health: 0.8
On military: 3.0
On education: 0.4

FACTORS AFFECTING HUMAN RIGHTS

After the precipitate withdrawal in 1960 of the Belgian colonial power, the country was
subjected to a prolonged civil war along tribal lines. A period of instability followed
until the present president with the support of the military forces seized power in 1965.
Since then his rule has been absolute, with authority to rule by decree, and there has
been little respect for human rights. Although having ratified the major covenants of the
United Nations, there have been many arbitrary killings, atrocities, torture, rape, and
incommunicado detentions by ill-disciplined, corrupt, and underpaid police and
security forces. Recent riots and violent protests against oppressive rule have, however,
forced the president to concede a form of multiparty rule but, to date, this has neither
been fully established nor free of his dominant influence.

	FREEDOM TO:		COMMENTS
1	Travel in own country	yes	Many roadblocks and checkpoints where underpaid police and military impose on-the-spot fines and accept bribes
2	Travel outside own country	yes	Exit visas required. Passports may be refused on "security" grounds. But bribes may facilitate clearance
3	Peacefully associate and assemble	no	Violent demonstrations by many opposing factions and rioting soldiers erupt periodically, with many murders, looting, and arson. In the circumstances, the law against assemblies protesting against the president is academic
4	Teach ideas and receive information	yes	Position improving, perhaps temporarily, as the democracy movement progresses. Student protests in most universities are harshly suppressed by security forces
5	Monitor human rights violations	yes	National and international monitors, while not being obstructed, are not being encouraged

FREEDOM TO:		COMMENTS
6	Publish and educate in ethnic language	**yes**	But certain tribal groups have not developed a written language

FREEDOM FROM:		COMMENTS
7	Serfdom, slavery, forced or child labor	**yes**	Pattern of child labor follows regional customs
8	Extrajudicial killings or "disappearances"	**NO**	Numbers of killings vary, with national human rights groups claiming "hundreds." Many during suppression of student protests and in arbitrary actions by the military in remote areas of the country
9	Torture or coercion by the state	**NO**	Well-documented evidence of torture, maltreatment, rape of female prisoners, etc.
10	Compulsory work permits or conscription of labor	**YES**	Rights respected
11	Capital punishment by the state	**NO**	Hanging, shooting by firing squad
12	Court sentences of corporal punishment	**no**	Flogging
13	Indefinite detention without charge	**NO**	Many held for long periods. Some incommunicado. Others exiled to remote areas
14	Compulsory membership of state organizations or parties	**yes**	Membership of the ruling MPR party an advantage in gaining government employment and socially
15	Compulsory religion or state ideology in schools	**YES**	Rights respected
16	Deliberate state policies to control artistic works	**yes**	The unstable security and political situation makes prudence necessary in the performing arts. Songs to be recorded should be passed by censor
17	Political censorship of press	**no**	The prospect of a more democratic political system has meant some relaxation in the previous strict censorship. But some arrests of journalists who take "freedom" too far
18	Censorship of mail or telephone tapping	**no**	General surveillance – though not officially admitted

FREEDOM FOR OR RIGHTS TO:		COMMENTS
19	Peaceful political opposition	**no**	Demonstrations have forced the president to introduce a form of political opposition along party lines but its effectiveness is nullified by his arbitrary domination

FREEDOM FOR OR RIGHTS TO: COMMENTS

20	Multiparty elections by secret and universal ballot	no	The president, with the support of the military, has appointed an opposition prime minister. Promises of multiparty elections and a new constitution are made against a background of riots, social anarchy, and unrestrained violence by the military against the civilian population (October 1991)
21	Political and legal equality for women	no	Some improvement in the place of women in the professions but little presence at senior government level
22	Social and economic equality for women	no	Discrimination continues along traditional lines. Most agricultural labor done by women
23	Social and economic equality for ethnic minorities	no	In practice, the treatment of the hundreds of different tribes depends on local factors. The discrimination particularly affects Pygmies, an exploited minority
24	Independent newspapers	yes	Prudence required to ensure independence. Journalists need an approved union press card
25	Independent book publishing	yes	Degree of self-censorship practiced. Occasional government pressures
26	Independent radio and television networks	**NO**	State owned and controlled
27	All courts to total independence	**NO**	President has wide powers to influence courts, depending on the need to use them. Judges augment low pay by taking bribes
28	Independent trade unions	no	The president's erratic moves to liberalize the previous party control have still to make significant changes

LEGAL RIGHTS: COMMENTS

29	From deprivation of nationality	no	Possible deprivation under a loose interpretation of "Zairean authenticity" ruling
30	To be considered innocent until proved guilty	no	A profitable area for ill-disciplined and low-paid police. Innocence proved by the payment of "release money"
31	To free legal aid when necessary and counsel of own choice	no	Means test but no choice of counsel. In remote areas, however, defense counsel may not be available
32	From civilian trials in secret	**NO**	A decision for the security forces – protected by presidential decree
33	To be brought promptly before a judge or court	no	Within 48 hours by law but this is frequently ignored. Apart from arbitrary police powers in practice, system overburdened

	LEGAL RIGHTS:		COMMENTS
34	From police searches of home without a warrant	no	Legal requirements often disregarded by police and military. Motives vary, one being to supplement low pay
35	From arbitrary seizure of personal property	no	As question 34. In remote areas, looting, arbitrary seizures

	PERSONAL RIGHTS:		COMMENTS
36	To interracial, interreligious, or civil marriage	YES	Rights respected though the complexity of hundreds of tribes with different customary laws may affect situation
37	Equality of sexes during marriage and for divorce proceedings	no	Inequalities throughout the system. Wife needs husband's permission to work outside the home, to obtain a passport, to open a bank account, etc.
38	To practice any religion	yes	Some minority sects such as Jehovah's Witnesses banned
39	To use contraceptive pills and devices	YES	Limited state support
40	To noninterference by state in strictly private affairs	yes	Occasional presidential decrees on dress and speech requirements, though recently rescinded. Harassment by ill-disciplined police

Zambia

Human rights rating: 57%

YES 11 **yes** 13 **no** 15 NO 1

Population: 8,500,000
Life expectancy: 54.4
Infant mortality (0–5 years)
  per 1,000 births: 125
United Nations covenants ratified:
Civil and Political Rights, Economic,
Social and Cultural Rights, Convention
on Equality for Women

Form of government: Multiparty
democracy
GNP per capita (US$): 290
% of GNP spent by state:
On health: 1.2
On military: 3.2
On education: 2.2

FACTORS AFFECTING HUMAN RIGHTS
In November 1991 the first multiparty election in 23 years resulted in the end of one-party government, which had ruled for most of the period since the country gained its independence from Britain in 1964. The new Movement for Multiparty Democracy has just been installed and the information below applies to the situation as of the election of the new government. If the transition to a democracy is successful, there will be major changes with respect to human rights.

	FREEDOM TO:		COMMENTS
1	Travel in own country	yes	Police checks on roads and in towns are part of wide crime control measures
2	Travel outside own country	no	Passports of political dissidents withheld
3	Peacefully associate and assemble	no	Permits usually refused for opposition meetings though some relaxation with prospects of the first multiparty election at end-October 1991
4	Teach ideas and receive information	yes	Understood guidelines. Academics may lose posts if criticism of president or party too extreme
5	Monitor human rights violations	yes	Some reluctance by government to cooperate with international organizations
6	Publish and educate in ethnic language	YES	Rights respected

	FREEDOM FROM:		COMMENTS
7	Serfdom, slavery, forced or child labor	yes	Regional pattern of child labor. Allegations of forced labor by prison inmates
8	Extrajudicial killings or "disappearances"	no	Cases of deaths caused by overzealous police "antirobbery squads." Also when held in custody

FREEDOM FROM:		COMMENTS
9	Torture or coercion by the state	no	Many abuses by police and security forces though some are charged with the offenses and face trial
10	Compulsory work permits or conscription of labor	YES	Rights respected
11	Capital punishment by the state	NO	Hanging for treason, murder, theft with violence, etc.
12	Court sentences of corporal punishment	no	Caning for theft, burglary, and minor crimes
13	Indefinite detention without charge	no	State of emergency expired September 1991. Assumed that this will mean some releases of political detainees
14	Compulsory membership of state organizations or parties	yes	Membership in the ruling party an advantage when applying for government posts or favors
15	Compulsory religion or state ideology in schools	YES	Rights respected
16	Deliberate state policies to control artistic works	YES	Rights respected
17	Political censorship of press	no	The two major newspapers are party and government owned. With the announcement of elections late in 1991, opposition views are being given more space
18	Censorship of mail or telephone tapping	yes	Occasionally of dissidents – system limited and incompetent

FREEDOM FOR OR RIGHTS TO:		COMMENTS
19	Peaceful political opposition	YES	The first election on a multiparty basis in 23 years was described by international observers as fair and "mature"
20	Multiparty elections by secret and universal ballot	YES	The new Movement for Multiparty Democracy was voted into power at the end of October by a fair and secret ballot
21	Political and legal equality for women	yes	Limited representation on the Central Committee of the governing party and in the professions
22	Social and economic equality for women	no	Some improvements in women's status but tradition perpetuates inequalities. Women provide much of agricultural labor
23	Social and economic equality for ethnic minorities	yes	Asian traders protest at discrimination and disadvantages, particularly in political representation
24	Independent newspapers	no	The two major newspapers are state owned. Smaller independent papers occasionally banned or seized

FREEDOM FOR OR RIGHTS TO:		COMMENTS
25	Independent book publishing	no	Nothing that is "not in the interest of the people." Provocative works prohibited
26	Independent radio and television networks	no	State owned and controlled but currently less biased with prospects of first multiparty election
27	All courts to total independence	yes	All courts subject to president's decree powers – which are rarely used
28	Independent trade unions	yes	The new Industrial Relations Act of 1990 permits independent unions. Strike controls usually ineffective

LEGAL RIGHTS:		COMMENTS
29	From deprivation of nationality	no	The president retains powers to deprive, deport, etc.
30	To be considered innocent until proved guilty	no	Senior officials have authority to hold suspects on police premises for 28 days without warrant
31	To free legal aid when necessary and counsel of own choice	yes	Means test. Court provides counsel
32	From civilian trials in secret	no	The announcement of end of the state of emergency (September 1991) too recent to ensure that secret trials have definitely ended
33	To be brought promptly before a judge or court	yes	Within 24 hours but many instances of long delays
34	From police searches of home without a warrant	yes	End of state-of-emergency provisions should mean end of searches without warrants
35	From arbitrary seizure of personal property	YES	Rights respected

PERSONAL RIGHTS:		COMMENTS
36	To interracial, interreligious, or civil marriage	YES	Legal at 18 years for both sexes
37	Equality of sexes during marriage and for divorce proceedings	no	The worst inequalities are against women in tribal areas where traditional practices continue – and in marriage, divorce, inheritance, and status
38	To practice any religion	YES	Rights respected
39	To use contraceptive pills and devices	YES	State support
40	To noninterference by state in strictly private affairs	YES	Rights respected

Zimbabwe

Human rights rating: 65%

YES 18 **yes** 11 **no** 9 NO 2

Population: 9,700,000
Life expectancy: 59.6
Infant mortality (0–5 years)
  per 1,000 births: 90
United Nations covenants ratified:
None

Form of government: Parliamentary system
GNP per capita (US$): 650
% of GNP spent by state:
On health: 2.9
On military: 5.0
On education: 8.5

FACTORS AFFECTING HUMAN RIGHTS

The elections of 1980, after almost 15 years of unconstitutional white rule, resulted in the first all-party Parliament by a democratic vote. But the governing Zimbabwe African National Union, with the stated objective of introducing one-party rule, united with the main opposition party and, after the 1990 elections, dominated pparliament with 117 of 120 seats. The president, who has been in control since independence, has been in favor of altering the constitution to permit absolute one-party rule, but public opposition and that of many of his colleagues have deterred such a move. The state of emergency was lifted in 1990 after 25 years, but government control of most of the media continues as does surveillance of the limited political opposition and dissident tribal elements.

	FREEDOM TO:		COMMENTS
1	Travel in own country	YES	Rights respected
2	Travel outside own country	YES	Rights respected. The only restraint is problem of obtaining foreign currency
3	Peacefully associate and assemble	yes	Government has reserve powers which, since the end of the state of emergency, are seldom used
4	Teach ideas and receive information	yes	Circumspection necessary on a few sensitive subjects. Protesting students occasionally arrested, university briefly closed
5	Monitor human rights violations	yes	But major international monitors such as Amnesty International, described as "enemy of the people," denied entry into country
6	Publish and educate in ethnic language	YES	Rights respected

FREEDOM FROM: | COMMENTS

#			Comments
7	Serfdom, slavery, forced or child labor	yes	Regional pattern of child labor. Tribal practice of settling interfamily disputes by offering a young daughter still survives
8	Extrajudicial killings or "disappearances"	**YES**	With the return of a stable society, the arbitrary killings of previous years have ceased
9	Torture or coercion by the state	no	By ill-disciplined security forces. Beatings and maltreatment
10	Compulsory work permits or conscription of labor	**YES**	Rights respected
11	Capital punishment by the state	**NO**	Resumed after a 2-year respite in mid-1980s
12	Court sentences of corporal punishment	no	Caning
13	Indefinite detention without charge	**YES**	Rights respected
14	Compulsory membership of state organizations or parties	**YES**	Rights respected
15	Compulsory religion or state ideology in schools	**YES**	Rights respected
16	Deliberate state policies to control artistic works	**YES**	Rights respected
17	Political censorship of press	yes	Position improved but a degree of "self-censorship" in practice. Major newspaper government owned
18	Censorship of mail or telephone tapping	no	Surveillance continues despite end of state of emergency

FREEDOM FOR OR RIGHTS TO: | COMMENTS

#			Comments
19	Peaceful political opposition	no	In practice the dominant ruling party, with 96% of parliamentary seats after the 1990 election, was able to intimidate the smaller parties of the opposition
20	Multiparty elections by secret and universal ballot	yes	Many violations of fair democratic elections, mostly by the dominant ZANU party
21	Political and legal equality for women	yes	Legal equality but discrimination at all levels, particularly among the poorly educated
22	Social and economic equality for women	no	Traditional practices defy modern legislation. Brides still bought for cattle in customary marriages
23	Social and economic equality for ethnic minorities	yes	Intertribal rivalries and hostilities lead to discrimination, particularly for government jobs. Shona majority over 60% of population

FREEDOM FOR OR RIGHTS TO:		COMMENTS
24	Independent newspapers	no	Mostly government controlled or party affiliated. Smaller independents need to display prudence as well as commercial acumen
25	Independent book publishing	yes	A few bannings of subversive works. Fewer restrictions on imported books
26	Independent radio and television networks	NO	State owned and controlled
27	All courts to total independence	YES	Rights respected
28	Independent trade unions	YES	Strike restrictions removed with end of state of emergency

LEGAL RIGHTS:		COMMENTS
29	From deprivation of nationality	no	Reserve powers under the seldom-used Law and Order Maintenance Act
30	To be considered innocent until proved guilty	YES	Rights respected
31	To free legal aid when necessary and counsel of own choice	no	Only for murder charges and similar. Court ultimately chooses counsel
32	From civilian trials in secret	YES	Rights respected
33	To be brought promptly before a judge or court	yes	Within 96 hours but in practice many delays
34	From police searches of home without a warrant	yes	Occasional abuses
35	From arbitrary seizure of personal property	YES	Rights respected

PERSONAL RIGHTS:		COMMENTS
36	To interracial, interreligious, or civil marriage	YES	Rights respected
37	Equality of sexes during marriage and for divorce proceedings	no	Large section of population still follows customary laws which treat wife as husband's "property." New Marriage Act has failed to remedy this
38	To practice any religion	YES	Rights respected
39	To use contraceptive pills and devices	YES	State support
40	To noninterference by state in strictly private affairs	YES	Rights respected

Universal Declaration of Human Rights

Adopted by the UN General Assembly in 1948 without a dissenting vote

Article 1
All human beings are born free and equal in dignity and rights. They are endowed with reason and conscience and should act towards one another in a spirit of brotherhood.

Article 2
Everyone is entitled to all the rights and freedoms set forth in this Declaration, without distinction of any kind, such as race, colour, sex, language, religion, political or other opinion, national or social origin, property, birth or other status.

Furthermore, no distinction shall be made on the basis of the political, jurisdictional or international status of the country or territory to which a person belongs, whether it be independent, trust, non-self-governing or under any other limitation of sovereignty.

Article 3
Everyone has the right to life, liberty and the security of person.

Article 4
No one shall be held in slavery or servitude; slavery and the slave trade shall be prohibited in all their forms.

Article 5
No one shall be subjected to torture or to cruel, inhuman or degrading treatment or punishment.

Article 6
Everyone has the right to recognition everywhere as a person before the law.

Article 7
All are equal before the law and are entitled without any discrimination to equal protection of the law. All are entitled to equal protection against any discrimination in violation of this Declaration and against any incitement to such discrimination.

Article 8
Everyone has the right to an effective remedy by the competent national tribunals for acts violating the fundamental rights granted him by the constitution or by law.

Article 9
No one shall be subjected to arbitrary arrest, detention or exile.

Article 10
Everyone is entitled in full equality to a fair and public hearing by an independent and impartial tribunal, in the determination of his rights and obligations and of any criminal charge against him.

Article 11
1. Everyone charged with a penal offence has the right to be presumed innocent until proved guilty according to law in a public trial at which he has had all the guarantees necessary for his defence.
2. No one shall be held guilty of any penal offence on account of any act or omission which did not constitute a penal offence, under national or international law, at the time when it was committed. Nor shall a heavier penalty be imposed than the one that was applicable at the time the penal offence was committed.

Article 12
No one shall be subjected to arbitrary interference with his privacy, family, home or correspondence, nor to attacks upon his honour and reputation. Everyone has the right to the protection of the law against such interference or attacks.

Article 13
1. Everyone has the right to freedom of movement and residence within the borders of each state.
2. Everyone has the right to leave any country, including his own, and to return to his country.

Article 14
1. Everyone has the right to seek and to enjoy in other countries asylum from persecution.
2. This right may not be invoked in the case of prosecutions genuinely arising from non-political crimes or from acts contrary to the purposes and principles of the United Nations.

Article 15
1. Everyone has the right to a nationality.
2. No one shall be arbitrarily deprived of his nationality nor denied the right to change his nationality.

Article 16
1. Men and women of full age, without any limitation due to race, nationality or religion, have the right to marry and to found a family. They are entitled to equal rights as to marriage, during marriage and at its dissolution.
2. Marriage shall be entered into only with the free and full consent of the intending spouses.
3. The family is the natural and fundamental group unit of society and is entitled to protection by society and the State.

Article 17
1. Everyone has the right to own property alone as well as in association with others.
2. No one shall be arbitrarily deprived of his property.

Article 18
Everyone has the right to freedom of thought, conscience and religion; this right includes freedom to change his religion or belief, and freedom, either alone or in community with others and in public or private, to manifest his religion or belief in teaching, practice, worship and observance.

Article 19
Everyone has the right to freedom of opinion and expression; this right includes freedom to hold opinions without interference and to seek, receive and impart information and ideas through any media and regardless of frontiers.

Article 20
1. Everyone has the right to freedom of peaceful assembly and association.
2. No one may be compelled to belong to an association.

Article 21
1. Everyone has the right to take part in the government of his country, directly or through freely chosen representatives.
2. Everyone has the right of equal access to public service in his country.
3. The will of the people shall be the basis of the authority of government; this will shall be expressed in periodic and genuine elections which shall be by universal and equal suffrage and shall be held by secret vote or by equivalent free voting procedures.

Article 22
Everyone, as a member of society, has the right to social security and is entitled to realization, through national effort and international co-operation and in accordance with the organization and resources of each State, of the economic, social and cultural rights indispensable for his dignity and the free development of his personality.

Article 23

1. Everyone has the right to work, to free choice of employment, to just and favourable conditions of work and to protection against unemployment.
2. Everyone, without any discrimination, has the right to equal pay for equal work.
3. Everyone who works has the right to just and favourable renumeration ensuring for himself and his family an existence worthy of human dignity, and supplemented, if necessary, by other means of social protection.
4. Everyone has the right to form and to join trade unions for the protection of his interests.

Article 24

Everyone has the right to rest and leisure, including reasonable limitation of working hours and periodic holidays with pay.

Article 25

1. Everyone has the right to a standard of living adequate for the health and well-being of himself and of his family, including food, clothing, housing and medical care and necessary social services, and the right to security in the event of unemployment, sickness, disability, widowhood, old age or other lack of livelihood in circumstances beyond his control.
2. Motherhood and childhood are entitled to special care and assistance. All children, whether born in or out of wedlock, shall enjoy the same social protection.

Article 26

1. Everyone has the right to education. Education shall be free, at least in the elementary and fundamental stages. Elementary education shall be compulsory. Technical and professional education shall be made generally available and higher education shall be equally accessible to all on the basis of merit.
2. Education shall be directed to the full development of the human personality and to the strengthening of respect for human rights and fundamental freedoms. It shall promote understanding, tolerance and friendship among all nations, racial or religious groups, and shall further the activities of the United Nations for the maintenance of peace.
3. Parents have a prior right to choose the kind of education that shall be given to their children.

Article 27

1. Everyone has the right freely to participate in the cultural life of the community, to enjoy the arts and to share in scientific advancement and its benefits.
2. Everyone has the right to the protection of the moral and material interests resulting from any scientific, literary, or artistic production of which he is the author.

Article 28

Everyone is entitled to a social and international order in which the rights and freedoms set forth in this Declaration can be fully realized.

Article 29

1. Everyone has duties to the community in which alone the free and full development of his personality is possible.
2. In the exercise of his rights and freedoms, everyone shall be subject only to such limitations as are determined by law solely for the purpose of securing due recognition and respect for the rights and freedoms of others and of meeting the just requirements of morality, public order and the general welfare in a democratic society.
3. These rights and freedoms may in no case be exercised contrary to the purposes and principles of the United Nations.

Article 30

Nothing in this Declaration may be interpreted as implying for any State, group or person any right to engage in any activity or to perform any act aimed at the destruction of any of the rights and freedoms set forth herein.

International Covenant on Civil and Political Rights

In 1948 the Universal Declaration of Human Rights was adopted by the General Assembly of the United Nations. It did not have the force of law, however, and in 1966 the General Assembly adopted the International Covenant on Civil and Political Rights (ICCPR) and the International Covenant on Economic, Social and Cultural Rights, which transformed the original principles into treaty provisions. Not all the member-states ratified the covenants, which established legal obligations to honor their articles. At the beginning of 1991 the covenants had been ratified by 92 of the 164 members.

An optional protocol to the ICCPR was also introduced. This enables individuals to seek redress by filing a complaint against the offending state party. But only 51 of the countries ratifying the covenant have agreed to submit to this further obligation.

The articles of the ICCPR most relevant to this work are set out below.

PART I

Article 1

1. All peoples have the right to self-determination. By virtue of that right they freely determine their political status and freely pursue their economic, social and cultural development.
2. The States Parties to the present Covenant, including those having responsibility for the administration of Non-Self-Governing and Trust Territories, shall promote the realization of the right of self-determination, and shall respect that right, in conformity with the provisions of the United Nations Charter.

PART II

Article 2

1. Each State Party to the present Covenant undertakes to respect and to ensure to all individuals within its territory and subject to its jurisdiction the rights recognized in the present Covenant, without distinction of any kind, such as race, colour, sex, language, religion, political or other opinion, national or social origin, property, birth or other status.
2. When not already provided for by existing legislative or other measures, each State Party to the present Covenant undertakes to take the necessary steps, in accordance with its constitutional processes and with the provisions of the present covenant, to adopt such legislative or other measures as may be necessary to give effect to the rights recognized in the present Covenant.
3. Each State Party to the present Covenant undertakes:
   (a) To ensure that any person whose rights or freedoms as herein recognized are violated shall have an effective remedy notwithstanding that the violation has been committed by persons acting in an official capacity;
   (b) To ensure that any person claiming such a remedy shall have his right thereto determined by competent judicial, administrative or legislative authorities, or by any other competent authority provided for by the legal system of the State, and to develop the possibilities of judicial remedy;
   (c) To ensure that the competent authorities shall enforce such remedies when granted.

Article 3

The States Parties to the present Covenant undertake to ensure the equal right of men and women to the enjoyment of all civil and political rights set forth in the present Covenant.

PART III

Article 6

1. Every human being has the inherent right to life. This right shall be protected by law. No one shall be arbitrarily deprived of his life.

2. In countries which have not abolished the death penalty, sentence of death may be imposed only for the most serious crimes in accordance with the law in force at the time of the commission of the crime and not contrary to the provisions of the present Covenant and to the Convention on the Prevention and Punishment of the Crime of Genocide. This penalty can only be carried out pursuant to a final judgement rendered by a competent court.

3. When deprivation of life constitutes the crime of genocide, it is understood that nothing in this article shall authorize any State Party to the present Covenant to derogate in any way from any obligation assumed under the provisions of the Convention on the Prevention and Punishment of the Crime of Genocide.

4. Anyone sentenced to death shall have the right to seek pardon or commutation of the sentence. Amnesty, pardon or commutation of the sentence of death may be granted in all cases.

5. Sentence of death shall not be imposed for crimes committed by persons below eighteen years of age and shall not be carried out on pregnant women.

6. Nothing in this article shall be invoked to delay or to prevent the abolition of capital punishment by any State Party to the present Covenant.

Article 7

No one shall be subjected to torture or to cruel, inhuman or degrading treatment or punishment. In particular, no one shall be subjected without his free consent to medical or scientific experimentation.

Article 8

1. No one shall be held in slavery; slavery and the slave-trade in all their forms shall be prohibited.

2. No one shall be held in servitude.

Article 9

1. Everyone has the right to liberty and security of person. No one shall be subjected to arbitrary arrest or detention. No one shall be deprived of his liberty except on such grounds and in accordance with such procedures as are established by law.

2. Anyone who is arrested shall be informed, at the time of arrest, of the reasons for his arrest and shall be promptly informed of any charges against him.

3. Anyone arrested or detained on a criminal charge shall be brought promptly before a judge or other officer authorized by law to exercise judicial power and shall be entitled to trial within a reasonable time or to release. It shall not be the general rule that persons awaiting trial shall be detained in custody, but release may be subject to guarantees to appear for trial, at any other stage of the judicial proceedings, and, should occasion arise, for execution of the judgement.

4. Anyone who is deprived of his liberty by arrest or detention shall be entitled to take proceedings before a court, in order that that court may decide without delay on the lawfulness of his detention and order his release if the detention is not lawful.

5. Anyone who has been the victim of unlawful arrest or detention shall have an enforceable right to compensation.

Article 12

1. Everyone lawfully within the territory of a State shall, within that territory, have the right to liberty of movement and freedom to choose his residence.

2. Everyone shall be free to leave any country, including his own.

3. The above-mentioned rights shall not be subject to any restrictions except those which are provided by law, are necessary to protect national security, public order (ordre public), public health or morals or the rights and freedoms of others, and are consistent with the other rights recognized in the present Covenant.

4. No one shall be arbitrarily deprived of the right to enter his own country.

Article 14

1. All persons shall be equal before the courts and tribunals. In the determination of any criminal charge against him, or of his rights and obligations in a suit at law, everyone shall be entitled to a fair and public hearing by a competent, independent and impartial tribunal established by law. The Press and the public may be excluded from all or part

of a trial for reasons of morals, public order (*ordre public*) or national security in a democratic society, or when the interests of the private lives of the parties so requires, or to the extent strictly necessary in the opinion of the court in special circumstances where publicity would prejudice the interests of justice; but any judgement rendered in a criminal case or in a suit at law shall be made public except where the interest of juvenile persons otherwise requires or the proceedings concern matrimonial disputes or the guardianship of children.

2. Everyone charged with a criminal offence shall have the right to be presumed innocent until proved guilty according to law.
3. In the determination of any criminal charge against him, everyone shall be entitled to the following minimum guarantees, in full equality:
    (a) To be informed promptly and in detail in a language which he understands of the nature and cause of the charge against him;
    (b) To have adequate time and facilities for the preparation of his defence and to communicate with counsel of his own choosing;
    (c) To be tried without undue delay;
    (d) To be tried in his presence, and to defend himself in person or through legal assistance of his own choosing; to be informed, if he does not have legal assitance, of this right; and to have legal assistance assigned to him, in any case where the interests of justice so require, and without payment by him in any such case if he does not have sufficient means to pay for it;
    (e) To examine, or have examined, the witnesses against him and to obtain the attendance and examination of witnesses on his behalf under the same conditions as witnesses against him;
    (f) To have the free assistance of an interpreter if he cannot understand or speak the language used in court;
    (g) Not to be compelled to testify against himself or to confess guilt.

Article 16
Everyone shall have the right to recognition everywhere as a person before the law.

Article 17
1. No one shall be subjected to arbitrary or unlawful interference with his privacy, family, home or correspondence, nor to unlawful attacks on his honour and reputation.
2. Everyone has the right to the protection of the law against such interference or attacks.

Article 18
1. Everyone shall have the right of freedom of thought, conscience and religion. This right shall include freedom to have or to adopt a religion or belief of his choice, and freedom, either individually or in community with others and in public or private, to manifest his religion or belief in worship, observance, practice and teaching.
2. No one shall be subject to coercion which would impair his freedom to have or to adopt a religion or belief of his choice.
3. Freedom to manifest one's religion or beliefs may be subject only to such limitations as are prescribed by law and are necessary to protect public safety, order, health, or morals or the fundamental rights and freedoms of others.
4. The States Parties to the present Covenant undertake to have respect for the liberty of parents and, when applicable, legal guardians to ensure the religious and moral education of their children in conformity with their own convictions.

Article 19
1. Everyone shall have the right to hold opinions without interference.
2. Everyone shall have the right to freedom of expression; this right shall include freedom to seek, receive and impart information and ideas of all kinds, regardless of frontiers, either orally, in writing or in print, in the form of art, or through any other media of his choice.

Article 20
1. Any propaganda for war shall be prohibited by law.
2. Any advocacy of national, racial or religious hatred that constitutes incitement to discrimination, hostility or violence shall be prohibited by law.

Article 21

The right of peaceful assembly shall be recognized. No restrictions may be placed on the exercise of this right other than those imposed in conformity with the law and which are necessary in a democratic society in the interests of national security or public safety, public order (*ordre public*), the protection of public health or morals or the protection of the rights and freedoms of others.

Article 23

1. The family is the natural and fundamental group unit of society and is entitled to protection by society and the State.
2. The right of men and women of marriageable age to marry and to found a family shall be recognized.
3. No marriage shall be entered into without the free and full consent of the intending spouses.
4. States Parties to the present Covenant shall take appropriate steps to ensure equality of right and responsibilities of spouses as to marriage, during marriage and at its dissolution. In the case of dissolution, provision shall be made for the necessary protection of any children.

Article 24

1. Every child shall have, without any discrimination as to race, colour, sex, language, religion, national or social origin, property or birth, the right to such measures of protection as are required by his status as a minor, on the part of his family, the society and the State.
2. Every child shall be registered immediately after birth and shall have a name.
3. Every child has the right to acquire a nationality.

Article 25

Every citizen shall have the right and the opportunity, without any of the distinctions mentioned in Article 2 and without unreasonable restrictions:
(a) To take part in the conduct of public affairs, directly or through freely chosen representatives;
(b) To vote and to be elected at genuine periodic elections which shall be by universal and equal suffrage and shall be held by secret ballot, guaranteeing the free expression of the will of the electors;
(c) To have access, on general terms of equality, to public service in his country.

Article 26

All persons are equal before the law and are entitled without any discrimination to equal protection of the law. In this respect the law shall prohibit any discrimination and guarantee to all persons equal and effective protection against discrimination on any ground such as race, colour, sex, language, religion, political or other opinion, national or social origin, birth or other status.

Article 27

In those States in which ethnic, religious or linguistic minorities exist, persons belonging to such minorities shall not be denied the right, in community with the other members of their group, to enjoy their own culture, to profess and practise their own religion, or to use their own language.

International Covenant on Economic, Social and Cultural Rights

---

The basis of the questionnaire uses only three of the articles of this covenant. The reason for this is explained in the Introduction (see section on the human rights covenants).

America, South

GUYANA
SURINAME
FRENCH
GUIANA

VENEZUELA

COLOMBIA

ECUADOR

PERU

BRAZIL

BOLIVIA

PARAGUAY

URUGUAY

ARGENTINA

CHILE

Most human rights respected

Many human rights denied

Most human rights denied

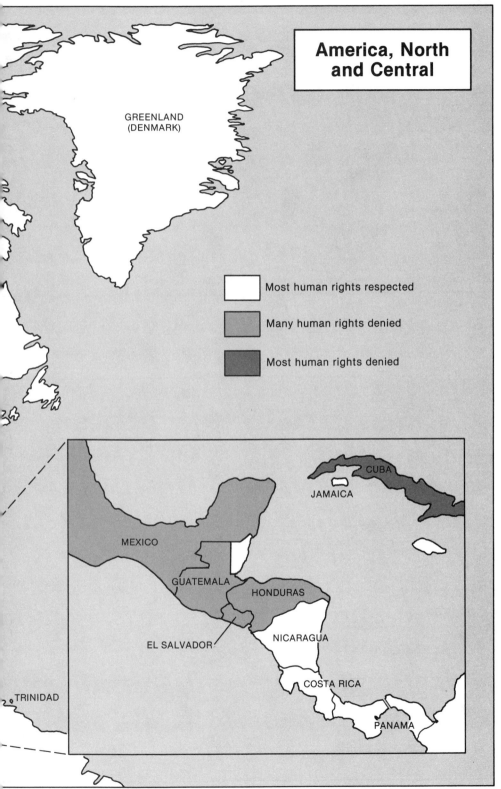

America, North and Central

GREENLAND
(DENMARK)

Most human rights respected

Many human rights denied

Most human rights denied

CUBA

JAMAICA

MEXICO

GUATEMALA

HONDURAS

EL SALVADOR

NICARAGUA

COSTA RICA

TRINIDAD

PANAMA

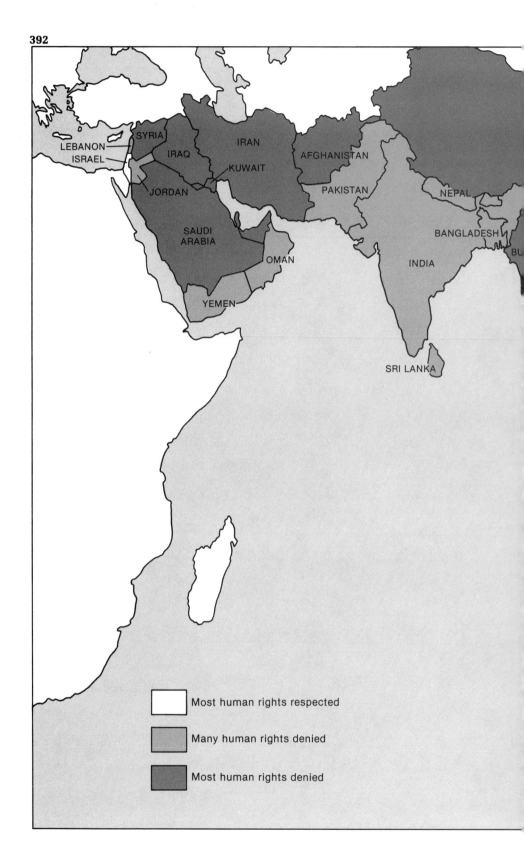

Most human rights respected

Many human rights denied

Most human rights denied

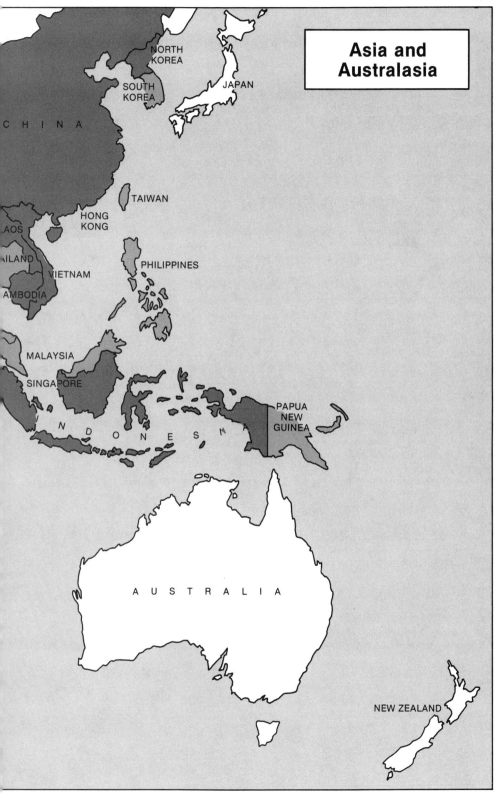

Asia and Australasia

CHINA

NORTH KOREA

SOUTH KOREA

JAPAN

TAIWAN

HONG KONG

AOS

ILAND

VIETNAM

AMBODIA

PHILIPPINES

MALAYSIA

SINGAPORE

INDONESIA

PAPUA NEW GUINEA

AUSTRALIA

NEW ZEALAND

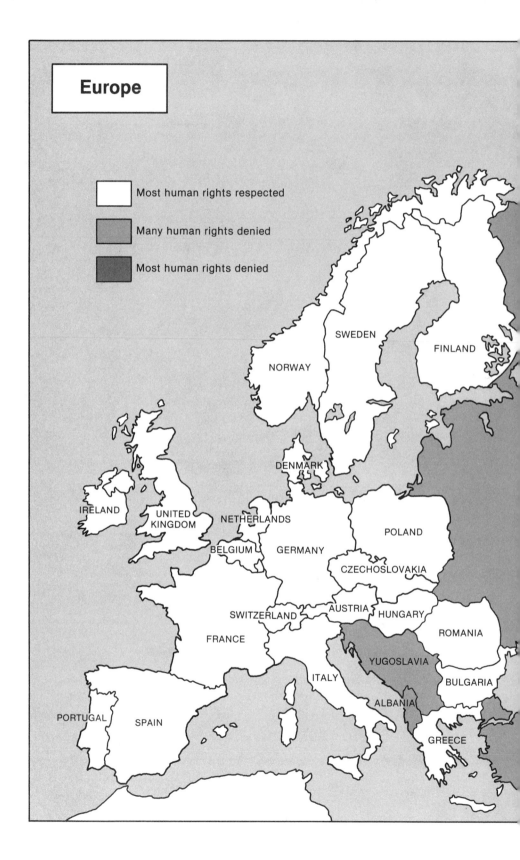

Europe

- ☐ Most human rights respected
- ▨ Many human rights denied
- ▓ Most human rights denied

NORWAY

SWEDEN

FINLAND

DENMARK

IRELAND

UNITED KINGDOM

NETHERLANDS

POLAND

BELGIUM

GERMANY

CZECHOSLOVAKIA

AUSTRIA

HUNGARY

ROMANIA

SWITZERLAND

YUGOSLAVIA

FRANCE

ITALY

BULGARIA

ALBANIA

PORTUGAL

SPAIN

GREECE

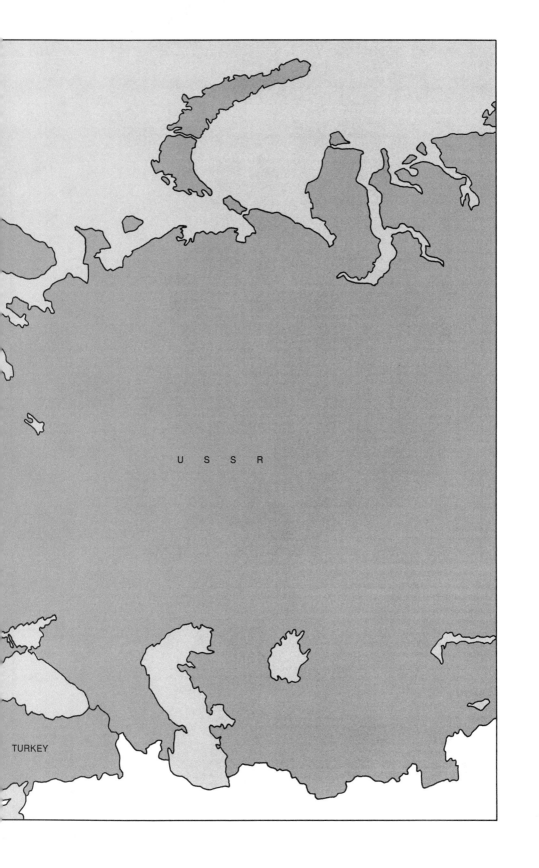

U   S   S   R

TURKEY

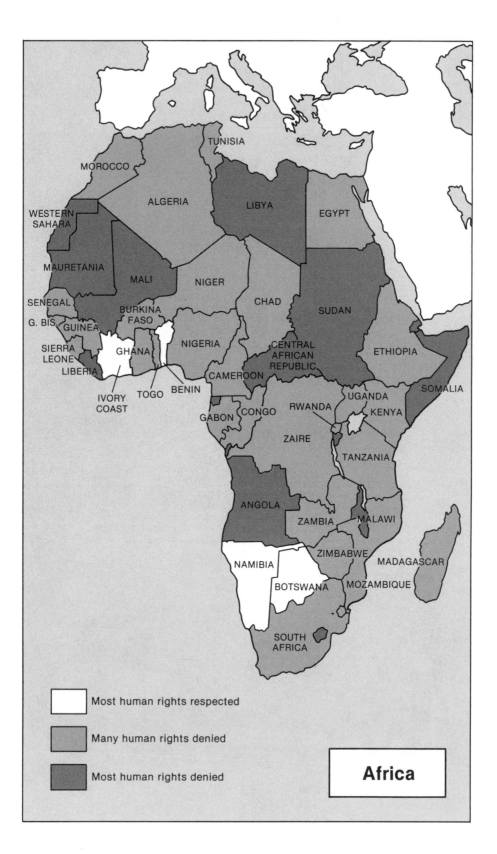

Most human rights respected

Many human rights denied

Most human rights denied

Africa